Bacteria: Microbiology and Molecular Genetics

Bacteria: Microbiology and Molecular Genetics

Edited by **Henry Evans**

SYRAWOOD
PUBLISHING HOUSE

New York

Published by Syrawood Publishing House,
750 Third Avenue, 9th Floor,
New York, NY 10017, USA
www.syrawoodpublishinghouse.com

Bacteria: Microbiology and Molecular Genetics
Edited by Henry Evans

International Standard Book Number: 978-1-68286-062-5 (Hardback)

Printed in the United States of America.

Contents

Preface

The main aim of this book is to educate learners and enhance their research focus by presenting diverse topics covering this vast field. This is an advanced book which compiles significant studies by distinguished experts. This book addresses successive solutions to the challenges arising in the area of application, along with it; the book provides scope for future developments.

The advancements and discoveries in the fields of microbiology and molecular genetics have immensely benefitted mankind with their applications in pharmaceuticals, bioengineering, environmental science, etc. This book brings forth some of the crucial concepts and developments in the study of bacteria and their applications in microbial processes. It is a compilation of some important topics in the field of bacteriology and molecular genetics like bacterial physiology, bacterial endotoxins, cell signalling, etc. Scientists and students actively engaged in this field will find this book full of crucial and unexplored concepts.

It was a great honour to edit this book, though there were challenges, as it involved a lot of communication and networking between me and the editorial team. However, the end result was this all-inclusive book covering diverse themes in the field.

Finally, it is important to acknowledge the efforts of the contributors for their excellent chapters, through which a wide variety of issues have been addressed. I would also like to thank my colleagues for their valuable feedback during the making of this book.

Editor

Prevalence and seasonality of fowl typhoid disease in Zaria-Kaduna State, Nigeria

I. J. Mbuko, M. A. Raji*, J. Ameh, L. Saidu, W. I. Musa and P. A. Abdul

Faculty of Veterinary Medicine, Ahmadu Bello University, P. O. Box 1044, Samaru-Zaria, Kaduna State-Nigeria.

A five years retrospective study (2003 - 2007) of the prevalence of fowl typhoid (FT) and other poultry diseases diagnosed at the avian unit of the Veterinary Teaching Hospital (VTH), Ahmadu Bello University, Zaria-Kaduna State, Nigeria was conducted. A prevalence rate of 18.4% (129 cases) was recorded for FT out of 700 cases of poultry disease. The highest number of outbreaks of FT was recorded during the rainy season (July - September). 48 cases (29.1%) of FT were recorded in birds 15 weeks and above, 124 cases (18.4%) in layers and only two cases (16.7%) in turkeys. Outbreaks of FT were closely associated with age, type and poultry species of birds (P < 0.05). The outbreaks of FT were also observed to be 3.1 times more likely to occur in December, 2.4 and 1.3 times more likely to occur in birds of 15 weeks and above and 1 - 5 weeks of age respectively. It was concluded from the study that FT is more commonly reported in the chicken than other poultry species and adult birds (>15 weeks) are more susceptible to the disease in Zaria, Nigeria. This study recommends that poultry farmers should be encouraged to practice prompt and regular vaccination of layers against fowl typhoid disease.

Key words: Salmonella enterica serovars Gallinarium, chicken, Nigeria.

INTRODUCTION

Fowl typhoid (FT) is an acute infectious enteritis (Okwori et al., 2007) causing heavy mortality in growers or adult birds although chicks can be affected (Jordan and Pattison, 1992). It is caused by the bacterium *Salmonella enterica* serovars Gallinarium (Jordan and Pattison, 1992; Aiello, 1998; Abdu, 2007), a member of the family enterobacteriaceae which is widely distributed throughout the world (Roa, 2000). *S. enterica* serovars Gallinarium is highly adapted and seldom causes significant problems in hosts other than chickens, turkeys and pheasants (Jordan and Pattison, 1992; Aiello, 1998; Mdegela et al., 2000; Okwori et al., 2007). It was formerly known as *Shigella gallinarum* when first isolated in 1889 by Klein in England (Okwori et al., 2007). It was named fowl typhoid (FT) in 1902 (Jordan and Pattison, 1992).

The disease has been eradicated from commercial poultry in many developed countries of Western Europe, USA, Canada, Australia and Japan with an intensive poultry industry (OIE, 2005). In Africa, FT has been reported in many countries including Nigeria (Sa'idu et al., 1994), Tanzania (Mtie and Msami, 1996), Uganda (Okoj, 1993), Zambia (Sharma et al., 1991), Libya (Hamid and Sharma, 1990), and Senegal (Arbelot et al., 1992). FT infection usually follows the ingestion of food or water contaminated by the fecal material. The clinically infected birds are carriers of FT which can also be transmitted by attendants through hands, feet, clothes and rodents (Jordan and Pattison, 1992; Aiello, 1998; Abdu, 2007; Okwori et al., 2007) or via eggs laid by vaccinated birds (Jordan and Pattison, 1992).

A definite diagnosis of FT requires the isolation and identification of *S. enterica Serovars Gallinarium*. However, a tentative diagnosis can be made based on the flock history, clinical signs, mortality and lesions (Barrow et al., 1992). The aim of this study was to review all the FT cases and other non-FT poultry diseases diagnosed at the Poultry Health Clinic, Ahmadu Bello University, Veterinary Teaching Hospital (ABUVTH) and confirmed in the Microbiology Laboratory from 2003 - 2007 and to determine if any association exists between prevalence of the disease and factors of the host and environment for example, breed, poultry species, types of bird and

*Corresponding author. E-mail: rajmash2002@yahoo.com.

Table 1. Yearly distribution of fowl typhoid (FT) and other poultry diseases in Zaria, Nigeria.

Year	FT cases	Non FT cases	Year specific rate (%)	OR	95% CI on OR
2003	14	97	12.6	0.595	0.33 - 1.08
2004	15	88	14.6	0.722	0.40 - 1.30
*2005	17	126	11.9	0.536	0.31 - 0.93
2006	23	148	13.5	0.620	0.38 - 1.01
*2007	60	112	34.9	3.624	2.42 - 5.43
Total	129	571			

*significant at $p < 0.05$.

season of the year.

MATERIALS AND METHODS

Study area

Zaria is located between 11°07'N and 7°44'E within the Sudan Savannah Zone.The dominated vegetation is trees and grasses (Sa'idu et al., 2004). The average rainfall ranges from 1000 - 1250 mm and the average daily temperature ranges from 17 - 33°C (Sa'idu et al., 2004).

Samples

The samples submitted to the Veterinary Microbiology laboratory for investigation were intestine, liver, gall bladder, spleen, heart, fecal content and ovarian follicles collected from sick commercially reared and local chickens from Zaria and environs. The samples were immediately transported to the laboratories in a cool thermos and were processed for culture.

Cultures

Salmonella was isolated according to standard methods (ISO 6579, 1993). The intestine, liver, gall bladder, spleen and ovarian follicles were aseptically added to 9 ml of the pre-enrichment medium tetra-thionate broth. All samples were incubated at 37°C for 24 h. One loopful from each of the enriched broths was streaked onto plates of *Salmonella Shigella* (SS) agar (Difco) and xylose lysine deoxycholate (XLD) agar (Difco) and incubated at 37°C for 24 h. The plates were examined for the presence of typical colonies of *Salmonella*, that is, transparent colonies with black centres on SS agar and red colonies with black centres on XLD agar (Antunes et al., 2003). Suspected colonies were confirmed positive by conventional biochemical methods (Lautrop et al., 1979; Nissen, 1984).

Data collection

The data on age, breed, type and poultry species were collected from the avian unit of ABUVTH, Zaria, Nigeria from January 2003 to December 2007. Information on FT was extracted from the clinic records and the non-FT cases were considered together as a group. A case was defined as a farm that reported an outbreak of a disease and diagnosis based on history, clinical signs, post mortem findings and laboratory results in ABUVTH, Zaria, Nigeria. The age of the birds were categorized as follows; (i) 1-5 weeks old, (ii) 6-10 weeks old, (iii) 11-15 weeks old, (iv) greater than 15 weeks old and (v) unknown age. The birds were categorized according to the purpose of keeping the birds i.e. Breeders, Layers, Broilers, Cockerels,

and mixed breed, species that is, chicken and turkey, and breed, that is, local and improved breed (Halle et al., 1999).

The seasons in Zaria and environ were categorized as follows: (i) Dry season (January - March). (ii) Pre-rainy season (April - June). (iii) Rainy season (July-August). (iv) Pre-dry season (October – December) (Abdu et al., 1992). The monthly variation in the distribution of FT for the period of 2003 - 2007 was determined by reducing the 5 year data to one year using the 12 months ratios to moving average method (Saidu et al., 2007). The graph of the isolated monthly index (ISMI) was plotted.

Data analysis

The data were analyzed statistically using Genstat and SPSS computer based programs.

RESULTS

A total of 700 cases of poultry disease were documented with 129 (18.4%) of the diseases diagnosed positive for FT. The highest prevalence rate of 34.9% and lowest prevalence of 11.9% were recorded in 2007 and 2005 respectively (Table 1).

The month specific rate (MSR) for FT was highest in December (39.1%) and lowest in May (8.7%). However, only the odds ratios for July (1.29), August (1.09), October (2.15), November (1.08) and December (1.15) were significant at the 95% CI (Table 2). This means that FT was 1.29, 1.09, 2.15, 1.08 and 3.15 times more likely to occur in July, August, October, November and December respectively when all months were compared (Table 2). The isolated monthly index (ISMI) showed that FT peaked in July, August, October and December (Figure 1). The highest peak was in August and the lowest peaks were in Febuary, April and May. 50 (29.9%) cases were recorded in the rainy season and 16 cases (12.3%) were recorded in the pre-rainy season (Table 3). This means that FT was 2.42 times more likely to occur in the rainy season (July - September).

The age specific rate of FT showed that birds >15 weeks old had the highest rate (29.1%), followed by bird's 1 - 5 weeks old (21.6%) while the lowest (13.3%) was recorded in birds within the age 11 - 15 weeks old. This means that FT was 2.4 times more likely to occur in adult birds (Table 4).

The breed distribution of FT disease compared to other

Table 2. Monthly distributions of FT and other poultry diseases in Zaria, Nigeria.

Month	FT cases	Non FT cases	Monthly specific rate (%)	OR	95% CI on OR
Jan	6	38	13.6	0.684	0.28 - 1.66
Feb	4	38	9.5	0.305	0.11 - 0.87
Mar	8	36	18.2	0.993	0.45 - 2.19
Apr	4	41	8.9	0.414	0.15 - 1,18
May	4	42	8.7	0.492	0.17 - 1.40
Jun	8	68	10.5	0.489	0.23 - 1.05
Jul	17	60	22.1	1.293	0.73 - 2.30
Aug	19	78	19.6	1.092	0.64 - 1.88
Sept.	14	65	17.7	0.948	0.51 - 1.75
* Oct	18	40	31.0	2.153	1.19 - 3.89
Nov	9	37	19.6	1.082	0.51 - 2.30
* Dec	18	28	39.1	3.145	1.68 - 5.88
Total	129	571			

*significant at p < 0.05.

Table3. Seasonal distribution of FT and other poultry diseases in Zaria, Nigeria.

Season	FT cases	Non FT cases	Season specific rate (%)	OR	95% CI on OR
*Dry (Jan-March)	18	132	12.0	0.539	0.316 - 1.921
*Pre-rainy (Apr-Jun)	16	114	12.3	0.568	0.323 - 0.996
*Rainy (Jul-Sept)	50	117	29.9	2.420	1.610 - 3.638
Pre dry (Oct-Dec	45	208	17.8	0.935	0.627 - 1.395
Total	129	571			

*significant at p < 0.05.

poultry disease revealed that the breed specific rate was high in improved breeds (18.5%) (Table 5).

The poultry species specific rate for FT showed that chicken had the highest rate (18.5%) compared to turkeys (16.7%) (Table 6).

The chicken type specific rate for FT showed that broilers had the highest rate of 25% followed by layers 18.4%. This means that FT was 1.5 and 0.9 times more likely to occur in broilers and layers, respectively (Table 7).

DISCUSSION

The prevalence of FT in this study was 18.4% which was higher than the 8.4% reported in Kaduna by Salami et al. (1989) and 9.4% reported in Jos by Okwori et al. (2007) with 2007 having the highest year specific rate (34.9%). These may be due to increases in the number of back-yard poultry farmers in Zaria that are compounding feeding by themselves in the period under study.

In this work, FT was observed to occur mostly from July to December, the ISMI showed that the disease has three peaks; one continuous peak from June to July followed by July to August another one from October to December. It is important to note that these months fall within

the rainy season in most parts of Nigeria including the study area, this is similar to what was reported by other workers (Calnek, 1995). These months are characterized by wet weather and it is the period of high egg production (Calnek, 1995). Roa (2000) also reported that outbreaks of FT were seen in summer particularly when the weather is wet and moisture is persistant in the air.

The most susceptible age groups to FT in this study were birds >15 weeks, followed by 1 - 5 weeks and 6 - 15 weeks. This observation is similar to the work of Calnek, (1995) who reported that FT is a disease of adult and growing chickens but mortalities have been recorded in young chicks. The outbreak of FT in young chicks may be associated with vaccination against FT practiced by most breeders which leads to vertical transmission of the disease (Jordan and Pattison, 1992; Roa, 2000). In this work, the least susceptible age groups to FT are 6 - 10 weeks and 11 - 15 weeks, this may be due to the fact that in this environment, layers are usually vaccinated against FT at 6 weeks. This may protect them throughout their growing age. It is also important to note that infected chicks remain carriers of the disease (Falade and Ehizokhale, 1981; OIE, 2005) and any stress may trigger an outbreak of FT (Aiello, 1998).

The layers and broilers were at the highest risk of FT than other type of birds; this may be due to the fact that

Table 4. Age distribution of FT and other poultry diseases in Zaria, Nigeria.

Age group	FT cases	Non FT cases	Age group specific rate (%)	OR	95% CI on OR
1-5	8	29	21.6	261	0.56 - 2.83
6-10	2	12	14.3	0.748	0.17 - 3.38
11-15	2	13	13.3	0.689	0.15 - 3.09
* >15	48	117	29.1	2.368	1.57 - 3.58
Unknown	67	402	14.3	0.475	0.31 -0.70
Total	129	571			

*significant at p < 0.05

Table 5. Breed distribution of FT and other poultry diseases in Zaria, Nigeria.

Breed	FT cases	Non FT cases	Specific rate (%)	OR	95% CI on OR
Improved	129	569	18.5	-	-
Local	0	2	0	-	-
Total	129	571			

*significant at p < 0.05.

Table 6. Distribution of FT and other poultry diseases in Zaria, Nigeria based on poultry species.

Poultry species	FT cases	Non FT cases	Specific rate (%)	OR	95% CI on OR
Chicken	127	561	18.5	0.903	0.25 - 3.23
Turkey	2	10	16.7	0.883	0.19 - 4.08
Total	129	571			

*significant at p < 0.05.

Table 7. Distribution of FT and other poultry diseases in Zaria, Nigeria based on type of birds.

Type of bird	FT cases	Non FT cases	Specific rate (%)	OR	95% CI on OR
Layers	124	551	18.4	0.900	0.33 - 2.45
Broilers	4	6	25.0	1.483	0.30 - 7.43
Mixed breed	0	2	0	-	-
*Breeder	1	0	100	-	-
Cockerel	0	2	0	-	-
Total	129	571			

*significant at p <0.05

most backyard and commercial farmers in Zaria raise layers and the stress of egg production depresses their ability to resist infection (Saidu et al., 2004). The broilers are at high risk because most backyard poultry farmers that keep these types of birds for meat do not vaccinate them against FT.

In this study, 98.4% of all the FT cases were recorded in chickens and 1.6% was recorded in turkeys. This observation lends support to the work by Kaupp and Dearstyne, (1924) cited by Calnek, (1995) who reported that turkeys are less susceptible than chickens. The small number of backyard and commercial farmers in Zaria that raise turkeys may be the reason why a low number of FT outbreaks were reported in turkeys. In conclusion this study observed that FT was common in improved breed of birds and occurred in adult birds, and this may be due to the stress of egg production. This study recommends that poultry farmers should be encouraged to practice prompt and regular vaccination of layers. The breeders should not be vaccinated instead regular blood testing should be done to cull and replace infected birds after two consecutive testings.

ACKNOWLEDGEMENT

We thank those who contributed to this study in the

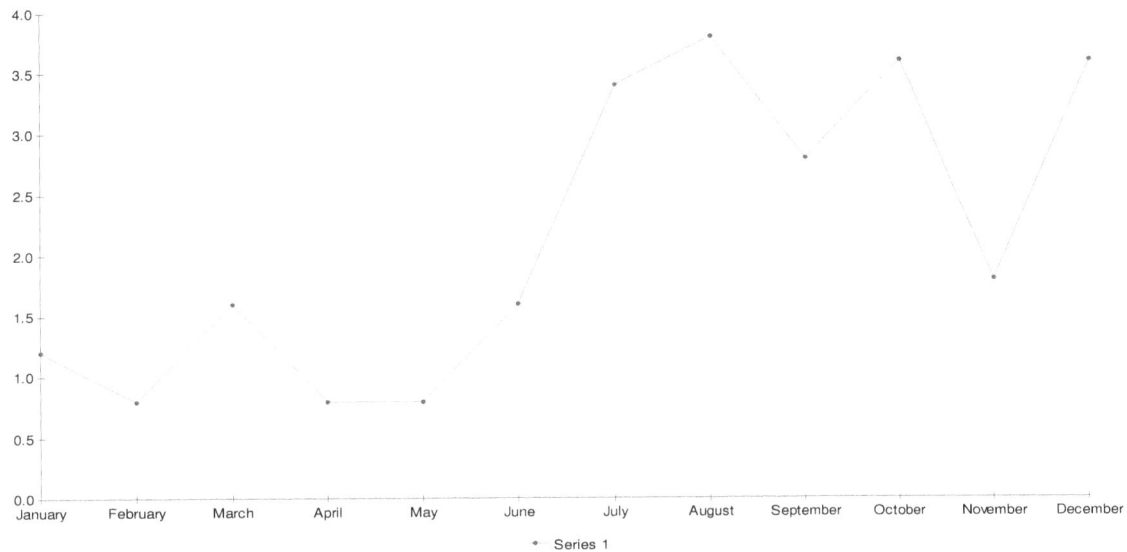

Figure 1. Monthly distribution of fowl typhoid in Zaria, Nigeria.

Faculty of Veterinary Medicine, Ahmadu Bello University, Zaria for their immense support in collection of the data and in the preparation of the manuscript.

REFERENCES

Abdu PA (2007). Manual of Important Poultry disease in Nigeria 2nd Edition. MacChin Multimedia Designers, Zaria pp. 42-47

AbduBDU PA, Mera UM, Sa'idu L (1992). A study of chicken mortality in Zaria, Nigeria, In: Proceeding of National Workshop on Livestock and Veterinary services, Held at National Veterinary Research Institution, Vom, on 11-14 August, 1992 pp. 51-55.

Aiello SE (Ed) (1998). The Merck Veterinary Manual, 8th Edition, Merck & Co., Inc, USA pp. 1995-1996.

Antunes PC, Reu JC, Sousa LP, Pestana N (2003). Incidence of Salmonella from poultry products and their susceptibility to antimicrobial agents. Int. J. Food Microbiol. 82: 97-103.

Arbelot B, Dayon JF, Mamis D, Gueye JC, Tall F, Samb H (1997). Sero survey of dominant avian diseases in Senegal: mycoplasmosis, fowl typhoid and pullorum disease, Newcastle, infectious bursal and infectious bronchitis diseases. Revue d'Elevage et de Medicine Veterinarier des Pays Tropicaux. 50: 197– 203.

Barrow PA, Berchieri JR, Al-haddad OA (1992). The Serological Response of Chickens to Salmonella Gallinarum-Salmoneka Diseases. 36: 227–236.

Calnek BW (Ed) (1995). The Merck Veterinary Manual. 8th Edition. MacChin Multimedia Designer, Zaria pp. 42-47.

Falade S, Ehizokhale M (1981). Salmonella and Escherichia coli strains isolated from poultry in Ibadan, Nigeria. Bulletin Animal Health Production in Africa 29:99.

Halle PD, Umoh JU, Saidu L,Abdu PA (1999). Prevalence and Seasonality of Newcastle disease in Zaria, Nigeria. Tropical Vet. 17: 53-62.Hamid MA, Sharma RN (1990). Epidemiology of fowl typhoid in and around Tropoli (Libya). Bulletin of Animal Health and Production of Animal in Africa. 38: 315-317.

ISO 6579, (1993). Microbiology General Guidance on Methods for the Detection of Salmonella. International Organization of Standardization, Geneva, Switzerland.

Jordan FTW, Pattison M (1992). Poultry Disease 4th Edition. W.B. Sauder Company Ltd London pp 169-171.

Lautrop H, Høiby N, Bremmelgaard A, Korsager B (1979). Bakteriologiske undersøgelsesmetoder (Bacteriological investigations), in Danish. FADL's Forlag, Copenhagen, Denmark.

Mdegela RH, Yongolo MGS, Minga UM, Olsen JE (2000). Molecular epidemiology of Salmonella gallinarum in chickens in Tanzania. Avian Pathology. 29: 5:457-463

Mtei BJ, Msami HM (1996). Reflections on current animal health status in Tanzania. Tanzan.ia Vet.erinary Journal. 16: 37– 49.

Nissen B (1984). Microtest for rapid identification of Enterobacteriaceae. Acta Pathologica, Microbiologica et Immunologica Scandinavica. 92: 239-245.

OIE (2005). Fowl Typhoid. Http:// www.cfsph.iastate.edu. 15/10/08 8.30 pm.

Okoj L (1993). Diseases as important factors affecting increased poultry production in Uganda. Der Tropenlandwirt, Zeitschrift in dentropen und Subtropen Jahrgang. 94: S37–S44.

Okwori AE, Hasimil GA, Adetunji JA, Akaka IO, Junards SA (2007). Serological survey of Salmonella gallinarum antibodies in chicken around Jos, Plateau Sate. Nigerian Online Journal Health Allied.6 (2) www.ojhas.org/issue22/2007-2-2htm, accessed 15/10/2008 5.30 pm.

Roa G (2000). A Comprehensive Textbook on Poultry Pathology. Medical publisher ltd pp. 7-10.

Sa'idu L, Abdu PA, Tekdet LB, Umoh JU (2004). Retrospective Pullorum Detected by Enzyme - Linked Immunosorbent Assay. Avian study of Newcastle disease cases in Zaria, Nigeria. Nigeria Vet.erinary and Journal. 27 (3): 53-62

Saidu L, Abdu PA, Umoh JU, Abdullahi US (1994). Diseases of Indigenous chickens. Bulletin Animal Health Production in Africa 42 :19-23.

Indigenous chickens. Bulletin Animal Health Production in Africa 42: 19-23

Salami JO, Egbulem BN, Kwaga JKP, Yusufu HT, Abdu PA (1989). Disease diagnosed in poultry in Kaduna, Nigeria (1981-1985). Bulletin Animal Health Production in Africa. 37: 109-114.

Sharma RN, Pandey GS, Khan AQ (1991). Salmonella isolation from poultry in the Republic of Zambia. Bulletin of Animal in Health and Production in Africa. 39: 173-175.

Diarrheagenic *Escherichia coli* (DEC): prevalence among in and ambulatory patients and susceptibility to antimicrobial chemotherapeutic agents

O. J. Akinjogunla[1]*, N. O. Eghafona[2], and O. H. Ekoi[1]

[1]Department of Microbiology, Faculty of Sciences, University of Uyo, P.M.B 1017, Uyo, Akwa Ibom State, Nigeria.
[2]Department of Microbiology, Faculty of Life Sciences, University of Benin, Benin City, Edo State, Nigeria.

The prevalence of diarrheagenic *Escherichia coli* both in an ambulatory patients passing out loose stools with or without blood and/or mucus in Anua General Hospital, University of Uyo Teaching Hospital and University of Uyo Health Centre from June to September, 2008 were determined using standard microbiological techniques. Susceptibility to seven different conventional and commonly available chemotherapeutic drugs/antibiotics: ampicillin, chloramphenicol, ciprofloxacin, gentamycin, tetracycline, cephalothin and ofloxacin were assessed using a disc diffusion technique (DDT). The macroscopic analysis of the stool samples showed that 31 of the 100 cases (31%) were diarrhea bloody and 33% mucoid. Sixty-nine diarrheagenic *E. coli* were isolated from 100 stool samples collected and were more prevalent in females (69.4%) than in males (30.6%). The observed percentage prevalence of diarrheagenic *E. coli* among the age groups (in years) 1 -15, 16 - 30, 31 - 45, 46 - 60 and 61 and above were 95, 80, 55, 70 and 45%, respectively. The results of antibiotic susceptibility showed that the *E. coli* were highly resistant to ampicillin (73.9%), tetracycline (75.4%) and gentamycin (68.1%), and moderately resistant to chloramphenicol (46.4 %) and cephalothin (43.5%), but highly sensitive to ciprofloxacin (71.0%) and ofloxacin (66.7%). The findings of this study showed ciprofloxacin and ofloxacin to be drugs of choice for the treatment of diarrheagenic *E. coli,* while ampicillin, tetracycline and gentamycin should not be used without first performing culture and sensitivity tests.

Key words: Diarrheagenic, *Escherichia coli,* prevalence, chemotherapy, susceptibility.

INTRODUCTION

Escherichia coli are common members of the normal flora of the human intestine (Nataro and Kaper, 1998; Yah et al., 2006). Strains of *E. coli* that acquire bacteriophage, plasmid DNA encoding enterotoxin or invasion factors become virulent. This virulence increases the ability of *E. coli* to adapt to new niches to cause a broad spectrum of diseases such as urinary tract infections and nosocomial infections resulting in either a plain, watery diarrhea or inflammatory dysentery. *E. coli* are prominent members of Enterobacteriaceae and are widely distributed in nature, they are present in the intestinal tract of man and animals, and in water and soil (Nataro et al., 1987; Smith et al., 2003). Diarrhea caused by *E. coli*

infection is one of the major public health concerns in many developing countries and has contributed exceedingly to morbidity and mortality, and also the associated increase in health costs (Adachi et al., 2001; Ogata et al., 2002; Robins-Browne and Hartland, 2002). *E. coli* have also been reported to be the leading cause of diarrhea-causing diseases in addition to bacterial pathogens such as *Salmonella* spp, *Shigella* spp, *Yersinia* spp, *Vibrio* spp, *Campylobacter* spp, *Enterobacter* spp, *Citrobacter* spp, *Proteus* spp, and parasitic pathogens such as *Entamoeba histolytica,* and *Giardia lamblia* in developing countries (Su and Brandt 1995; Smith et al., 2003; Prescott et al., 2008). Individuals who are debilitated or have other predisposing factors are at a much higher risk of infection than healthy persons. Strains of *E. coli* can be classified as commensal, intestinal pathogenic or extra intestinal pathogenic *E. coli* (ExPEC). *E. coli* pathotypes responsi-

*Corresponding author. E-mail: papajyde2000@yahoo.com.

responsible for intestinal infections are enteroaggregative *E. coli* (EAEC), enterohaemorrhagic *E. coli* (EHEC), enteroinvasive *E. coli (EIEC)*, enteropathogenic *E. coli* (EPEC), or enterotoxigenic *E. coli* (ETEC) (Rademaker et al., 1993; Paton and Paton 1998; Yah et al., 2006). Consumption of faecally contaminated water is an important route of transmission of diarrheagenic pathogens especially *E. coli,* in many regions of the world lacking infrastructure to guarantee water quality and safe management of human waste (Swerdlow et al., 1992). *E. coli* is an important opportunistic pathogen that has shown an increasing antimicrobial resistance to most antibiotics (Winokur et al., 2001; Miranda et al., 2004; Poppe et al., 2005). Antimicrobial resistance of *E. coli* has played an important role in clinical infectious diseases (Winokur et al., 2001). Thus, the aim of this investigation was to determine the prevalence of diarrheagenic *E. coli* both in ambulatory patients in Uyo City and assess their susceptibility to different conventional and commonly available chemotherapeutic agents/antibiotics.

MATERIALS AND METHODS

Collection and processing of samples

Stool samples from diarrheagenic patients (patients passing out at least three loose stools in a 24 h period accompanied by symptoms such as nausea and/or abdominal cramp and/or fever (>38°C) were collected between July and September, 2008 for a prospective study of three different hospitals in Uyo City: Anua General Hospital, University of Uyo Teaching Hospital and University of Uyo Health Centre. Stool samples from patients who had not received antibiotic treatment at the time of investigation were collected aseptically using clean, sterile wide-mouth containers and taken to the Microbiology Laboratory of University of Uyo for bacterial analyses within 1 - 4 h of collection. Stool samples that could not be analyzed immediately were refrigerated at 4°C for less than 24 h. Characterization and identification of *E. coli* cultures were made on the basis of morphology, cultural characteristics, and biochemical reactions. All the stool samples were cultured into MacConkey agar (MCA) for primary isolation of common intestinal pathogens and incubated at 37°C for 24 h. All colonies on MacConkey agar plates suspected to be *E. coli* (lactose fermenter, non-mucoid, 2 - 3 mm diameter, circular, smooth and convex) were streaked on Eosin Methylene Blue (EMB) agar and incubated at 37°C for 24 h. Green metallic sheen colonies positive for *E. coli* were further sub-cultured onto nutrient agar and incubated for another 24 h. The cultures on nutrient agar plates were subjected to Gram's-staining, motility, urease production, glucose, oxidase, sucrose, mannitol, lactose, indole, Voges proskauer, and citrate utilization tests. All Gram-negative, rod-shaped, motile, indole-negative, urease-negative isolates that produced acid on Triple Sugar Iron agar slants were identified as species of the genus *E. coli* with reference to Cowan and Steel (1985); Fawole and Oso (1988); Cheesbrough (2004).

Antibiotic sensitivity testing

In vitro susceptibility of the *E. coli* to seven different antibiotics was determined using a disk-diffusion technique (NCCLS, 2004). Sterile Petri dishes of Mueller Hinton agar were prepared according to the manufacturer's specification. 0.1 ml of *E. coli* was seeded into each of the Petri dishes containing Mueller-Hinton agar and were allowed to stand for 45 min to enable the inoculated organisms to pre-

diffuse. The commercially available discs containing the following antibiotics: gentamycin (Gen, 10 □g), ofloxacin (Ofl, 30 □g), ampicillin (Amp, 10 □g), tetracycline (Tet, 30 □g), cephalothin (Cep, 30 □g), ciprofloxacin (Cip, 5 □g), chloramphenicol (Crp, 30 □g) (Oxoid, UK) were aseptically placed on the surfaces of the sensitivity agar plates and these were incubated for 18 - 24 h at 37°C. Zones of inhibition after incubation were observed and the diameters of inhibition zones were measured in millimeters. The interpretation of the measurement as sensitive, intermediate and resistant was made according to the manufacturer's standard zone size interpretive manual which were as follows: ofloxacin (S ≥ 21, I =16 - 20 and R ≤ 15), ciprofloxacin (S ≥ 21, I = 16 - 20 and R ≤ 15), gentamicin (S ≥ 15, I = 13 - 14 and R ≤ 12), ampicillin (S ≥ 17, I = 14 - 16 and R ≤ 13), chloramphenicol (S ≤ 18, I = 13 - 17 and R ≥ 12), cephalothin (S ≥ 18, I =15 - 17 and R ≤ 14) and tetracycline (S ≥ 19, I = 15 - 18 and R ≤ 4) where S = sensitivity, I = intermediate and R = resistance.

The intermediate readings were considered as sensitive for the assessment of the data. The choice of the above antibiotics used was based on local availability.

RESULTS

A total of sixty-nine (69) diarrheagenic *E. coli* were isolated from 100 stool samples collected from three different hospitals which translated to 69% of all samples being positive for *E. coli* during the study period (Tables 1 and 2). The macroscopic analysis of the stool samples showed that 31 of the 100 cases (31%) were diarrhea bloody and 33% mucoid. Forty-four of the subjects were male and 56% were female. Table 3 shows that diarrheagenic *E. coli* were more prevalent in females (69.6%) than in males (30.4%) and the observed percentage prevalence of diarrheagenic *E. coli* among the age groups (in years) 1 - 15, 16 - 30, 31 - 45, 46 - 60 and 61 and above were 95, 80, 55, 70 and 45%, respectively (Table 2).

The antimicrobial sensitivity tests of diarrheagenic *E. coli* to seven antibiotics by a disc diffusion method is shown in Tables 4 and 5. The results showed that 66.7 - 71.0% of diarrheagenic *E. coli* were found to be highly sensitive against ofloxacin (≥15 mm diameter) and ciprofloxacin (≥16 mm diameter), while 43.5 and 46.4% were found to be moderately resistant against cephalothin (≤18 mm diameter), and chloramphenicol (≤12 mm diameter), respectively. The results of the antibiotic susceptibility also showed most of the isolates to be highly resistant to the antibiotics ampicillin (73.9%), tetracycline (75.4%) and gentamycin (68.1%) with inhibitory zones ranging from 0 to ≥10 mm. Multidrug resistance (≥2 - 6) to antibiotics was observed in 49 cases (71.0%) and the major resistance profile was ampicillin-gentamycin- tetracycline.

DISCUSSION

E. coli infection is one of the major public health problems in many developing countries and has contributed exceedingly to morbidity, mortality and increased health costs (Adachi et al., 2001; Ogata et al., 2002; Robins-

Table 1. Macroscopic and diarrhea status of *Escherichia coli.*

Nature of Stool Sample	No. Positive for *E. coli*	No. Negative for *E. coli*	Total
Watery	33	16	49
Watery + Bloody	11	07	18
Watery + Mucoid	16	04	20
Watery + Bloody + Mucoid	09	04	13
Total	69	31	100

Table 2. Age-specific prevalence of diarrheagenic *Escherichia coli.*

Age (Years)	No. of Samples	Occurrence of *E. coli* Isolated	Percentage of Isolates
0-15	20	19	95
16-30	20	16	80
31-45	20	11	55
46-60	20	14	70
61 and above	20	09	45
Total	100	69	69

Table 3. Sex-specific prevalence of diarrheagenic *Escherichia coli.*

Sex	No. of Samples	No. of Isolates	Percentage of Isolates
Male	44	21	30.4
Female	56	48	69.6
Total	100	69	100.0

Browne and Hartland, 2002)

Our results reveal the prevalence of diarrheagenic *E. coli* among the age groups (in years) 1 - 15, 16 - 30, 31 - 45, 46 - 60 and 61 and above to be 95, 80, 55, 70 and 45%, respectively. The prevalence of diarrheagenic *E. coli* among females was greater than that of males and these results are in conformity with those obtained by Diame et al. (1990) and Lothar et al. (1998).

In recent years, antibiotic resistance of diarrheagenic pathogens has reached alarming proportions worldwide. The misuse of antibiotics has been found to be the most important selecting force in bacterial antibiotic resistance (Okeke et al., 1999; Yah and Eghafona, 2007; Akinjogunla et al., 2008). Antibiotic resistance of *E. coli* to ampicillin, gentamycin and chloramphenicol recorded in this study are similar to those obtained by Okeke et al. (2000); Okoli et al. (2002). The resistance of some diarrheagenic *E. coli* in this study to at least one of the seven antibiotics tested especially tetracycline, ampicillin and gentamycin could be a result of the routine and uncontrolled use in patients. Also, regarding the sensitivity pattern, this study showed that ofloxacin and ciprofloxacin were effective against diarrheagenic *E. coli* and is in agreement with previous findings (Yah and Eghafona, 2007). The low level of resistance of these quinolones may be because they are relatively new antibiotics and

are also more expensive than tetracycline, ampicillin and gentamycin. There is evidence indicating that tetracycline survives longer in the environment than other antibiotics which may be critical in maintaining the level of tetracycline resistance at a high level (Yah and Eghafona, 2007)

Monitoring drug resistance patterns of *E. coli* will give vital clues to clinicians regarding therapeutic regimes to be adopted against individual cases and will be an important tool to devise a comprehensive chemo-prophylaxis.

Conclusion

The development of new antibiotics may offer a short-term solution to the problem of resistance among diarrheagenic bacteria especially *E. coli* but more effective measures, such as health education and further research on the prevention of infections through quality sanitation, should be emphasized.

ACKNOWLEDGEMENTS

The authors gratefully acknowledge the expertise of staff of the Department of Medical Microbiology and Parasitology, University of Uyo Teaching Hospital, University of Uyo Health Centre and Anua General Hospital for their immense co- operation and assistance in the collection of

Table 4. Occurrence and percentage of antibiotic resistant diarrheagenic *Escherichia coli*.

Hospital	No. of Samples Collected	No. of E. coli Isolated	No. (%) Resistant to CIP	No. (%) Resistant to GEN	No. (%) Resistant to CRP	No. (%) Resistant to AMP	No. (%) Resistant to OFL	No. (%) Resistant to TET	No. (%) Resistant to CEP
AGH	33	22	6 (27.3)	14 (63.6)	10 (45.5)	16 (72.7)	7 (31.8)	17 (77.3)	11 (50.0)
UUTH	40	29	9 (31.0)	20 (68.9)	14 (48.3)	22 (75.9)	11 (37.9)	20 (69.0)	11 (37.9)
UUHC	27	18	5 (27.8)	13 (72.2)	8 (44.4)	13 (72.2)	5 (27.8)	15 (83.3)	8 (44.4)
Total	100	69	20 (30.0)	47 (68.1)	32 (46.4)	51 (73.9)	23 (33.3)	52 (75.4)	30 (43.5)

Keys: AGH: Anua General Hospital; UUTH: University of Uyo Teaching Hospital; UUHC: University of Uyo Health Centre; CIP: Ciprofloxacin; GEN: Gentamycin; CRP: Chloramphenicol; AMP: Ampicillin; OFL: Ofloxacin; TET: Tetracycline; CEP: Cephalothin.

Table 5. Occurrence and percentage of antibiotic sensitive diarrheagenic *Escherichia coli*.

Hospital	No. of Samples Collected	No. of E. coli Isolated	No. (%) Sensitive to CIP	No. (%) Sensitive to GEN	No. (%) Sensitive to CRP	No. (%) Sensitive to AMP	No. (%) Sensitive to OFL	No. (%) Sensitive to TET	No. (%) Sensitive to CEP
AGH	33	22	16 (72.7)	8 (36.4)	12 (54.5)	6 (27.3)	15 (68.2)	5 (22.7)	11 (50.0)
UUTH	40	29	20 (69.0)	9 (31.0)	15 (51.7)	7 (24.1)	18 (62.1)	9 (31.0)	18 (62.1)
UUHC	27	18	13 (72.2)	5 (27.8)	10 (55.6)	5 (27.8)	13 (72.2)	3 (16.7)	10 (55.6)
Total	100	69	49 (71.0)	22 (31.9)	37 (53.6)	18 (26.1)	46 (66.7)	17 (24.6)	39 (56.5)

Keys: AGH: Anua General Hospital; UUTH: University of Uyo Teaching Hospital; UUHC: University of Uyo Health Centre; CIP: Ciprofloxacin; GEN: Gentamycin; CRP: Chloramphenicol; AMP: Ampicillin; OFL: Ofloxacin; TET: Tetracycline; CEP: Cephalothin.

of specimens

REFERENCES

Adachi JA, Jiang ZD, Mathewson JJ, Verenkar MP, Thompson S, Martinez-Sandoval F, DuPont HL (2001).Enteroaggregative *Escherichia coli* as a major etiologic agent in traveler's diarrhea in 3 regions of the world. Clin. Infect. Dis. 32: 1706–1709.

Akinjogunla OJ, Inyang CU, Ekoi OH, Ntinya AJ (2008).Prevalence and Antibiogram of *Salmonella typhi and Salmonella paratyphi* in Stools of Symptomatic Typhoid Patients. Int. J. Biotechnol. Allied Sci. 3(1):377-382

Cheesbrough M (2004). District Laboratory Practice in Tropical Countries (Part II). Cambridge University pp. 50-120

Cowan ST, Steel KJ (1985). Manual for the Identification of Medical Bacteria. (4th Edition). Cambridge University Press. London. p.217

Diame EM, Ndiaye S, Airey P (1990). Diarrhoeal Morbidity Among Young Children: Findings from the DHS survey of Senegal 1986, in Hill Allen G. (2nd edition .Determinants of health and mortality in Africa, DHS Surveys, Further Analysis Series pp. 47-72.

Fawole MO, Oso BA (1988). Laboratory Manual for Microbiology (1st Edition), Spectrum Book Ltd, Ibadan. pp 22-45.

Lothar B, Sonja Z, Kerstin G (1998). Human Infections with Shiga Toxin- Producing *Escherichia coli* Other Than Serogroup O157 in Germany. Emerg. Infect. Dis.4 (4).1-10

Miranda S, David MG, Peter JC (2004). Evolution of multi-resistance plasmids in Australia clinical isolates of *Escheri-chia coli*. Microbiol. 150:1539-1546.

Nataro JP, Kaper JB, Robins-Browne R, Prado V, Vial P, Levine MM (1987). Patterns of adherence of diarrheagenic *Escherichia coli* to HEp-2 cells. Pediatr. Infect. Dis J. 6: 829-831

Nataro JP, Kaper JB. (1998). Diarrheagenic *Escherichia coli*. Clin. Microbiol. Rev. 112: 142-201.

NCCLS – National Committee for Clinical Laboratory Standards (2004).Performance standards for Antimicrobial susceptibility testing. Fourteenth informational supplemented. M100-S14, Wayne, PA, USA.

Ogata K, Kato R, Ito K, Yamada S (2002). Prevalence of *Escherichia coli* possessing the eaeA gene of enteropatho-genic E. coli (EPEC) or the aggR gene of enteroaggregative E. coli (EAggEC) in traveler's diarrhoea diagnosed in those returning to Tama, Tokyo from other Asian countries. Jpn J.

Infect. Dis. 55: 14–18.

Okeke IN, Fayinka ST, Lamikanra A (2000). Antibiotic Resistance in *E. coli* from Nigerian Students, 1986-1998. Emerg. Infect. Dis. 6:393-396.

Okeke IN, Lamikanra A, Edelman R (1999). Socio-economic and behavioral factors leading to acquired bacterial resistance to antibiotics in developing countries. Emerg. Infect. Dis. 5: 13-27.

Okoli IC, Okeudo NJ, Onwuchekwa CI (2002). New trends in antibiotic resistance among *E. coli* isolates from southern Nigeria. In Book *of Abstracts for the 39th Annual National Congress*, Nigerian Veterinary Medical. Association 27th to 31st, October, Sokoto, Nigeria. p. 16.

Paton JC, Paton AW (1998). Pathogenesis and diagnosis of shiga-toxin producing *Escherichia coli* infections. Clin. Microbiol. Rev. 11: 450-479.

Poppe C, Martin LC, Gyles CL, Reid-Smith R, Boerlin P, McEwen SA, Prescott JF, Forward KR (2005). Acquisition of resistance to extended spectrum cephalosporin by *Salmonella enterica* subsp. *Enterica serovar Newport* and *Escherichia coli* in Turkey Poultry Intestinal Tract. Appl. Environ. Microbiol. 71: 1184- 1192.

Prescott LM, Harley JP, Klein DA. (2008). In: Microbiology 7th edition, pp. 1-12

Rademaker CMA, Martinez-Martinez L, Perea EJ, Jansze M, Fluit AC, Glerum JH, Verhoef J. (1993). Detection of enterovirulent *Escherichia coli* associated with diarrhea in Seville, southern Spain, with non-radioactive DNA probes. J. Med. Microbiol. 38: 87-89

Robins-Browne RM (1987). Traditional enteropathogenic *Escherichia coli* of infantile diarrhoea. Rev. Infect. Dis. 9: 28–53

Smith SI, Aboaba, OO, Odeigha P, Shodyo K, Adeyeye JA, Ibrahim A, Adebiyi T, Onibokum H, Odunukwe NN (2003). Plasmids profiles *E*

coli from apparently healthy animals. Afr. J. Biotechnol. 2(9): 322-324.

Su C, Brandt LJ (1995). *Escherichia coli* O157:H7 infection in humans. Ann. Intern. Med. 123:698-714

Swerdlow DL, Woodruff BA, Brady RC, Griffin PM, Tippen S, Donnell HD, Geldreich E, Payne BJ, Meyer A, Wells JG (1992). A waterborne outbreak in Missouri of *Escherichia coli* O157:H7 associated with bloody diarrhoea and death. Ann. Intern. Med. 117:812-819.

Winokur PL, Vonstein DL, Hoffman LJ, Uhlenhopp EK, Doern GV (2001). Evidence for transfer of CMY-2 AmpC- lactamase plasmids between *E. coli* and Salmonella isolates food, animals and humans. Antimicrobial Agents and Chemotherapy. 45(10): 2716-2722.

Yah CS, Eghafona NO (2007).Plasmids: A Vehicle For Rapid Transfer of Antibiotic Resistance Markers Of Salmonella Species In Animals. J. Am. Sci. 3(4) 86-92

Yah SC, Eghafona NO, Enabulele IO, Aluyi HAS (2006). Ampicillin Usage and Ampicillin Resistant (Ampr) Plasmids Mediated *Escherichia Coli* Isolated from Diarrheagenic Patients Attending Some Teaching Hospital in Nigeria. Shiraz E-Medical J. 7(4):1-12

Intestinal helminth infection and anaemia during pregnancy: A community based study in Ghana

S. E. Baidoo[1,2], S. C. K. Tay[1], K. Obiri-Danso[3] and H. H. Abruquah[4]*

[1]Department of Clinical Microbiology, School of Medical Sciences, Kwame Nkrumah University of Science and Technology, Kumasi, Ghana.
[2]Department of Laboratory Technology, University of Cape Coast, Cape Coast, Ghana.
[3]Department of Theoretical and Applied Biology, Kwame Nkrumah University of Science and Technology, Kumasi, Ghana.
[4]Division of Microbiology and Infectious Diseases, University Hospital, Kwame Nkrumah University of Science and Technology, Kumasi, Ghana.

Intestinal helminths are among the most common and widespread of human infections, contributing to poor nutritional status, anaemia and impaired growth. Anaemia and iron deficiency in pregnancy is a major public health problem in developing countries, but their causes are not always known. The objective of this study was to assess the prevalence and severity of anaemia and iron deficiency and their association with helminths, among pregnant women in the Ashanti region of Ghana. A cohort study was carried out in the Sekyere, East district of the Ashanti region of Ghana. One hundred and eight (108) pregnant women were followed until 5-10 weeks postpartum, during the period of (December 2005 - November 2006). Haemoglobin and total serum iron concentrations were evaluated in venous blood samples, and helminths infections evaluated in stool samples in each trimester using standard methods. Most of the 108 pregnant women, 54.9% were found to be anaemic. The highest prevalence of anaemia and low iron stores (57.4 and 32.4%, respectively) were found in the second trimester. Only 17.6% had evidence of helminths infection, with *Necator americanus* (hookworm) being the commonest (13.9%). There was a significant association between hookworm infection and low iron stores. The study concluded that hookworm infection is a strong predictor of iron status. These findings reinforce the need to provide anthelminthic therapy to infected women before conception as a public health strategy in reducing the prevalence of hookworm infection, and in addition to providing nutritional and iron supplements to effectively control anaemia in pregnancy.

Key words: Intestinal helminth, anaemia, paragravids, primigravids.

INTRODUCTION

Intestinal helminthes are among the most common and widespread of human infections. They contribute to poor nutritional status, anaemia and impaired growth in children of school going age (Dickson et al., 2000. Epidemiological surveys have revealed that, poor sanitation and inappropriate environmental conditions coupled with indiscriminate defaecation, geophagy, and contamination of water bodies are the most important predisposing factors to intestinal worm infestation (Brooker et al., 2008). The prevalence and intensity of infection is especially high in developing countries, particularly among populations with poor environmental sanitation (van Eijk et al., 2009). Other practices such as hand washing, disposal of refuse, personal hygiene, wearing of shoes and others, when not done properly may contribute to the infection or picking of these worms from the environment (Stoltzfus et al.,1997).

*Corresponding author. E-mail: kowabruquah@yahoo.co.uk.

Anaemia denotes the complex of signs and symptoms associated with low haemoglobin levels. Globally, the most common cause of anaemia is believed to be iron deficiency due to inadequate dietary iron intake, physiologic demands of pregnancy and rapid growth and iron losses due to parasitic infections (Ayoya et al., 2006; Dreyfuss et al., 2000). However, iron deficiency is not the only cause of anaemia. Other prevalent causes of anaemia include malaria, chronic infections and nutritional deficiencies of vitamin A, folate and vitamin B-12 (Dreyfuss et al., 2000). The relative contributions of these causes of anaemia and iron deficiency varies by sex, age and population and are not well described in many populations (Dreyfuss et al., 2000).

Iron deficiency is responsible for 95% of cases of anaemia during pregnancy (Porter and Kaplan, 1999) which is often due to inadequate dietary intake (especially in teenage girls), to a previous pregnancy or to the normal loss of iron in menstruation (Porter and Kaplan, 1999).

Impact of intestinal helminth infections on anaemia during pregnancy is aggravated by low nutritional status of subjects whose staple foods, such as rice, cassava, and maize are poor sources of folate, and iron (Ayoya et al., 2006; Pasricha et al., 2008). The most important cause of pathological chronic loss of blood and iron in the tropics is hookworm and other soil-transmitted helminthes (Brooker et al., 2008), and malaria in pregnancy (Fleming, 1982). At a hospital in Kathmandu, Nepal, hookworm infection was associated with severe but not moderate anaemia among women receiving antenatal care (Bondevik et al., 2000). Data on the epidemiology of iron deficiency, anaemia in East Africa and elsewhere point to the important contribution of hookworms to this condition (Ayoya et al., 2006; Stoltzfus et al., 1997). Hookworm infection has been established as a strong predictor of iron deficiency and anaemia in other populations (Ayoya et al., 2006; Stoltzfus et al., 1997), but few studies have examined their relationships in pregnant women (WHO, 1993).

The occurrence of helminth infection at high rates among pregnant women is mostly indicative of faecal pollution of soil and domestic water supply around homes due to poor sanitation and improper sewage disposal (Bundy et al., 1995; van Eijk et al., 2009). Pregnant women are at high risk of infection because of their close relationship with children (Bundy et al., 1995; van Eijk et al., 2009). Most of these worms are transmitted through the soil whilst, the practice of soil- eating (geophagy) is common amongst pregnant women in many communities in developing countries (Brooker et al., 2008). It is in this vain that, this study investigates the prevalence and some clinical effects of anaemia in pregnant women infected with intestinal helminthes in the Sekyere East district of the Ashanti region of Ghana. This would help largely in the management of intestinal worm burden interfering with iron stores in pregnant mothers.

MATERIALS AND METHODS

Setting

Subjects were recruited from the Sekyere East district of the Ashanti region with the Effiduase hospital at the base laboratory. The examination center was centrally located in the study area and represented the general characteristics of the district. This analysis includes data collected from December, 2005 - November, 2006 from pregnant women attending antenatal clinics in the clinic sub study area.

Subject selection

The study population involved pregnant women of ages between 12 - 45 years who participated in a randomized community intervention. Three hundred eligible pregnant women who visited the clinic had pregnancies confirmed with a β-human chorionic gonadotropin urine test. Persons were excluded from the study if the initial ANC screening detected that they had any haemoglobinopathy such as G6PD or sickle cell diseases.

Sample collection and analysis

Blood was collected via venipuncture. Anaemia and Iron status were assessed with haemoglobin (Hb) and total serum iron (TSI) or ferritin concentrations. Morphological study of red cells was carried out as a confirmatory test for iron deficiency anaemia (Leishman staining). Hb was measured with a Haemocue hemoglobinometer (Mission Viejo, CA). Serum ferritin (TSI) was assessed with a fluorometric immunoassay (Delfia System; Wallac, Gaithersburg, MD). The formol-ether concentration method was used in the preparation of stool samples for microscopy and detection of helminthes.

Data analysis

Firstly, estimates of the severity of anaemia, defined by Hb concentration, and of iron deficiency, defined by TSI or ferritin concentration were examined as a cause of anaemia. Anaemia was defined as Hb <10 g/dl and differentiated as Hb <10 g/dl – 'severe anaemia'; Hb <11 – 11.5 g/dL – 'moderate anaemia' and Hb 12 – 15.2 g/dL - 'normal'. Persons with TSI or serum ferritin <60 μg/dL were also described as 'anaemic' (Porter and Kaplan, 1999). In the morphological examination of iron deficiency anaemia, the morphology of the red cells should be microcytic, hypochromic with anisocytosis and poikilocytosis (Porter and Kaplan, 1999). Gestation age and gravidity were also retained in all models because iron status and haemoglobin concentration is known to be strongly associated with gestation age and gravidity (Porter and Kaplan, 1999). Data was analyzed using SPSS (version 11.0; SPSS Inc, Chicago), while statistical significance was defined as a p-value of <0.05. The Pearson Chi – square test was used to examine the association between both total serum iron and haemoglobin concentrations and helminth infection as indicators of anaemia.

RESULTS

Most of the 108 women recruited into the study, 66 (61.1%) were paragravids and 42 (39.9%) primigravids. Clinical data obtained within the first month of recruitment show that 27.8% of these women were in their first

Table 1. Total serum iron (TSI) concentration of study subjects as distributed over the trimesters of pregnancy. Using TSI 60 persons were categorized as being anaemic as compared with 59 using Hb levels

| Gestation | Serum iron concentration | | Total |
	< 60 ug/dl (Anaemic, %)	> 60 ug/dl (Normal)	
1st Trimester	17 (15.7)	13	30
2ndTrimester	35 (32.4)	27	62
3rd Trimester	8 (7.4)	8	16
Total	60 (55.6)	48	108

trimester, while 57.4 and 14.8% were in their second and third trimesters respectively. Of the paragravids 15 (13.8%), 39 (36.1%), and 12 (11.1%) were in their first, second and third trimester, respectively; and 15 (13.8%), 23 (21.3%) and 4 (3.7%) of primigravids were also in their first, second and third trimesters, respectively.

Prevalence of anaemia and iron deficiency

Anaemia was present in more than 50% of the pregnant women studied (Table 1). Iron deficiency appeared to be the dominant cause of anaemia, especially, moderate to severe anaemia in these communities with 60 subjects having TSI concentrations of <60 μg/dl while 59 persons had Hb levels < 12 g/dl (55.6 and 54.6%, respectively, Figure 1).

Helminth infection

Most of the 108 pregnant women, only 19 or 17.6% were infected with helminthes; the commonest being *Necator americanus* (hookworm) infection (15 or 13.9%) followed by *Strongloides stercoralis* (2 or 1.9%), *Ascaris lumbricoides* (1 or 0.9%) and *T. trichiura* (1 or 0.9%)

All 19 women infected with at least one species of pathogenic intestinal helminthes were found to be moderate to severe anaemia. All 15 pregnant women who were infected with hookworm were found to be anaemic (Hb <10.0 g/dL and/or TSI < 20 μg/dL), with 13 of them being severely anaemic (Figure 2). Hookworm infection had a strong association with all two indicators of anaemia and iron deficiency (p= 0.001 and 0.00 for TSI and Hb respectively).

DISCUSSION

In developing countries, most pregnant women generally become anaemic and this is presumed to be primarily, as a result of iron deficiency (WHO, 1996). However, the definition and identification of iron deficiency has been a problem especially in situations where there are multifactorial causes of anaemia. Nevertheless,

hookworm has been established as a major predictor for iron deficiency anaemia (van den Broek, 1996; van den Broek et al., 1998). It is also known that poverty, ignorance, geophagy, promiscuous defaecation and poor personal hygiene and environmental sanitation also predispose humans to hookworm infection (Larry and Janovy Jnr, 1996).

The 108 pregnant women studied, 55.6% of them were classified as anaemic, which although is on the high side, it is not uncharacteristic of the community setting of which the study was undertaken (Jans et al., 2008; Pasricha et al., 2008). In a recent study in Netherlands, the prevalence of anaemia in pregnancy was found to be 3.4% (Jans et al., 2008). Despite the discrepancies in the number of study subjects and the study areas of the two studies, ethnicity played a major role in defining the risk of being anaemic in pregnancy, with women of non-northern European descent, there are more at risk than their northern European counterparts (Jans et al., 2008). In similar studies in India and Nepal, the percentages of pregnant women who were anaemic were 88 and 81, respectively, (Agarwal et al., 1987; WHO, 1998). This affirms the predominance of predisposing factors of anaemia within rural settings of developing countries. In our part of the world, anaemia may also be caused by malaria, HIV and poor nutrition (Ayoya et al., 2006; Muhangi et al., 2007).

Iron deficiency appeared to be the dominant cause of anaemia in these communities as defined by TSI concentration and Hb levels (55.6 and 54.6%, respectively) and although only 19 of the 60 anaemic women (31.7%) had helminthes infection, iron deficiency was found to be significantly associated with hookworm infection (p = 0.001).

It was encouraging to note that, only 17.6% were infected with helminths in this study, the commonest hookworm (13.9%), compared with a similar study in western Kenya (van Eijk et al., 2009). However, since this study is an epidemiological one and the subjects were asymptomatic, any helminth ova or larvae present would be in very low intensity and possibly undetectable (Ayoya et al., 2006; Bundy et al., 1995; van Eijk et al., 2009).

Hookworm infection has been established as a strong predictor of iron deficiency and anaemia in other populations (Bundy et al., 1995; Hopkins et al., 1997;

Distribution of Helminth Infestation

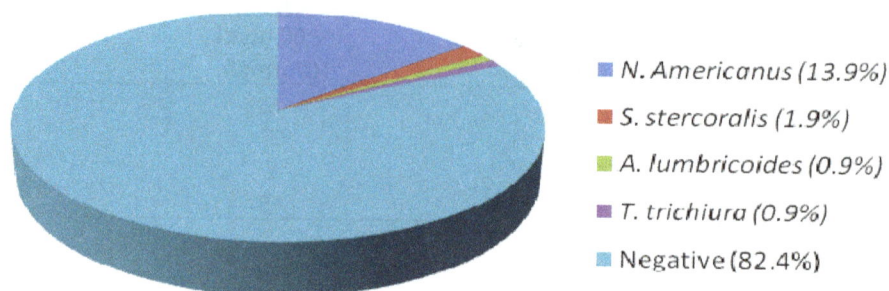

Figure 1. Different species of helminthes identified among the pregnant women studied with *Necator americanus* (hookworm) being the most common.

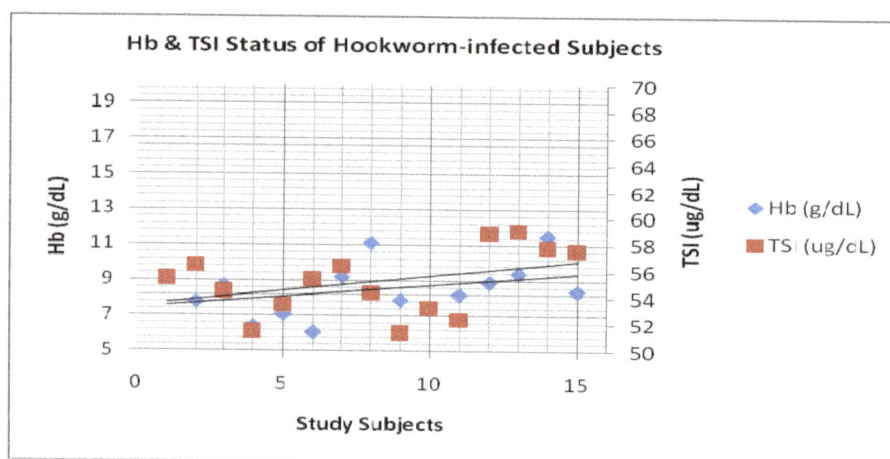

Figure 2. Haemoglobin and TSI levels in 15 pregnant women who had hookworm infection. All 15 were found to have Hb levels and TSI concentrations below normal, with 13 of them being severely anaemic.

Stoltzfus et al., 1997), and few studies have examined these relationships in pregnant women. In this study, hookworm infection was associated with severe anaemia (95% CI 0.2-0.8) which conformed to published data (Ayoya et al., 2006; Bondevik et al., 2000; Brooker et al., 2008; van Eijk et al., 2009).

A single course of anthelminthic therapy in addition to iron-folate supplementation significantly increased hemoglobin concentrations and improved iron status (serum ferritin and EP) in pregnant Sri Lankan plantation workers, suggesting that hookworm infection caused iron deficiency anaemia in that population (Atukorala et al., 1994). However, allocation to anthelminthic therapy was nonrandom and the prevalence and intensity of hookworm infection were not assessed.

The pathenogenicity of hookworm infection shows that the disease manifests in three main phases, with the intestinal phase representing the most important period.

A moderate hookworm infection according to studies, will gradually produce iron deficiency anaemia as the body reserves of iron are used up, with the severity depending on the worm load and the dietary intake of iron (Larry and Janovy Jnr, 1996).

In sub-Saharan Africa, up to 24 million women may become pregnant each year (Bundy et al., 1995). This high rate is often related to an increase in the susceptibility of pregnant women to infections because pregnancy is a time of high hormone activity which may exert immuno-suppressive effects on the child bearing woman (Beer and Billingham, 1978; McGregor et al., 1983). Information also shows that, iron deficiency is responsible for about 95% cases of anaemia during pregnancy usually due to inadequate dietary intake (especially in teenage girls), to previous pregnancies or to normal loss of iron in blood with menses to the interference of iron stores by parasites (Porter and

Kaplan, 1999).

Findings in this study provide a population based picture of iron status during pregnancy among women in the Sekyere East district, who live in conditions of poverty, malnutrition and endemic infections. After evaluating all the commonly available variables for measuring iron status, van der Broek and Letsky, 1998 estimated TSI (ferritin) as the best predictor of iron deficiency (van den Broek et al., 1998).

Given that women in these communities are low income generating people, with majority of study subjects between the ages of 20 - 29, malnutrition is a possibility which could account for the high endemicity of anaemia.

Conclusion

The study concludes that anaemia is prevalent within rural communities of the Sekyere East district of the Ashanti region and that hookworm infection is a strong predictor of iron status.

Recommendation

Anthelmintic therapy is inexpensive and safe during pregnancy after the first trimester, therefore, it should be part of the antenatal programme since malaria diagnosis and treatment is also part of the antenatal programme (Larocque and Gyorkos, 2006). The world health organization recommends anthelmintic therapy for women to control hookworm infection in areas, in which the prevalence of infection is high and anaemia is prevalent (WHO, 1996).

Anthelminthic therapy and improvement in the amount and quality of dietary iron intake among pregnant women should be encouraged as an important long-term goal, and if well established, it can meet the iron needs of pregnant women with hookworm and other intestinal helminth infections, particularly among populations who consume diets of low iron bioavailability.

Also, other causes of anaemia should be looked at in the Sekyere East district to arrive at a comprehensive approach in solving the problem of anaemia.

REFERENCES

Agarwal DK, Agarwal KN, Tripathi AM (1987). Nutritional status in rural pregnant women of Bihar and Uttar Pradesh. Indian Pediatr. 24: 119-125.

Atukorala TM, Silva LD, Dechering WH, Dassenaeike TS, Perera RS (1994). Evaluation of effectiveness of iron-folate supplementation and anthelminthic therapy against anemia in pregnancy--a study in the plantation sector of Sri Lanka. Am. J. Clin. Nutr. 60: 286-292.

Ayoya MA, Spiekermann-Brouwer GM, Traore AK, Stoltzfus RJ, Garza C (2006). Determinants of anemia among pregnant women in Mali. Food Nutr. Bull. 27: 3-11.

Beer AE, Billingham RE (1978). Maternal immunological recognition mechanisms during pregnancy. Ciba Found Symp. pp. 293-322.

Bondevik GT, Eskeland B, Ulvik RJ, Ulstein M, Lie RT, Schneede J, Kvale G (2000). Anaemia in pregnancy: possible causes and risk factors in Nepali women. Eur. J. Clin. Nutr. 54: 3-8.

Brooker S, Hotez PJ, Bundy DA (2008). Hookworm-Related Anaemia among Pregnant Women: A Systematic Review. PLoS Negl Trop. Dis. 2: e291.

Bundy DA, Chan MS, Savioli L (1995). Hookworm infection in pregnancy. Trans. R Soc. Trop. Med. Hyg. 89: 521-522.

Dickson R, Awasthi S, Demellweek C, Williamson P (2000). Anthelmintic drugs for treating worms in children: effects on growth and cognitive performance. Cochrane Database Syst. Rev. CD000371.

Dreyfuss ML, Stoltzfus RJ, Shrestha JB (2000). Hookworms, malaria and vitamin A deficiency contribute to anemia and iron deficiency among pregnant women in the plains of Nepal. J. Nutr. 130: 2527-2536.

Fleming AF (1982). Iron deficiency in the tropics. Clin. Haematol. 11: 365-388.

Hopkins RM, Gracey MS, Hobbs RP, Spargo RM, Yates M, Thompson RC (1997). The prevalence of hookworm infection, iron deficiency and anaemia in an aboriginal community in North-west Australia. Med. J. Aust. 166: 241-244.

Jans SM, Daemers DO, Vos R, Lagro-Jansen AL (2008). Are pregnant women of non-Northern European descent more anaemic than women of Northern European descent? A study into the prevalence of anaemia in pregnant women in Amsterdam. Midwifery.

Larocque R, Gyorkos TW (2006). Should deworming be included in antenatal packages in hookworm-endemic areas of developing countries? Can. J. Public Health 97: 222-224.

Larry SR, Janovy Jnr J (1996). Introduction to Parasitology. In Foundations Parasitology, pp. 1 – 7, 410-415.: The McGraw Hill Companies, Inc.

McGregor IA, Wilson ME, Billewicz WZ (1983). Malaria infection of the placenta in The Gambia, West Africa; its incidence and relationship to stillbirth, birthweight and placental weight. Trans. R Soc. Trop. Med. Hyg. 77: 232-244.

Muhangi L, Woodburn P, Omara M (2007). Associations between mild-to-moderate anaemia in pregnancy and helminth, malaria and HIV infection in Entebbe, Uganda. Trans. R Soc. Trop. Med. Hyg. 101: 899-907.

Pasricha SR, Caruana SR, Phuc TQ (2008). Anemia, iron deficiency, meat consumption, and hookworm infection in women of reproductive age in Northwest Vietnam. Am. J. Trop. Med. Hyg. 78: 375-381.

Porter RS, Kaplan JL (1999). The Merck Manual of diagnosis and therapy, 17th Ed edn. White house station, N.J.

Stoltzfus RJ, Chwaya HM, Tielsch JM, Schulze KJ, Albonico M, Savioli L (1997). Epidemiology of iron deficiency anemia in Zanzibari schoolchildren: the importance of hookworms. Am. J. Clin. Nutr. 65: 153-159.

Van den Broek N (1996). The aetiology of anaemia in pregnancy in West Africa. Trop. Doct. 26: 5-7.

Van den Broek NR, Letsky EA, White SA, Shenkin A (1998). Iron status in pregnant women: which measurements are valid? Br. J. Haematol. 103: 817-824.

Van Eijk AM, Lindblade KA, Odhiambo F (2009). Geohelminth Infections among Pregnant Women in Rural Western Kenya; a Cross-Sectional Study. PLoS Negl. Trop. Dis. 3: e370.

WHO (1993). Prevention and management of severe anaemia in pregnancy: report of a technical working group. Geneva. WHO/FHE/MSM/ 93.

WHO (1996). Report of the WHO Informal Consultation on Hookworm Infection and Anaemia in Girls and Women World Health Organization Geneva. WHO/CTD/SIP/ 96.

WHO (1998). UNICEF and UNU Iron Deficiency: Indicators for Assessment and Strategies for Prevention: Geneva.

4

Skin diseases among preschool children

Allma Koçinaj[1]*, Dardan Koçinaj[1] and Merita Berisha[2]

[1]University Clinical Center of Kosova, Prishtina, Kosova.
[2]National Institute of Public Health of Kosova, Prishtina, Kosova.

In developing countries skin diseases can affect more than 60% of the general population and usually are not well managed. Skin diseases are common in children. Epidemiologic studies of the general population, however, are still limited and missing for the region of Kosova. The aim of this study was to investigate disease prevalence in children under 6 years old of out- and in-patients at the department of Dermatological clinic - a tertiary health care center, a cross-sectional study was carried out during a period of one year. A total of 1,998 children 0 - 6 years old were examined. In the out-patients scabies presented in 18% of overall dermatoses, atopic dermatitis 7.9%, urticaria 7.9%, pyodermia 7.8%, tinea superificialis 5.9%, staphylodermia 5.5%, dermatitis amoniacalis 4.7%, eczema infantum 4%, impetigo contagiosa 3.5% and exanthema toxo-allergica 2.7%. While in the in-patients, acute urticaria presented in 26.1% of the overall dermatoses, scabies 20.7% and atopic dermatitis 10.8%, epidemiologic data are necessary for the monitoring of skin changes in school children and provides the basis of training programs for medical professionals in primary health care with the aim to reduce long-term morbidity and socioeconomic impact.

Key words: Skin disease, prevalence, children.

INTRODUCTION

In developing countries skin diseases can affect more than 60% of the general population and usually are not well managed (Hay et al., 1994; Ogunbiyi et al., 2004). Skin of infants, children, adults and the elderly, with its anatomic and physiologic characteristics, acts as a barrier for different environmental insults, and undergoes certain changes in each period during human life. Skin diseases are common in children (Nanda et al., 1999). Epidemiologic studies of the general population, however, are limited and missing for the region of Kosova. Some skin diseases in children, although not life-threatening, may be particularly distressing and chronic skin disease may have a severe psychological impact (Chen et al., 2008). Kosova is a state which is spread on 10908 km^2 land, including >2 million inhabitants living in 30 municipalities. It is characterized by a progressive type of population, with a slightly male gender domination (51.6%) and a relatively young, under 25 years old (57.85%), population. This population may be ideal for an epidemiologic prevalence study in younger people. From the beginning of human life skin changes occur and as a child grows the skin is exposed to several irritating and infectious agents (Fung and Lo, 2000). Genetic related disorders of the skin play an important role in manifesting different skin disorders. Children represent an individual population group and the history data often are gathered by parents or care providers. Successful treatment is achieved due to good relationships and cooperation between parents, children and dermatologists (Ricci et al., 2009). Pediatric dermatology is a relatively new field so there are only a few prospective studies in the literature (Massa et al., 2000) mainly in developed countries (Shibeshi, 2000; Wenk and Itin, 2003; Ogunbiyi et al., 2005; Wu et al., 2000; Popescu et al., 1999). The aim of this study was to investigate disease prevalence in children below 6 years old of out- and in-patients at the department of Dermatological clinic - a tertiary health care center.

MATERIALS AND METHODS

Patients

This was a cross-sectional study carried out during a period of 1 year, June 2003 – 2004 including in- and out-patients at the

Table 1. Structure of out- and in-patients according to the group of dermatoses and hospitalization rate.

Dermatoses Type	Out-patients		In-patients		Hospitalization Rate
	N	%	N	%	(%)
Examined	1525	100.0	111	100.0	7.3
Seborrhoicdermatosesp>0.05	45	3.0	5	4.5	11.1
Infective dermatosesp>0.05	874	57.3	50	45.0	5.7
Allergic dermatosesp < 0.01	488	32.0	54	48.6	11.1
other	163	10.7	2	1.8	1.2
			P < 0.01		

department of Dermatology clinic, University clinical center of Kosova, Prishtina. All patients in the study population were examined by a dermatologist. A total of 1,998 children 0 - 6 years old (with male gender domination) were examined.

Assessment

An assessment of the whole body including head was conducted and confirmed by at least 2 dermatologists. Dermatoses were divided into 4 main groups: infectious, seborrhoic, allergic and other dermatoses. Apart from a clinical assessment based on recognized criteria, diagnoses were also by laboratory examination (microscopic examination of fungal slides of the scales scraped from a lesion) and microbial culture isolation. The collected data were registered in a relevant database. Children in the out-patient department who required further management were hospitalized.

Statistical analysis

The statistical analysis of the data was by chi square test, p < 0.01 was considered statistically significant.

RESULTS

During 2003 - 2004, the total number of patients referred at the clinic was 13,749; 93.59% were out-patients and of these 14.66% were under 6 years old. There were 881 in-patients, 12.6% of whom were children <6 years old. The mean age of children that were in-patients was 3.0 ± 1.9) and for out-patients was 2.5 ± 1.8).

Structure of the out- and in-patients according to different groups of dermatoses and the hospitalization rate

Seborrhoic dermatoses were more evident for in-patients with a hospitalization rate of 11.1%; allergic dermatoses were also highly presented in in-patients compared to out-patients (48.6% vs. 32.0%, p < 0.01) with the same hospitalization rate (11.1%) (Table 1 and Figure 1). This could be due to different clinical features and outcomes of allergic diseases. Infectious dermatoses were higher in out-patients (57.3%) compared to in-patients (45%) with a hospitalization rate of 5.7%.

Common dermatoses of out-patients

The presence of scabies 18%, atopic dermatitis 7.9%, urticaria 7.9%, pyodermia 7.8%, tinea superificialis 5.9%, staphylodermia 5.5%, dermatitis amoniacalis 4.7%, eczema infantum 4%, impetigo contagiosa 3.5% and exanthema toxo-allergica 2.7% were the most common dermatoses of out-patients (Table 2 and Figure 2).

DISCUSSION

In a 2 year prospective study (Nnoruka, 2004), with a total of 1,019 patients aged 4 weeks, 57 years in more than half (51.3%) atopic dermatitis was presented before the age of 10. In infants, the earliest age of this disease appearance was 6 weeks (12.7%). According to Shafer et al. (2000), the prevalence of this disease was 10.4% in preschool (5 - 6 years old) children. Another cross-sectional study (Foley et al., 2001) showed that aotpic dermatitis was present in 30.8% of cases, with a decrease in prevalence after the third year of life. In an analysis of 1,760 children of age 1 - 5 years, Emerson et al. (1998) found that the prevalence of atopic dermatitis was 16.5%. According to our study, the mean age in in-patients with allergic dermatoses was 3.1 ± 2, while more than half of the out-patients diagnosed with allergic dermatoses were under 2 years old.

In a survey based study including preschool children 3 - 6 years old, consisting of 2,311 (52.8%) males and 2,062 (47.2%) females, allergic diseases were detected in 34.6% of cases (Wang et al., 1998). In this study the prevalence of atopic dermatitis was 6.6% and urticaria 6.8%, in males and females, with no significant difference according to gender (Wang et al., 1998). From another study of preschool children, 12.9% suffered atopic dermatitis (Schafer et al., 1996). In our study, in the group of allergic dermatoses of the out-patient population, atopic dermatitis and acute urticaria were around 25% each. In the overall population atopic dermatitis was present at 7.9% for out-patients and 10.8% for in-patients.

Seborrhoic dermatitis and Pityriasis capitis are common in early childhood. Foley et al. (2003) have determined the prevalence of these diseases among 1,116 preschool children of the age till 6 years and found that the

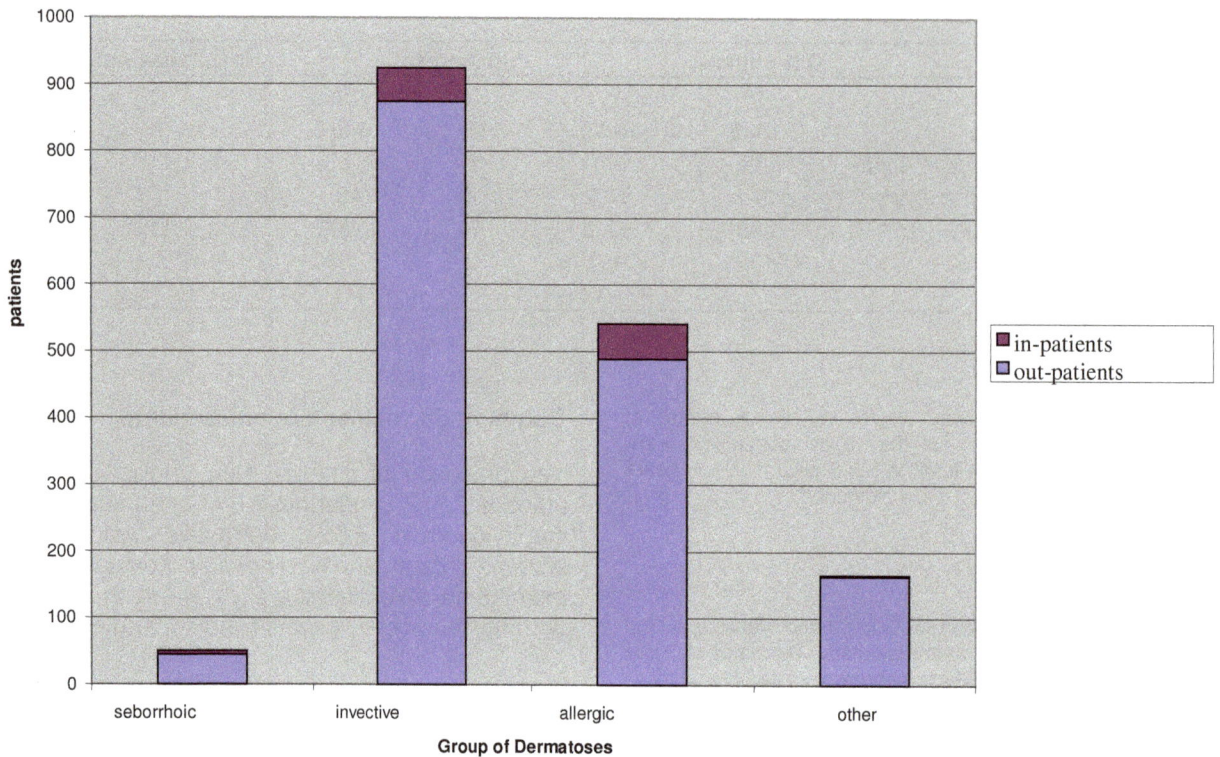

Figure 1. Structure of the out- and in-patients according to the group of dermatoses.

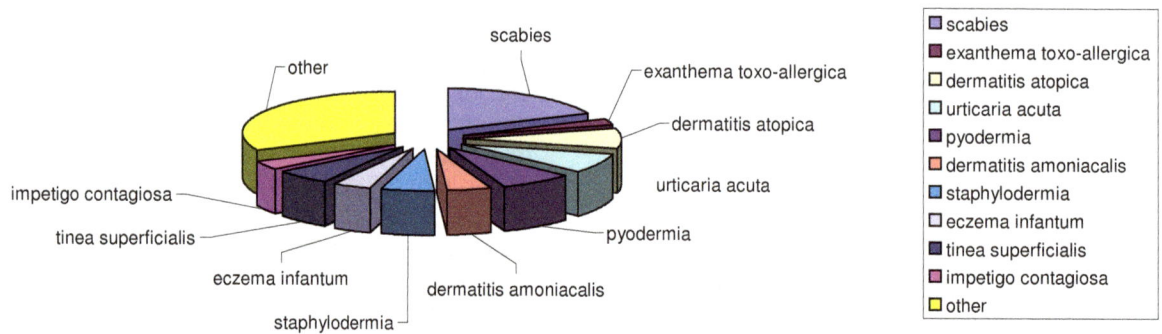

Figure 2. Common skin diseases in out-patients.

prevalence of seborrhoic dermatitis was 10% and was more common in males and during the first 3 months of life. In our study, dermatitis seborrhoica was present in 4.5% of in-patients with no cases above 5 years old and 3% of out-patients. As for out-patients in the group of seborrhoic dermatoses, seborrhoic dermatitis was present at 44.44%. This disease together with Leiner disease was present in the age group below years.

In another study (Masawe and Nsanzumuhire, 1975) of 532 preschool children, scabies was found to be the most common skin disease (31%), primary pyodermia was present in 7% cases, while fungal infections occurred in 2.4% cases. Dermatoses in our study presented at 57.3%

in out-patients and 45% in in-patients. Within the dermatoses group, analysis of out-patients showed that scabies was diagnosed in 31.4%, pyodermia 13.6% and tinea superficialis in 10.3% cases. For in-patients, again scabies was mostly presented in the dermatoses group (47.9%), with 14.6% pyoccocia, 12.5% staphylodermia, 8.3% Kerion Celsi, 8.3% pyodermia, 4.2% herpes simplex, 2.1% herpes Zoster and 2.1% tinea capitis cases. Mahe et al. (1995) reported a mean prevalence of 34% skin diseases in children, with pyodermia 12.3%, tinea capitis 9.5%, pediculosis capitis 4.7%, scabies 4.3% and molluscum contagiosa 3.6%.

Kumar et al. (2004) published a retrospective epidemio-

Table 2. Common skin diseases in out-patients.

Skin Disease	N	%
Scabies	338	18.01
Dermatitis atopica	149	7.94
Urticaria acuta	148	7.88
Pyodermia	146	7.78
Tinea superficialis	111	5.91
Staphylodermia	103	5.49
Dermatitis amoniacalis	89	4.74
Eczema infantum	76	4.05
Impetigo contagiosa	65	3.46
Exanthema toxo-allergica	50	2.66
Other	602	32.07
Total	1877	100.00

logical study including 419 children under the age of 14, in which males of age 6 - 10 years and females 10 - 14 years were mostly affected by psoriasis. In our study, cases of psoriasis were sporadic, presented with a low number of cases and unable to be grouped as a specific entity.

In a study population of 10,000 patients up to 18 years old, patients under 1 year were mostly presented (Ruiz-Maldonado et al., 1977). According to this study, parasitic dermatoses and viral and bacterial skin disorders were most common. Papullar urticaria was reported in 16.3%, atopic dermatitis 12.9%, scabies 10.4%, benign warts 8.4%, impetigo 6.8%, pityriasis alba 6.6% and other diseases around 2% of cases.

According to our study, in out-patients scabies presented in 18% of overall dermatoses, atopic dermatitis 7.9%, urticaria 7.9%, pyodermia 7.8%, tinea superificialis 5.9%, staphylodermia 5.5%, dermatitis amoniacalis 4.7%, eczema infantum 4%, impetigo contagiosa 3.5% and exanthema toxo-allergica 2.7%. While in in-patients, acute urticaria was presented in 26.1% of the overall dermatoses, scabies 20.7% and atopic dermatitis 10.8%. These data correspond with those already published (Ruiz-Maldonado et al., 1977).

Epidemiologic data are necessary for the monitoring of skin changes in school children and provides the basis of training programs for medical professionals in primary health care with the aim to reduce long-term morbidity and socioeconomic impact.

Competing interests

All authors declare that they have no competing interests.

ACKNOWLEDGEMENTS

Allma Koçinaj participated in the design and coordination of the study, analysis and interpretation of data, drafting the manuscript and revising it critically. Dardan Koçinaj helped to draft the manuscript and revised it critically. Merita Berisha performed the statistical analysis and revised the manuscript critically. All authors read and approved the final manuscript.

REFERENCES

Chen G, Cheng Y, Wang Ch, Hsu T, Hsu M, Yang P, Chen W (2008). Prevalence of skin diseases among schoolchildren in Magong, Penghu, Taiwan: A Community-based Clinical Survey. J. Formos. Med. Assoc. 107(1): 21-29.

Emerson RM, Williams HC, Allen BR (1998). Severity distribution of atopic dermatitis in the community and its relationship to secondary referral. Br. J. Dermatol. 139(1): 73-76.

Foley P, Zuo Y, Plunkett A, Marks R (2001). The frequency of common skin conditions in preschool-age children in Australia: atopic dermatitis. Arch. Dermatol. 137(3): 293-300.

Foley P, Zuo Y, Plunkett A, Merlin K, Marks R (2003). The frequency of common skin conditions in preschool-aged children in Australia: seborrheic dermatitis and pityriasis capitis (cradle cap). Arch. Dermatol. 139(3): 318-322.

Fung WK, Lo KK (2000). Prevalence of skin disease among school children and adolescents in a Student Health Service Center in Hong Kong. Pediatr. Dermatol. 17(6): 440-446.

Hay RJ, Estrada CR, Alarcon HH, Chavez LG, Lopez FLF, Paredes SS, Andersson N (1994). Wastage of family income on skin disease in Mexico. BMJ. 309(6958): 848.

Kumar B, Jain R, Sandhu K, Kaur I, Handa S (2004). Epidemiology of childhood psoriasis: a study of 419 patients from northern India. Int. J. Dermatol. 43(9): 654-658.

Mahe A, Prual A, Konate M, Bobin P (1995). Skin diseases of children in Mali: a public health problem. Trans R Soc Trop Med Hyg. 89(5): 467-470.

Masawe AE, Nsanzumuhire H (1975). Scabies and other skin diseases in pre-school children in Ujamaa villages in Tanzania. Trop Geogr Med. 27(3): 288-294.

Massa A, Alves R, Amado J, Matos E, Sanches M, Selores M, Santos C, Costa V, Velho G, Oliveira M, Ferreira E, Taveira M, Silva NS, Granado E, Lemos A, Calheiros JM (2000). Prevalence of cutaneous lesions in Freixo de Espada a Cinta. Acta Med Port. 13(5-6): 247-254.

Nanda A, Al-Hasawi F, Alsaleh QA (1999). A prospective survey of pediatric dermatology clinic patients in Kuwait: an analysis of 10,000 cases. Pediatr. Dermatol. 16: 6-11.

Nnoruka EN (2004). Current epidemiology of atopic dermatitis in south-eastern Nigeria. Int. J. Dermatol. 43(10): 739-744.

Ogunbiyi AO, Daramola OO, Alese OO (2004). Prevalence of skin diseases in Ibadan, Nigeria. Int. J. Dermatol. 43(1): 31-36.

Ogunbiyi AO, Owoaje E, Ndahi A (2005). Prevalence of skin disorders in school children in Ibadan, Nigeria. Pediatr. Dermatol. 22(1): 6-10.

Popescu R, Popescu CM, Williams HC, Forsea D (1999). The prevalence of skin conditions in Romanian school children. Br. J. Dermatol. 140(5): 891-896.

Ricci G, Bendandi B, Aiazzi R, Patrizi A, Masi M (2009). Three years of Italian experience of an educational program for parents of young children affected by atopic dermatitis: improving knowledge produces lower anxiety levels in parents of children with atopic dermatitis. Pediatr. Dermatol. 26(1): 1-5.

Ruiz-Maldonado R, Tamayo SL, Velazquez E (1977). Epidemiology of skin diseases in 10,000 patients of pediatric age. Bol. Med. Hosp. Infant. Mex. 34(1): 137-161.

Schafer T, Kramer U, Vieluf D, Abeck D, Behrendt H, Ring J (2000). The excess of atopic eczema in East Germany is related to the intrinsic type. Br. J. Dermatol. 143(5): 992-998.

Schafer T, Vieluf D, Behrendt H, Kramer U, Ring J (1996). Atopic eczema and other manifestations of atopy: results of a study in East and West Germany. Allergy. 51(8): 532-539.

Shibeshi D (2000). Pattern of Skin Disease at the Ethio-Swedish Pediatric Hospital, Addis Ababa, Ethiopia. Pediatr. Dermatol. 17(5):357.

Wang WC, Lue KH, Sheu JN (1998). Allergic diseases in preschool children in Taichung City. Zhonghua Min Guo Xiao Er Ke Yi Xue Hui Za Zhi. 39(5): 314-318.

Wenk Ch, Itin P (2003). Epidemiology of pediatric Dermatology and Allergology in the Region of Aargau, Switzerland. Pediatr. Dermatol. 20(6): 482.

Wu YH, Su HY, Hsieh YJ (2000). Survey of infectious skin diseases and skin infestations among primary school students of Taitung County, eastern Taiwan. J. Formos. Med. Assoc. 99(2): 128-134.

5

Induction of cinnamyl alcohol dehydrogenase in bacterial spot disease resistance of tomato

S. Umesha* and R. Kavitha

Department of Studies in Biotechnology, University of Mysore, Manasagangotri, Mysore 570 006, Karnataka, India.

Cinnamyl alcohol dehydrogenase (CAD) is known to be involved in the defense related responses in many host-pathogen systems. Induction of CAD enzyme in bacterial spot disease resistance in tomato (*Solanum lycopersicum* Mill.) was investigated in the present studies. Twenty different tomato cultivars were screened for resistance against bacterial spot disease incited by *Xanthomonas axonopodis* pv. *vesicatoria* under field conditions. Field screening was performed by artificially inoculating *X. axonopodis* pv. *vesicatoria* suspension to four-week-old tomato seedlings and observed for the typical symptoms of bacterial spot disease. They were categorized into highly resistant, resistant, susceptible and highly susceptible cultivars on the basis of disease incidence under field conditions. Tomato cultivars were subjected for estimating CAD - a defense-related enzyme. Temporal pattern of the enzyme was estimated using four cultivars representing each category of tomato cultivars by inoculating with six isolates of *X. a. vesicatoria*. Native PAGE analysis of CAD was carried out for the time course of enzyme activity and also by selecting three different tomato cultivars, after infecting with the pathogen. Based on the inducible amounts of the enzyme upon pathogen infection, the tomato cultivars were correlated with that of disease incidence under field conditions. A significant ($P \leq 0.05$) correlation was observed between the degree of host resistance and the enzyme level. In highly resistant tomato cultivar the enzymatic level was increased in comparison with highly susceptible tomato cultivars. Isoform analysis of CAD enzyme indicated a clear difference between the number of isoforms and also the intensity of each isoform after pathogen infection with the resistant and susceptible tomato cultivars. A possible regulation of CAD in imparting host resistance is discussed here.

Key words: Tomato, *Solanum lycopersicum*, bacterial spot, *Xanthomonas axonopodis* pv. *vesicatoria*, cinnamyl alcohol dehydrogenase, resistance, susceptibility.

INTRODUCTION

Seed- borne pathogens are the most serious disease causing agents in tomato (*Solanum lycopersicum* Mill.) among which *Xanthomonas axonopodis* pv.*vesicatoria* (Doidge) Dye (*Xav*) causes bacterial spot of tomato. When the environmental conditions are favorable for the pathogen, the disease is established in the field and it affects stem, leaves and fruit which finally leads to yield loss of up to 52% (Jones et al., 1998; El-Hendawy et al., 2005). Tomato is an important popular vegetable,

*Corresponding author. E-mail: pmumesh@gmail.com, mumesh@yahoo.com.

because of its high nutritive value, diversified use, nutritionally significant source of vitamin A and C. Red tomatoes have an important compound called lycopene, which has antioxidant properties and has the ability to cure human diseases like cancer and heart disease (Barone, 2003). The tomato production in India is 17.35 million tones with the productivity of 8.63 mth^{-1} and the area covered under production is 0.497 million hectares (FAO, 2006).

Plants have to develop a broad range of complex defense systems to struggle against pathogenic infections, because they have being confined to the place where they grow. The defense mechanism includes inducible response of pathogenesis related enzymes, pre existing physical and chemical barriers that have become marked by the action upon infection. The pathogen spread is controlled by the deposition of lignin at the point of infection by the plants when they are infected by the pathogen (Vance et al., 1980; Ride, 1975). Lignin plays an elementary role in higher plants by providing mechanical support, solute conductance and disease resistance (Barber and Mitchell, 1997). The effectual barrier to pathogen ingress and spread are the lignified cell walls (Ride, 1983). The composition and quantity of lignin varies between cell types and between tissues with the same plant (Whetten et al., 1998). The coordinated regulation of the three biosynthetic pathways, the Shikimate pathway, the general phenylpropanoid pathway and the lignin branch pathway leads to the biosynthesis of lignin. Among the enzymes involved in lignin biosynthesis, Cinnamyl alcohol dehydrogenase (CAD; E.C. 1.1.1.195) of phenylpropanoid metabolism specific for lignin synthesis has become the focus of a number of molecular analyses of lignification. Cinnamyl alcohol dehydrogenase catalyses the last step of monlignol biosynthesis, the reduction of the hydroxycinnamyl aldehydes to hydroxycinnamyl alcohol, which is considered to be highly specific marker for lignification (Mitchell and Barber, 1994). Cinnamyl alcohol dehydrogenase has been purified from tobacco (Halpin et al., 1992), *Eucalyptus gunii* (Goffner et al., 1992; Hawkins and Boudet, 1994), Wheat (Pillonel et al., 1992) *Lobolly pine* (O'Malley et al., 1992) and *Aralia cordata* (Hibino et al., 1993). Cinnamyl alcohol dehydrogenase deficiency causes drastic changes in the accumulation and nature of soluble phenolics; it also alters the structure of the lignin polymer that is deposited in the cell wall. The enzyme has a high affinity for coniferaldehyde and much lower affinity for sinapaldehyde (Sederoff et al., 1999).

Taking into consideration the role of CAD and lignin in imparting disease resistance to the plants, our study was aimed to investigate the possible role of CAD enzyme in imparting disease resistance to the plant, role of lignin deposition in disease resistance against bacterial spot in tomato and to correlate their activity with disease resistance of tomato to bacterial spot.

MATERIALS AND METHODS

Collection and screening of tomato seed samples

Twenty different tomato cultivars were procured from local seed agencies, Mysore, India. The collected seed samples were subjected to screening in the laboratory following direct plating method on semi-selective Tween B medium (Peptone 10 g/L, KBr 10 g/L, CaCl$_2$ 0.25 g/L, H$_3$BO$_3$ 0.30 g/L and Agar 15 g/L). After autoclaving; Tween 80 10 ml/L, Cyclohexamide 100 mg/L, Cephalexin 65 mg/L, 5-fluorouracil 12 mg/L and Tobramycin 0.4 mg/L was aseptically added (Mc Gurie et al., 1986). Tomato seed samples were plated directly on to the Tween B medium after surface sterilization with 70% ethyl alcohol, followed by repeated washing with sterile distilled water and blot dried. Plated seeds were incubated for 36-48 h at 28±2°C to observe typical yellow colonies with clear lipolytic zones around the seeds and sub-cultured onto the Tween B media. The number of seeds showing these typical *X. axonopodis vesicatoria* colonies was recorded. The experiments were conducted in four replicates of 100 seeds each and repeated thrice.

Field experiments

Field experiments were conducted in the experimental plot of the Department of Applied Botany and Biotechnology, University of Mysore, Manasagangotri, Mysore to evaluate the tolerance ability of all the twenty cultivars against *X. axonopodis vesicatoria* infection under field conditions. Four week–old seedlings were transplanted to well prepared field from raised beds (1x 4 m). The seedlings were planted leaving 0.5 m distance between plant to plant and 1.3 m between row to row. Each replicate consisted of single row, which was 10 m long with 18–20 plants per row. Plants were irrigated through furrow irrigation as and when required. A randomized complete block design was employed with four replicated rows.

Inoculation procedure

X. axonopodis pv. *vesicatoria* isolates were preserved on Tween B medium in the absence of light at 4°C. The inoculum was prepared by growing the bacteria in nutrient broth incubated at 28±2°C for 36 h on rotary shaker at 100 rpm (Janked and Kunkel, IKA Labortechnik, Germany). 36 h-old- culture was pelleted by centrifugation (thrice at 5000 rpm for 5 min) using bench top centrifuge (UniCen, 15 DR, Herolab GmbH, Germany). Inoculum was prepared by adjusting the bacterial concentration with sterile distilled water to 1 x 10^8 cfu/ml at A$_{610}$ nm using UV-visible spectrophotometer. Field plots were inoculated with 36 h-old *X. axonopodis vesicatoria* suspension (1x10^8 cfu/ml) at early morning and late evening as spray inoculation to completely run-off the plants, 48 h after transplantation (EPPO, 1998). All the normal agronomical practices were maintained throughout the experiment *viz.,* application of fertilizer once in every 20 days, weeding and irrigation as and when required etc.

Disease scoring

Disease assessment in field plots was done by counting infected plants at the intervals of 10 days upto 50 days starting from the appearance of first symptoms *viz.,* spots surrounded by yellow hallow, necrotic lesions that occur on leaves, stems and flower parts. The disease was confirmed by observing the infected leaf section for the bacterial ooze under compound microscope, plating the pieces of plant parts showing symptoms on to Tween B media and subjecting the isolated bacteria to various biochemical, physiological, hypersensitive response and pathogenicity tests. Individual genotypes were categorized into highly resistant (HiR), with no plants (0 %) showing any symptoms of bacterial spot disease; resistant (R), with 0.1 to 10.0% of plants showing slight marginal spots and 1- 20% of leaves becoming brown; susceptible (S), with 10.1 to 20.0% of plants showing sectorial spots and 20-40 % of leaves being brown; and highly susceptible (HS), with > 25% of the plants showing pronounced leaf collapse and more than 40% of leaves become brown. Experiments were conducted in two consecutive seasons and average disease incidence was calculated.

Determination of Cinnamyl alcohol dehydrogenase activity

Temporal pattern study of enzyme

To study the temporal pattern of CAD enzyme, the tomato seedlings of Safal, Rasi, PKM-1 and Golden seeds was selected from each category and were raised, by plating the seeds onto moist blotter discs inserted into the petridish (9 cm diameter), at the rate of 25 seeds per plate following standard procedures of International Seed Testing Association (ISTA, 2003). The plates were incubated at 28±2⁰C for 8 days until cotyledonary leaf completely opens. The eight-day -old- tomato seedlings were covered with polythene sheet, 2 h before inoculation to increase the humidity. The tomato seedlings were root dip and spray inoculated with all the six isolates of *X. axonopodis vesicatoria* (1×10^8 cfu/ml) followed by polythene sheet covering (EPPO, 1998; AVRDC, 1999). The seedlings were harvested at different time intervals *viz.,* 0, 3, 6, 9, 12, 15, 18, 21, 24, 27, 30 upto 72 h after pathogen inoculation and stored at −20⁰C until enzyme assay. Respective distilled water treated samples served as control.

Preparation of cell free extract

One gram tomato seedlings was macerated to a fine paste in a ice cold mortar with 1 g of acid purified sand in 100 mM Tris-HCl pH 7.5; 20 mM β-mercaptoethanol, 5% (w/v) Polyvinylpolypyrolidone, 2% (w/v) polyethelene glycol 6000, 5 mM DTT. The crude extract was clarified by centrifugation twice (10,000 rpm/ 10 min) at 4⁰C and the supernatant was used directly for enzyme assay (Sarni et al., 1984)

Cinnamyl alcohol dehydrogenase assay

Cinnamyl alcohol dehydrogenase activity was determined by measuring the increase in absorbance at 400 nm when coniferyl alcohol was oxidized to coniferaldehyde (Wyrambik and Grisebach, 1975). The assay was performed for 10 min at 30℃ in a total volume of 1 ml containing 100 mM Tris-HCl (pH 8.8), 100 µm coniferyl alcohol (Alfa Aesar, Johnson Matthey company, India), 200 µm NADP (Himedia, Laboratories limited Mumbai, India) and 200 µl enzyme extract. The values were expressed as a specific activity basis taking into account of proteins in the sample.

Lignin analysis

One gram of tomato seedlings was extracted with methanol (2 ml) with two changes for 24 h. The supernatant was discarded, the solids centrifuged at 10,000 rpm for 10 min were dried at 37℃ for 48 h and their dry weights were determined. The wall-bound phenolic acids from the dried samples were hydrolyzed by extracting in 2 ml of 0.2 M NaOH for 24 h. The incubated mixture was neutralized with 0.5 ml of 2 M HCL and the residue collected by centrifugation, washed twice with distilled water and resuspended in methanol (5 ml) and allowed to air dry. The air-dried pellet was resuspended in 5 ml of 2 M HCl to which 0.5 ml of thioglycolic acid was added. The tubes were sealed and placed in a 95℃ water bath for 4 h, then cooled on ice and the solids collected by centrifugation. The solids were washed twice with 2 ml of distilled water each time, followed by centrifugation. The solids were then suspended in 0.5 M NaOH (5 ml) overnight at 4℃ and again collected by centrifugation and washed twice with 2 ml water. The water washes were added to the NaOH supernatant, acidified with 1 ml of concentrated HCl, and the solution placed at 4℃ for 4 h to precipitate the lignothioglycolic acid (LTGA). The precipitate was collected by centrifugation and washed twice with 0.1 M HCl. The final pellet was dissolved in 0.5 M NaOH (3 ml) and insoluble material was removed by centrifugation. The absorbance of a 1 ml solution of LTGA was determined at 280 nm and the lignin content was expressed as absorbance at 280 nm ml^{-1} g^{-1} dry weight (Cahill and McComb, 1992).

Protein estimation

Protein contents of the extracts for the estimated enzyme was determined using the standard procedure of Bradford (1976) using BSA (Sigma, USA) as standard.

Nondenaturing PAGE and activity staining

Native PAGE (10% Polyacrylamide gel) was performed for the prepared enzyme extract (150 µg) and electrophoresis was performed as described by Laemmli (1970) but without SDS in buffers. Activity staining was performed as described by Mansell et al. (1974). Samples (150 µg) were loaded onto 10% (w/v) polyacrylamide gels with a vertical mini-gel electrophoresis unit (Biometra, Gottingen, Germany). The electrode buffer was Tris-base (3.0 g Tris-base, 7.2 g glycine and 1lt distilled water). Electrophoresis was performed at a constant voltage of 50 V initially

Table 1. Reaction of tomato cultivars for bacterial spot disease under field conditions.

Tomato cultivars	Bacterial spot incidence %	Categorization
Safal	0	HiR
Indam	8± 0.2[b]	R
Vignesh	6±0.3[a]	R
Rasi	5±0.3[a]	R
Pradhan	9±0.3[a]	R
Naveen	8±0.3[a]	R
Pioneer seeds	4 ±0.4[a]	R
Rukshita	11±0.6[c]	S
Marglobe	18±0.7 [c]	S
PKM-1	13±0.6[c]	S
Rohini	15±0.5[c]	S
SCL-4	14±0.5[c]	S
Utsav	17±0.6[d]	S
Leadbeter	13±0.6 [c]	S
Arka vikas	19±0.7[d]	S
Madanapalli	30±0.9[e]	HS
Heemsona	32±0.9[e]	HS
Vajra	38±0.8	HS
Amar	35±0.7 [e]	HS
Golden	32±0.9[e]	HS

Seeds of all the tomato cultivars were sown using Random Block Design and pathogen was sprayed 48 h after transplantation. Plants were observed for typical symptoms of bacterial spot at 10 days intervals upto 50 days. Tomato cultivars were categorized into highly resistant (HiR), resistant (R), susceptible (S) and highly susceptible (HS) based on the disease incidence. Values are the means ± S.E. of four replicates of 18-20 plants each. Values followed by same letters do not significantly differ at 5% level according to Fisher's least significant difference (*p*= 0.05).

for 1 h and 100 V to complete electrophoresis. The gels were incubated for 2 h at room temperature in darkness in 10 ml of a reaction mixture that contained 100 mM Tris-HCl (pH 8.8), 1.5 mg of nitrobluetetrazolium, 0.1 mg of phenazinemetho-sulphate, 2.5 mg of NADP and 2.5 mg of coniferyl alcohol.

Statistical analysis

All experiments were performed three times with similar results. The data from field experiments were analysed separately for each experiment and were subjected to two – way analysis of variance (ANOVA) using the statistical software SAS (version 9.0) for Microsoft windows. The means were compared for significance using Fisher's LSD. Significant effects of pathogen inoculation on enzyme activities were determined by the magnitude of the *F*-value (P≤0.05).

RESULTS

The reactions of the twenty tomato cultivars for bacterial spot disease recorded varied degree of disease

incidence under filed conditions. Cultivar Safal was exceptional which was completely free from the disease in both the seasons (Table 1). The bacterial spot incidence in the remaining cultivars ranged from 4 to 38%, accordingly tomato cultivars were categorized as Highly resistant (HiR), Resistant (R), Susceptible (S) and Highly susceptible (HS), based on the degree of host-resistance (Table 1).

Temporal pattern study of enzyme

The temporal activity of CAD enzyme in all the categorized tomato cultivars treated with *X. axonopodis* pv. *Vesicatoria*1 showed that in the HiR cultivar the CAD activity began to increase and reached the peak activity at 24 h with 126.6 Δ OD at 400 nm / min /mg protein where its control showed 24 Δ OD at 400 nm / min /mg protein. The R cultivar showed 113.7 Δ OD at 400 nm /

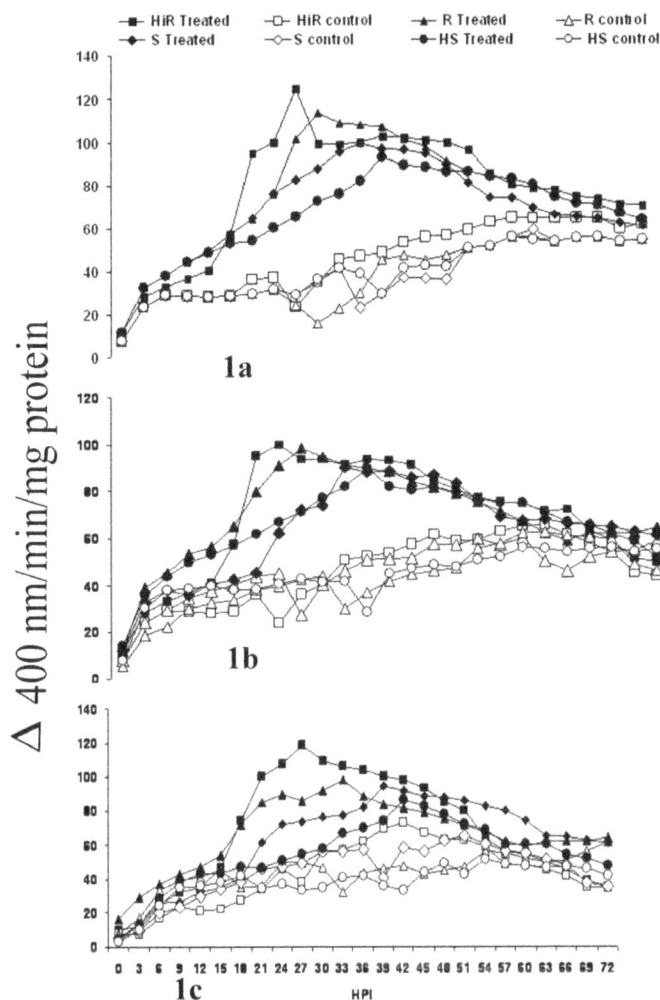

Figure 1a,b,c. The temporal pattern study of CAD activity in HiR, R, S, and HS tomato cultivars treated with Xav isolates 1, 2, and 3. The data are expressed as the averaged of three independent experiments with three replicates each and subjected to Fisher's least significant difference (p=0.05).

min /mg protein and its control showed 16 Δ OD at 400 nm / min /mg protein upon pathogen inoculation at 27 h. In the S cultivar the activity was 99.4 Δ OD at 400 nm / min /mg protein and its control showed 23.6 Δ OD at 400 nm / min /mg protein at 33 h after pathogen inoculation. The HS cultivars showed the maximum activity of CAD at 36 h with 93.2 Δ OD at 400 nm / min /mg protein and its control showed 30 Δ OD at 400 nm / min /mg protein (Figure 1a)

The HiR cultivar showed the maximum activity of CAD

at 21 h when treated with *Xav* 2, with 100 Δ OD at 400 nm / min /mg protein and its control showed 23.6 Δ OD at 400 nm / min /mg protein. Similarly the R cultivar showed 98.3 Δ OD at 400 nm / min /mg protein CAD activity which was maximum when compared to its control with 27.4 Δ OD at 400 nm / min /mg protein. The S cultivars also showed a maximum activity of CAD at 30 h with 90.5 Δ OD at 400 nm / min /mg protein and its control showed 30 Δ OD at 400 nm / min /mg protein. The HS cultivar showed the increase in CAD activity at 33 h with 89.9 Δ OD at 400 nm / min /mg protein and its control showed 28.6 Δ OD at 400 nm / min /mg protein (Figure 1b).

When the HiR was treated with the *Xav* 3 the CAD activity was found to be increased at 27 h with 118.7 Δ OD at 400 nm / min /mg protein and its respective control showed 38.1 Δ OD at 400 nm / min /mg protein. The R cultivar showed the CAD activity to be maximum at 33 h with 97.9 Δ OD at 400 nm / min /mg protein and its control showed 32.6 Δ OD at 400 nm / min /mg protein. The S cultivar also responded for the inoculation with *Xav* 3 by mounting CAD activity at 39 h with 94.5 Δ OD at 400 nm / min /mg protein and its control showed 40.6 Δ OD at 400 nm / min /mg protein. The HS cultivar showed the maximum CAD activity at 42 h with 86.3 Δ OD at 400 nm / min /mg protein in treated and its control showed 33.5 Δ OD at 400 nm / min /mg protein (Figure 1c).

The tomato cultivars showed varied responses to the infection of *Xav* 4 where HiR showed 120.6 Δ OD at 400 nm / min /mg protein when compared to its control where it showed 54.4 Δ OD at 400 nm / min /mg protein CAD activity at 24 h. The R cultivar showed CAD activity to be maximum at 30 h with 100.9 Δ OD at 400 nm / min /mg protein and its control showed 47.7 Δ OD at 400 nm / min /mg protein. The S cultivar showed the CAD activity to be maximum at 36 h with 95.5 Δ OD at 400 nm / min /mg protein and its control showed 32 Δ OD at 400 nm / min /mg protein. The HS cultivar Showed CAD activity at 39 h to be maximum with 90 Δ OD at 400 nm / min /mg protein and its control showed 24.6 Δ OD at 400 nm / protein (Figure 2a). The HiR cultivar showed the CAD activity to be maximum min /mg at 21 h with 115.3 Δ OD at 400 nm / min/mg protein and its control showed 27.5 Δ OD at 400 nm / min /mg protein when artificially inoculated with *Xav* 5. Similarly R cultivar showed maximum activity at 24 h with 103.6 Δ OD at 400 nm / min /mg protein and its control showed 35.4 Δ OD at 400 nm / min /mg protein. The S cultivar showed the maximum CAD activity at 33 h with 98.3 Δ OD at 400 nm / min /mg protein and its control showed 20 Δ OD at 400 nm / min /mg protein. The HS showed the maximum CAD activity at 42 h with 85.3 Δ OD at 400 nm / min /mg protein and its control showed 32.6 Δ

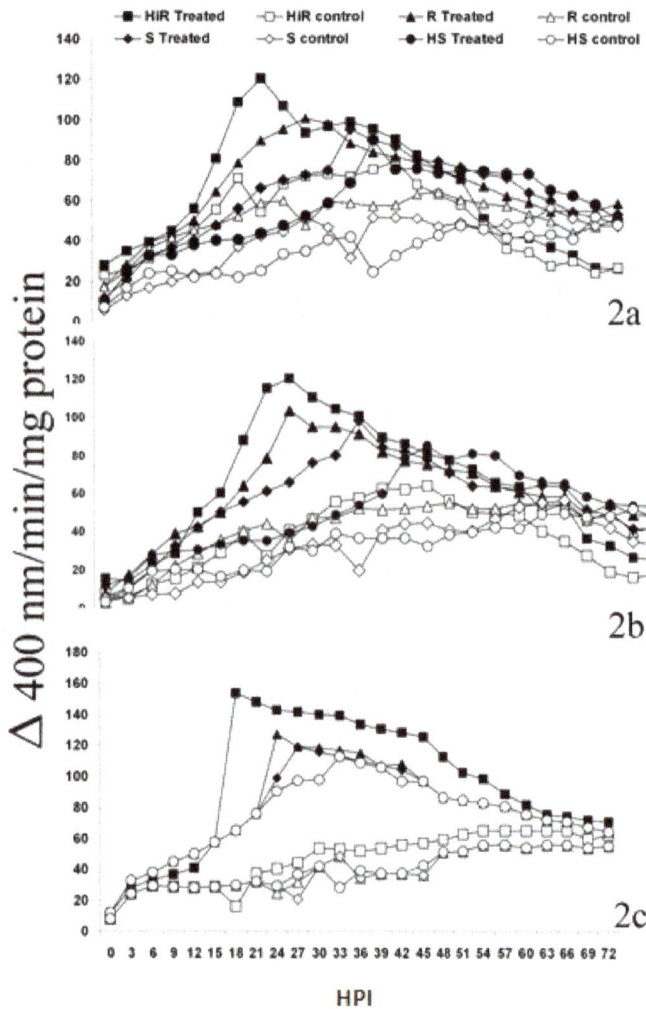

Figure 2a,b,c. The temporal pattern study of CAD activity in all the four category of tomato cultivars treated with Xav isolates 4, 5, and 6. The data are expressed as the averaged of three independent experiments with three replicates each and subjected to Fisher's least significant difference (p=0.05).

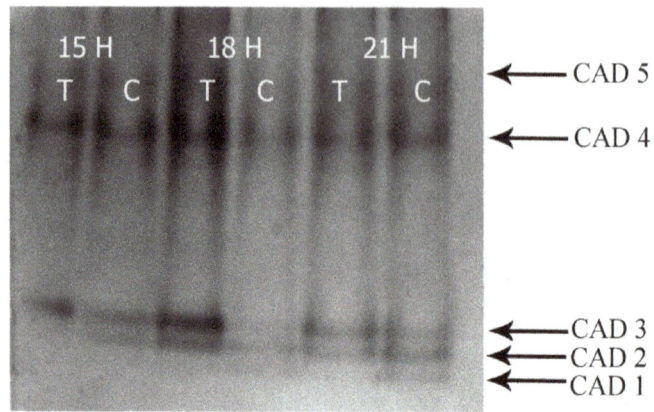

Figure 3. Native PAGE profile of CAD isoforms in tomato seedlings at different time intervals with and without pathogen inoculation. Total protein (150 µg) was loaded in each lane of 10% polyacrylamide gel. T= Treated; C = Control.

400 nm / min /mg protein. The HS cultivar showed 112.8 Δ OD at 400 nm / min /mg protein and its control showed 28 Δ OD at 400 nm / min /mg protein at 33 h after pathogen inoculation (Figure 2c). Our temporal pattern results show that the Xav 6 had maximum CAD activity in all the cultivars when compared to all the other isolates. So we selected the sixth isolate for confirming the results using Native PAGE analysis where the 18 h treated seedlings showed more intensified bands with maximum isoforms (5), when compared to its control where it showed 3 isoforms and also with less intensified bands. The 15 and 21 hpi treated and control lane showed 3, 4, 4 and 5 isoforms, respectively but not significant as that of 18 hpi treated lane (Figure 3).

CAD assay and host resistance of tomato cultivars to bacterial spot disease

Among the twenty tomato cultivars subjected for CAD assay, Safal being the HiR cultivar showed the maximum activity with 222.3 Δ OD at 400 nm / min /mg protein where its control showed 50.8 Δ OD at 400 nm / min /mg protein (Figure 4a). Among the resistant cv. Rasi showed CAD activity of 36.8 Δ OD at 400 nm / min/mg protein in control which significantly (P≤0.05) increased to 92.5 Δ OD at 400 nm / min / mg protein at 24 hpi (Figure 5a). Similarly in the susceptible cv. group PKM-1 showed 53.5 Δ OD at 400 nm / min /mg protein in treated seedlings where its control showed 24.7 Δ OD at 400 nm / min /mg

OD at 400 nm / min /mg protein (Figure 2b).

The first increase in CAD activity in HiR began at 3 h after inoculation with the 6th isolate of Xav, and peaked at 18 h with 153.8 Δ OD at 400 nm / min /mg protein where its respective control showed 16 Δ OD at 400 nm / min /mg protein. The R cultivar showed 127.1 Δ OD at 400 nm / min /mg protein and its control showed 24.7 Δ OD at 400 nm / min /mg protein at 24 h. The S cultivars showed maximum CAD activity at 27 h with 118.8 Δ OD at 400 nm / min /mg protein and its control was 20.6 Δ OD at

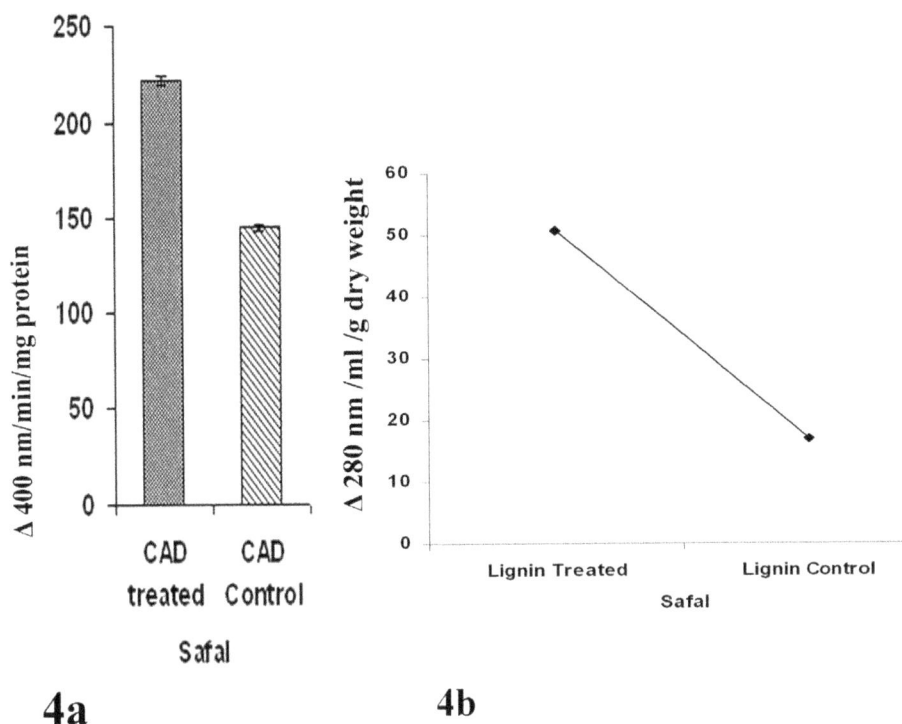

4a

4b

Figure 4a, b. CAD activity AND lignin content of HiR cultivar estimated spectrophotometrically in 8 day-old seedlings of HiR cultivar Safal in response to *X. axonopodis vesicatoria*. The data are expressed as the averaged of three independent experiments with three replicates each and subjected to Fisher's least significant difference (p=0.05). Bars indicate standard errors.

protein which was greater than the treated seedlings (Figure 6a). Among the highly susceptible category cultivar Golden showed 23.1 Δ OD at 400 nm / min /mgprotein where its control seedlings showed 20.2 Δ OD at 400 nm / min /mg protein (Figure 7a).

Native PAGE analysis

The spectrophotometric results were supported by the native PAGE analysis. The HiR treated seedlings showed 6 isoforms with thick and high intensified bands where its respective control showed only 3 isoforms. In resistant cultivars the treated seedlings showed 5 isoforms and its respective control showed 3 isoforms. Four isoforms were visualized in HS treated seedlings where its control expressed 3 isoforms (Figure 8). When compared to the HiR treated lane with all the other lanes it showed maximum isoforms with thickest bands, thus representing

the maximum enzyme activity.

Lignin analysis

Lignin content of the tomato cv. Safal known to be highly resistant showed 17 Δ OD at 280 nm ml^{-1} g^{-1} dry weights in the untreated control, which was significantly (P≤0.05) increased to 85.1 Δ OD at 280 nm ml^{-1} g^{-1} dry weight upon pathogen inoculation (Figure 4b). Cultivar Rasi showed 27.4 Δ OD at 280 nm ml^{-1} g^{-1} dry weight and 14.3 Δ OD at 280 nm ml^{-1} g^{-1} dry weight with and without pathogen inoculation respectively (Figure 5b). Susceptible cultivar PKM-1 showed 18.4 Δ OD at 280 nm ml^{-1} g^{-1} dry weight and its control showed 11 Δ OD at 280 nm ml^{-1} g^{-1} dry weight (Figure 6b). The cultivar Golden showed 10 Δ OD at 280 nm ml^{-1} g^{-1} dry weight in control whereas it remained unchanged upon pathogen inoculation with 10 Δ OD at 280 nm ml^{-1} g^{-1} dry weight

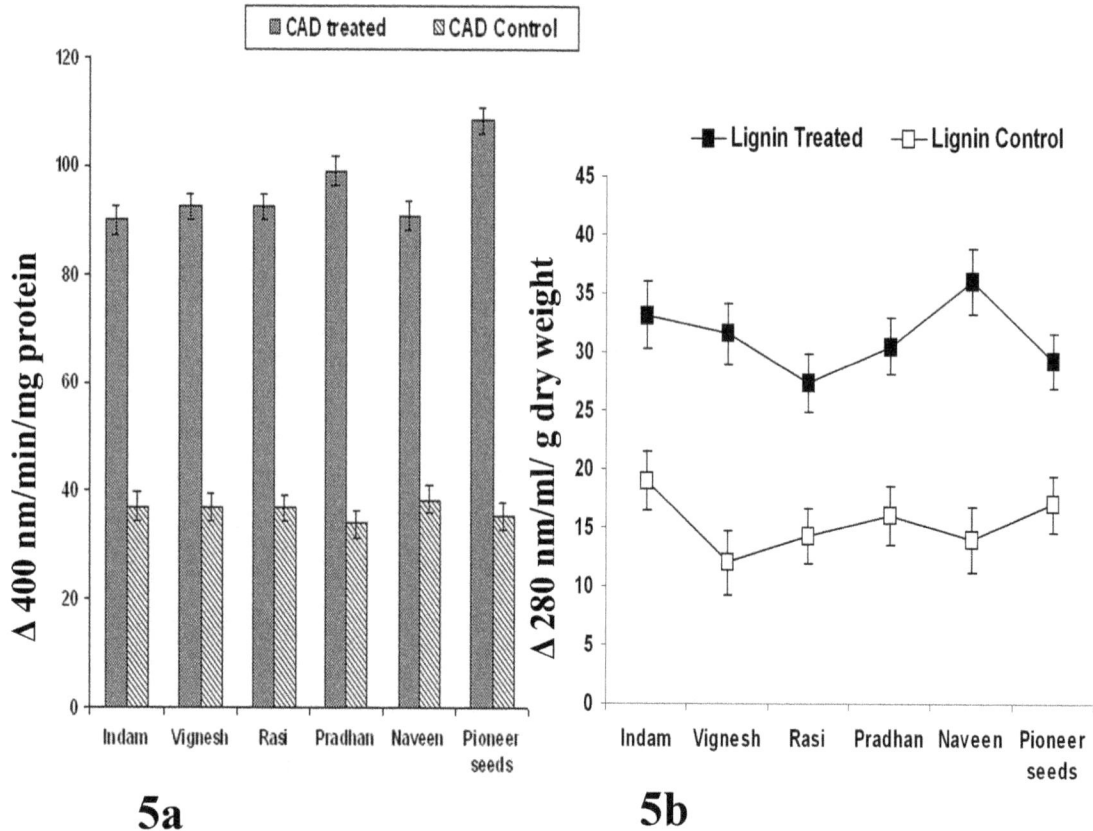

Figure 5a, b. CAD activity and lignin content of R cultivar in 8 day-old seedlings in response to *X. axonopodis vesicatoria* The data are expressed as the averaged of three independent experiments with three replicates each and subjected to Fisher's least significant difference (p=0.05). Bars indicate standard errors.

(Figure 7b).

DISCUSSION

The present work is an attempt to correlate the CAD enzyme activity in tomato to bacterial spot disease resistance. The twenty cultivars used in this study clearly varied in the degree of host resistance to the pathogen. All the resistant cultivars sampled in this study reacted to pathogen inoculation by mounting CAD enzyme activity along with lignin. In this study, we report the direct involvement of CAD during host-pathogen interaction and our results also suggest that CAD is involved in lignin synthesis.

Cinnamyl alcohol dehydrogenase induction has been studied in relatively few plant defense responses (De Sa

et al., 1992; Grand et al., 1987; Moerschbacher et al., 1988; Walter et al., 1988). Mitchell et al. (1994) first reported the CAD activity expression during the plant defense response involving lignin deposition.

In HiR, the CAD activity was maximum than R, S and HS cultivars. Similarly the lignin content of HiR was maximum than the other three grouped cultivars. The above findings supports that as the CAD activity increases upon pathogen attack the lignin accumulation is also increased. Lignin plays an important role in protection and defense and is synthesized in response to pathogen attack (Bostock and Stermer, 1989; Vance et al., 1980) and mechanical wounding (Hawkins and Boudet, 1996). Our findings also support this hypothesis, where we found that where the lignin content was maximum the disease expression was minimum and where lignin content was moderately deposited the

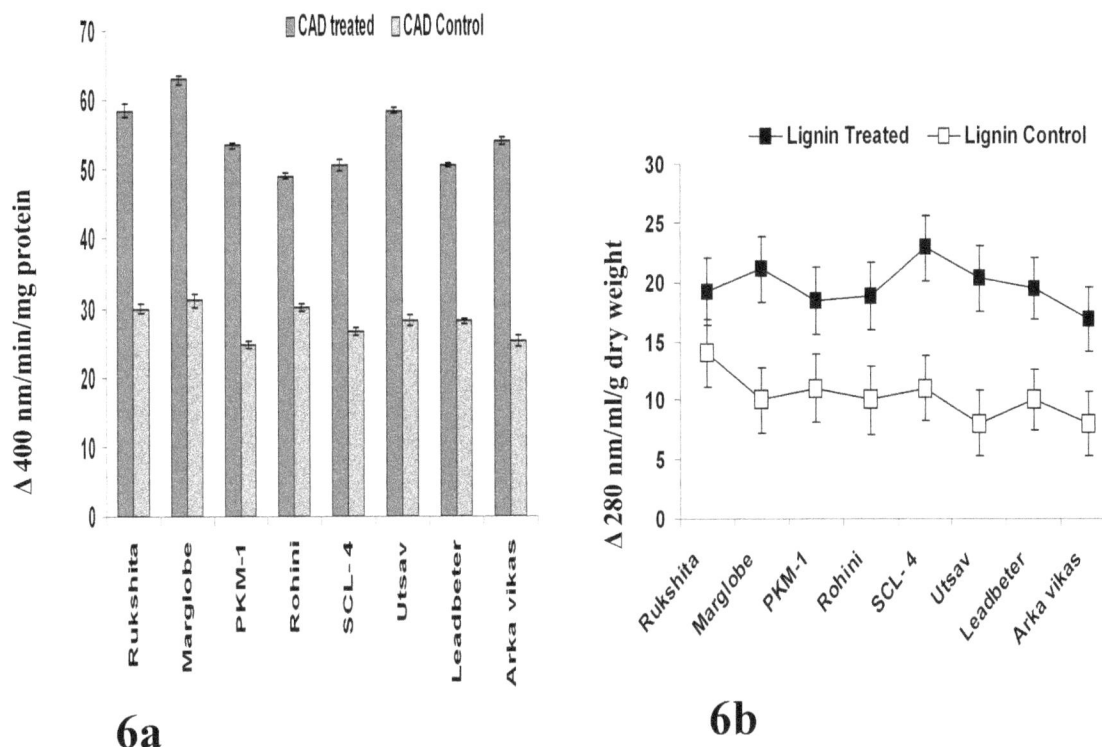

Figure 6a, b. CAD activity and lignin content of S cultivar estimated in 8 day-old seedlings by artificially inoculating *X. axonopodis vesicatoria* The data are expressed as the averaged of three independent experiments with three replicates each and subjected to Fisher's least significant difference (p=0.05). Bars indicate standard errors.

disease expression was moderate and in highly susceptible cultivar where its lignin content was minimum or very less the disease expression was maximum.

Tomato cultivars with maximum CAD activity and maximum lignin content showed high resistance against the *X. a. vesicatoria*, similarly cultivars which showed less CAD activity and less lignin content was found to be susceptible to the bacterial spot disease. This is supported by the findings of Hawkins et al. (1997) who revealed that plants down regulated in CAD activity could be more susceptible to pathogen attack, even though tAheir morphological development is apparently unaffected. Cinnamyl alcohol dehydrogenase is directly involved in lignification (Goffner et al., 1992; O'Malley et al., 1992). Our findings also support the above statement as we observed that CAD activity increased the lignin deposition increased in tomato seedlings. Our findings revealed that the HiR cultivar showed maximum lignin content upon pathogen infection when compared with its respective control and the R, S and HS cultivars. The

resistant cultivars showed moderate lignin content. The susceptible and highly susceptible cultivars showed minimal lignin content upon infection. We noticed that HiR cultivar showed maximum CAD activity and where there was maximum enzyme activity; the lignin deposition was also maximum, thus resisting the pathogen entry into the host. This support the findings of Mitchell et al. (1994) who found that the lignin deposition at wound margins in wheat leaves was consistent with the sequential induction of CAD.

Cinnamyl alcohol dehydrogenase isoforms have been demonstrated in a number of species including wheat, eucalyptus, soybean and bean (Mackay et al., 1995). Native PAGE analysis of CAD enzyme was preformed to visualize the isoforms expressed in HiR, R, S and HS cultivars. The results showed that HiR showed maximum isoforms upon infection when compared to R, S and HS cultivars. Differential expression of isoforms with different specifities has been put forth as a potential mechanism account for lignin heterogeneity (Mansell et al., 1976).

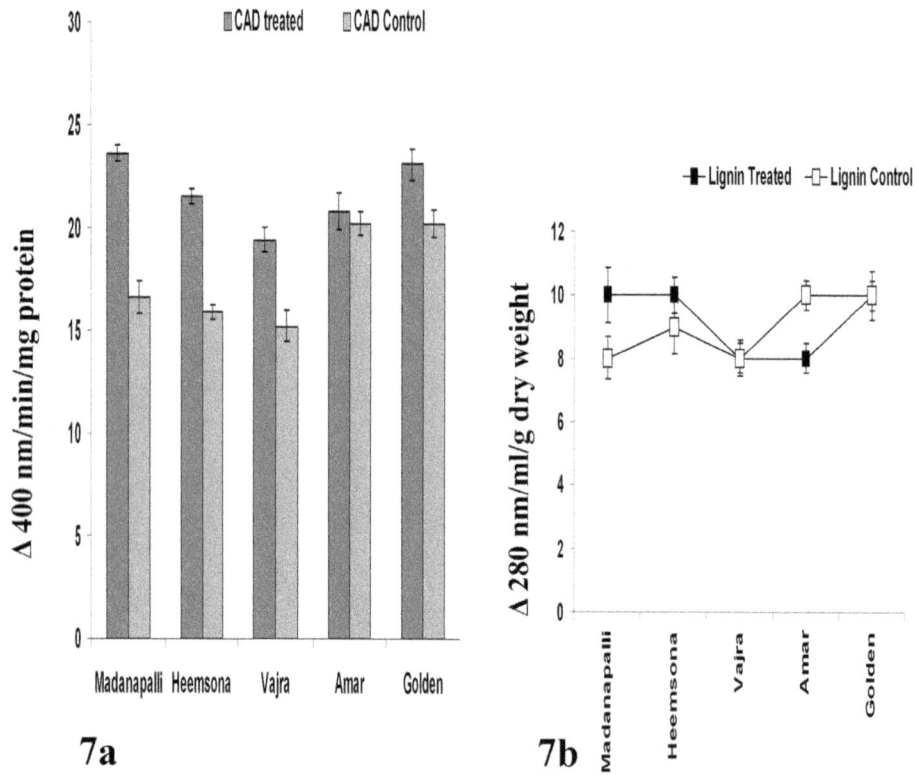

Figure 7a, b. CAD activity and lignin content of HS cultivar in 8 day-old seedlings in response to *X. axonopodis vesicatoria*. The data are expressed as the averaged of three independent experiments with three replicates each and subjected to Fisher's least significant difference (p=0.05). Bars indicate standard errors.

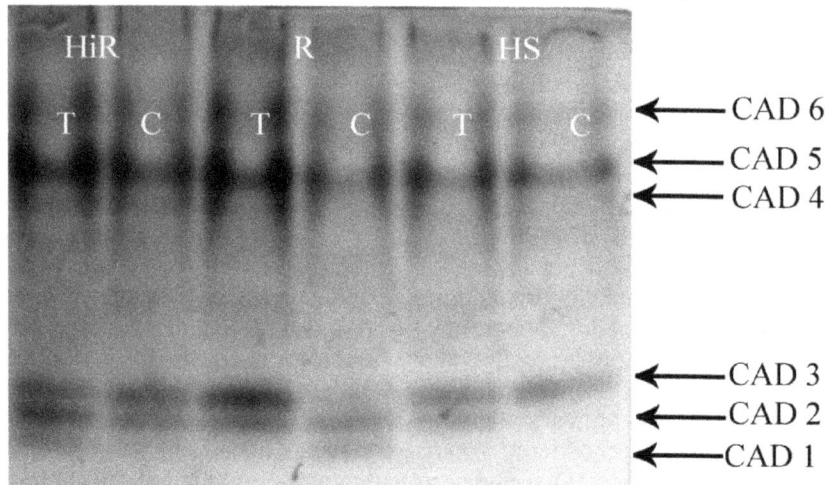

Figure 8. CAD isoforms visualized using Native PAGE of 10% in three tomato cultivars WITH and without *X. vesicatoria* inoculation. Total protein (150 μg) was loaded in each lane. HiR = Highly resistant, R= Resistant and HS = Highly susceptible

Some of these isoforms seems to be associated with the synthesis of defense lignin (Mitchell et al., 1994). Several studies have shown that CAD may be polymorphic with isoforms that differ not only on substrate affinity but on molecular mass also (Boudet et al., 1995; Goffner et al., 1992; Halpin et al., 1992; Luderitz and Griesebach, 1981; Mansell et al., 1974, 1976; Pillonel et al., 1992; Sarni et al., 1984). Dixon et al. (2001) and Higuchi et al. (1997) have found that in various species CAD isoforms are involved in monolignol biosynthesis. Two isoforms has been observed in Soybean (Wyrambik and Grisebach, 1975), *Eucalyptus gunii* (Goffner et al., 1992) and *Phaseolus vulgaris* (Pettenati et al., 1994).

It is clear from the present studies that the defense-related enzyme CAD is up-regulated in the highly resistant tomato cultivars along with lignin content, upon pathogen infection, whereas in susceptible tomato cultivars the level of defense enzyme and lignin content remains unchanged or down regulated. Hence we conclude that CAD enzyme is an enzyme, which prevents the plant from pathogen attack by inducing lignin deposition, where lignin is known to be a physical barrier, which prevents the pathogen entry into the host, and we have noticed that as CAD activity increases, lignin deposition also substantially increased. Thus we conclude that CAD is directly involved in lignification and helps the plant by imparting resistance to the plant against bacterial spot disease.

ACKNOWLEDGEMENT

The present work is the result of major research project entitled, "Investigations on bacterial spot of tomato and its management" awarded by University Grants Commission, Government of India, New Delhi, India, under 10th plan No. F.30-146/2004 (SR), dated 10th November' 2004. The authors wish to thank the Chairman, Department of Applied Botany, Seed Pathology and Biotechnology, University of Mysore, Manasagangotri, MYSORE 570 006, India.

REFERENCES

Barber MS, Mitchell HJ (1997). Regulation of phenylpropanoid metabolism in relation to lignin biosynthesis in plants. Inter. Rev of Cytol., 172: 243-293.

Barone A (2003). Molecular marker –Assisted selection for resistance to pathogens in tomato. In: Marker assisted selection in plants. MAS workshop of electronic forum on Biotechnology in food and agriculture, FAO during 17-18 Oct' 2003 at Turin, Italy.

Bostock RM, Stermer BA (1989). Perspectives on wound healing in resistance to pathogens. Ann. Rev. Phytopathol., 27: 343-371.

Boudet AM, Lapierre C, Grima-Pettenati J (1995). Biochemistry and Milecular biology of lignification. New Phytol. 129: 203-36.

Bradford MM (1976). A rapid and sensitive method for quantification of microgram quantities of protein utilizing the principle of protein due binding. Annal. Biochem., 72: 248- 254.

Cahill DM, McComb JA (1992). A comparison of changes in Phenyl alanine ammonia lyase activity, lignin and phenolic synthesis in the roots of *Eucalyptus calophylla* (Field resistant) and *E. marginata* (Susceptible) when infected with *Phytophthora cinnamomi*. Physiol. Mol. Plant Pathol., 46: 315- 332.

De Sa MM, Subramaniam R, Williams FE, Douglas CJ (1992). Rapid activation of phenylpropanoid metabolism in elicitor-treated hybrid popular (*Populus trichocarpa* torr and gray x *Populus deltoids* marsh) suspension- cultured cells. Plant Physiol. 98: 728-737.

Dixon RA, Chen F, Guo D, Parvathi K (2001). The biosynthesis of monolignols: a "metabolic grid". Or independent pathways to guaiacyl and syringyl units?. Phytochemistry, 57: 1069-1084.

El- Hendawy HH, Osman ME, Sorour NM (2005). Biological control of bacterial spot of tomato caused by *Xanthomonas campestris* pv. *Vesicatoria* by *Rahnella aquatilis*. Microbiol. Res., 160: 343-352.

FAO (2006). Food and Agriculture Organization of United Nations, Agriculture data. http:// www.Faostat.com.

Goffner D, Joffroy I, Grima-Pettenati J, Halpin C, Knight ME, Schuch W, Boudet AM (1992). Purification and characterization of isoforms of cinnamly alcohol dehydrogenase (CAD) eucalyptus xylem. Planta, 188: 48-53.

Grand C, Sarni F, Lamb CJ (1987). Rapid induction by fungal elicitor of the synthesis of cinnamly alcohol dehydrogenase a specific enzyme of lignin synthesis. Eur. J. Biochem., 169: 73-77.

Grima-Pettenati J, Campargue C, Boudet A, Boudet AM (1994). Purification and characterization of cinnamly alcohol dehydrogenase isoforms from *Phaseolus Vulgaris*. Phytochem., 37: 941-947.

Halpin C, Knight ME, Grima-Pettenati J, Goffner D, Boudet AM, Schuch W (1992). Purification and characterization of isoforms of cinnamly alcohol dehydrogenase from tobacco stem. Plant Physiol., 98: 12-16.

Hawkins S, Boudet A (1994). Purification and characterization of cinnamly alcohol dehydrogenase isoforms from the periderm of *Eucalyptus gunnii* Hook. Plant Physiol., 104: 75-84.

Hawkins S, Boudet A (1996). Wound-induced lignin and suberin deposition in a woody angiosperm (*Eucalyptus gunnii* Hook.): histochemisrty of early changes in young plants. Protoplas., 191: 96-104.

Hawkins S, Samaj J, Lauvergeat V, Boudet A, Grima-Pettenati J (1997). Cinnamly alcohol dehydrogenase: identification of new sites of promoter activity in transgenic popular. Plant Physiol., 113: 321-325.

Hibino T, Shibata D, Chen JQ, Higuchi T (1993). Cinnamly alcohol dehydrogenase from *Aralia cordata*: cloning of the cDNA and expression of the gene in lignified tissues. Plant Cell Physiol., 34: 659-665.

Higuchi T (1997). Biochemistry and Molecular Biology of Wood. *Springer-Verlag*, New York.

Jones JB, Bouzar H, Somodi GC, Stall RE, Pernezny K, El-Morsy G, Scott JW (1998). Evidence for the preemptive nature4 of tomato race 3 of *Xanthomonas campestris* pv. *vesicatoria* in Florida. Phytopathology, 88: 33-38.

Laemmli UK (1970). Cleavage of structural proteins during the assembly of the head of bacteriophage T4. Nature, 227: 680-685.

Luderitz T, Griesebach H (1981). Enzymic synthesis of lignin precursors. Comparison of cinnamoyl CoA reductase and cinnamyl alcohol: NADP dehydrogenase from spruce (*Picea abies* L.) and Soybean (*Glycine max.* L.). Eur. J. Biochem., 65: 119-24.

Mackay JJ, Liu W, Whetten RW, Sederoff RR, O'Malley DM (1995). Genetic analysis of cinnamyl alcohol dehydrogenase (*Cad*) in lobololly pine: single gene inheritance, molecular characterization and evolution. Mol. Gen. Genet., 247: 537-545.

Mansell RB, Babbel GR, Zenk MH (1976). Multiple forms and specificity of coniferyl alcohol dehydrogenase from cambial regions of higher plants. Phytochemistry, 15: 1849-1853.

Mansell RL, Babbel GR, Zenk MH (1976). Multiple forms and specificity of coniferyl alcohol dehydrogenase from cambial regions of higher plants. Phytochem. 15:1849-53.

Mansell RL, Gross C, Stockigt J, Franke H, Zenk MH (1974). Purification and properties of cinnamyl alcohol dehydrogenase from higher plants involved in lignin biosynthesis. Phytochemistry, 13: 2427-37.

Mitchell HJ, Barber MS (1994). Elicitor-induced cinnamly alcohol dehydrogenase activity in lignifying wheat (*Triticum aestivum* L.) leaves. Plant Physiol., 104: 551-556.

Moerschbacher BM, Noll UM, Flott BE, Reisener HJ (1988). Lignin biosynthesis enzymes in stem rust infected, resistant and susceptible near- isogenic wheat lines. Physiol. Mol. Plant Pathol., 33: 33-46.

O'Malley DM, Porter S, Sederoff RR (1992). Purification, characterization and cloning of cinnamyl alcohol dehydrogenase in loblolly pine (*Pinus taeda* L). Plant Physiol., 98: 1364-71.

Pillonel C, Hunziker P, Binder A (1992). Multiple forms of the constitutive Wheat cinnamyl alcohol dehydrogenase. J. Exp. Bot., 43: 299-305.

Ride JP (1975). Lignification in wounded wheat leaves in response to fungi and its possible role in resistance. Physiol. Mol. Plant Pathol., 5: 125-134.

Ride JP (1983). Cell walls and other structural barriers in defence. In: Biochemical Plant Pathology, ed. Callow, J.A. (Wiley, London), pp, 215-236.

Sarni F, Grand C, Boudet AM (1984). Purification and properties of cinnamly CoA reductase and cinnamly alcohol dehydrogenase from popular stems (*Populus x euramericana*). Eur. J. Biochem., 139: 259-65.

Vance CP, Kirk TK, Sherwood RT (1980). Lignification as a mechanism of disease resistance. Ann. Rev. Phytopathol., 18: 259-288.

Walter MH, Grima-Pettenati J, Grand C, Boudet AM, Lamb CJ(1988). Cinnamyl- alcohol dehydrogenase, a molecular marker specific for lignin synthesis: DNA cloning and mRNA induction by fungal elicitor. Proc Nat Acad of Sci USA, 85: 5546-5550.

Wyrambik D, Grisebach H (1975). Purification and properties of isoenzymes of cinnamly- alcohol dehydrogenase from soybean- cell- suspension cultures. *Eur. J. Biochem.*, 59: 9-15.

6

Isolation and molecular characterization of *Flavobacterium columnare* strains from fish in Brazil

F. A. Sebastião[1], F. Pilarski[1]* and M. V. F. Lemos[2]

[1]Aquaculture Center of São Paulo State University - CAUNESP, Rod. Paulo Donato Castellane, s/n Bairro Rural, Jaboticabal, SP CEP 14884-900, Brazil.
[2]Universidade Estadual Paulista, Department of Applied Biology for Agriculture, Rod. Paulo Donato Castellane, s/n Bairro Rural, Jaboticabal, SP CEP 14884-900, Brazil.

Flavobacterium columnare, the etiologic agent of columnaris disease, has a broad geographical distribution and accounts for a large number of mortalities in fish species. This study aimed to generate a faster method for diagnosis of columnaris through isolation and characterization of the *F. columnare* 16S rDNA gene from bacteria isolated from Nile tilapia *(Oreochromis niloticus)*, tambaqui *(Colossoma macropomum)* and matrinxã *(Brycon amazonicus)*. The bacteria were characterized biochemically and by PCR-RFLP. For isolation, rasping with "swab" was performed directly on the characteristic lesions and the cephalic kidney of the fish then transferred to culture medium suitable for *Flavobacterium*. DNA was extracted for PCR and digestion with restriction enzymes. Altogether, 37 isolates were obtained. Biochemical assays included testing of absorption of Congo red, production of flexirrubin, production of H_2S, nitrate reduction and motility. The results indicated that the isolates can be classified as *F. columnare*. The phylogram generated by the PCR-RFLP technique showed three main branches among of the *F. columnare* isolates. Therefore, the use of PCR-RFLP for identification of the bacteria was shown to be a more efficient and rapid tool than current biochemical techniques, which are time consuming and often inconclusive.

Key words: Fish, *Flavobacterium columnare*, PCR-RFLP, 16S rDNA.

INTRODUCTION

Flavobacterium columnare, which is responsible for columnaris disease in fish, is an opportunistic bacterium with a broad geographical distribution. It makes up part of the normal microbiotic of the water, skin and gills of a fish. It is commonly observed in fresh-water fish culture and is described as a causal agent of large numbers of mortalities in trout *(Oncorhynchus mykiss)*, salmon *(Salmo salar)*, carp *(Cyprinus carpio)*, tilapia (Oreo-chromis niloticus), perch *(Perca fluviatilis)* and catfish *(Ictalurus punctatus)* (Bernardet and Grimont, 1989).

Columnaris disease is characterized by grayish dots or yellowish areas of erosion that are usually surrounded by a hyperemic reddish zone on the head, body surface and gills. These sites display progressive necrosis involving the epidermis, dermis and muscles, and they often lead to systemic infection (Decostere et al., 1999).

In Brazil, many columnaris outbreaks have been observed, especially during the changing of seasons when daily water temperature variation is greater than 5°C and water quality is inadequate. During such times there is

*Corresponding author. E-mail: fabianap@caunesp.unesp.br.

often a reduction of the dissolved oxygen concentration, an excess of organic matter and inadequate management. Besides being a serious health problem, this disease reduces production due to high and rapid mortality of fingerlings, decimation of shoals and non-utilization of the product due to the lesions that it causes. Despite this fact, there are few studies in the Brazilian literature addressing columnaris disease (Pilarski et al., 2008).

Species that exhibit rapid growth, good feed conversion, easily induced reproduction and features appropriate for sport fishing have stimulated interest in fish culture. For this study the following species were selected: matrinxã *(Brycon amazonicus)*, a species of the Bryconinae subfamily from the Amazon Basin (Tavares-Dias et al., 1999); tambaqui *(Colossoma macropomum)*, a freshwater teleost fish belonging to the order Characiformes, family Serrasalmidae (Géry, 1977), which is native to the Amazon and Orinoco basins and their tributaries (Chagas and Val, 2003); and the Nile tilapia *(O. niloticus)*, which has become the most cultivated species in Brazil over the last decade, accounting for approximately 40% of the national aquaculture (Zimmermann and Hasper, 2003; Marengoni, 2006).

Molecular characterization has been used in epidemiological studies to identify different pathogen genotypes in a safe and highly efficient manner. Furthermore, these techniques can be performed faster than other methods. Recognition of the prevalence of a particular genotype and its patterns in an infection can lead to the development of more effective methods for ichtiopathology control (Darwish and Ismaiel, 2005; Olivares-Fuster et al., 2007b; Figueiredo and Leal, 2008).

Several studies describe the genetic variation observed by PCR and RFLP analysis of *F. columnare* strains found throughout the world (Song et al., 1998; Bernardet and Grimont 1989; Toyama et al., 1996).

Therefore, the objective of this study was to isolate and characterize strains of *F. columnare* using molecular and biochemical methods. The intraspecific variation of the 16S rDNA gene using the PCR-RFLP technique was used in cultured specimens of tilapia *(O. niloticus)*, matrinxã *(Brycon amazonicus)* and tambaqui *(C. macropomum)* collected from various regions in the Southeastern Brazil, with characteristic clinical signs of columnaris disease.

MATERIALS AND METHODS

The study was conducted at the Laboratory of Bacteria Genetics and Applied Biotechnology, belonging to the Department of Applied Biology, Faculty of Agriculture and Veterinary Sciences - UNESP, Jaboticabal Campus. Isolation of *F. columnare* strains occurred in the Ichtiopathology Laboratory of the Department of Veterinary Pathology, Faculty of Agriculture and Veterinary Sciences - UNESP, Jaboticabal Campus.

Bacterial isolation

Samples of *F. columnare* were taken from 18 specimens of tilapia *(O. niloticus)*, matrinxã *(B. amazonicus)* and tambaqui *(C. macropomum)* from different fish farms in the region of São Paulo, Brazil. The fish used in this study were clinically diagnosed as suffering from columnaris disease by the Pathology Laboratory of Aquatic Organisms (LAPOA-CAUNESP-Jaboticabal, SP).

For isolation of bacteria, a sterile swab was rasped in the characteristic lesions and in the kidneys of the fishes. The material was then immediately transferred to a culture medium (liquid and solid) favorable for growth of *F. columnare*, modified by Pilarski et al., (2008) and incubated at 30 °C for 48 h.

The bacterial isolates were grown on solid medium that was composed of 100 ml natural fish fillet, 5.0 mL yeast extract (yeast Fleishemann), 0.02 g sodium acetate and 2.5 g agar at pH 6.8. Media were autoclaved at 121 °C for 15 min in 100 ml, distributed in Petri dishes for plating and incubated for 48 h in a bacteriological incubator at 30 °C.

The characteristic colonies were inoculated in flasks containing 6.0 ml of liquid medium which consisted of 100 ml natural fish fillet, 5 ml yeast extract and 0.02 g sodium acetate at pH 6.8. This medium was autoclaved at 121 °C for 15 min and incubated at 30 °C for 48 h.

Biochemical characterization

The biochemical tests for identification of *F. columnare* isolates were performed using kits for bacterial biochemistry (LaborClin®), in which included assays for desamination of L-tryptophane, utilization of glucose as a sole carbon and energy source, gas and H_2S production, decarboxylation of ornithine, indole, rhamnose, citrate and lysine, motility, production of flexirrubin type of pigments, Congo Red adsorption, nitrate reduction, TSI (Triple Sugar Iron Agar), esculin, mannitol, inositol, arginine, sucrose, lysine, starch hydrolysis and catalase.

DNA extraction and PCR-RFLP

DNA was extracted according to the method described by Wilson (1987) with modifications including the addition of lysozyme for cell wall degradation, proteinase K to degrade proteins, a solution of CTAB/NaCl and treatment with RNase A at the end of the process for DNA purification.

DNA from the new isolates of *F. columnare* was analyzed by amplification of the conserved region of 16S rRNA using the PCR method. Two strains of *F. columnare* (referred to as 38 and 39), one strain of *Aeromonas hydrophila* and one strain of *Streptococcus agalactiae* belonging to the collection of the Pathology Laboratory of Aquatic Organisms (LAPOA-CAUNESP - Jaboticabal / SP) were used as controls.

The sequences of oligonucleotide primers used for amplification of the ribosomal subunit 16S rDNA were made according to Kuske et al. (1997) as follows: PAF 5'AGAGTTTAGTCCTGGCTCAG 3' (located in *E. coli* bases 8 - 27) and PC5B 5'TACCTTGTTACGACTT 3' (located in *E. coli* bases 1507 - 1492), with approximately 1500 bp.

The amplification reactions were conducted in a reaction volume of 20 µl containing 110 ng DNA to be amplified, 0.5 µl solution of dNTPs (10 mM), 0.6 µl MgCl₂ (1.25 mM), 2.0 µl buffer solution (10X) for the PCR reaction, 1.0 µl of each primer (10 pmol/µl), 5.0 U *Taq* polymerase and Milli Q distilled water (previously sterilized, q.s.p.). All amplification reactions were performed in sterile tubes containing no genetic material.

The cycle of amplification was performed in a MJ Research, Inc., PTC-100 TM model thermal cycler equipped with a "Hot Bonnet" circuit. The program initially consisted of the following steps: 94℃ for 2 min for denaturation, then 35 cycles at 94℃ for 30 s, 56℃ for 30 s and 72℃ for 1 min, followed by an extension step at 72℃ for 5 min. Samples were then maintained at 10℃ until removal of the different letter.

PCR products were submitted to electrophoresis in 1.5% agarose gels for 1 h, 30 min at 90 V and stained with 5 µg/ml ethidium bromides to visualize the band of 1500 bp. In all electrophoresis experiments, a DNA sample of known molecular weight was used (1 Kb ladder purchased from Fermentas®), which served as a reference for calculation of the molecular weights of the fragments obtained in the PCRs.

The amplified DNA fragments were separately cleaved with five restriction enzymes including *EcoRI, MboI, PstI, HhaI* and *HpaII* according to the manufacturer's instructions. To visualize the genetic profile generated by digestion with these restriction endonucleases, the reactions were subjected to electrophoresis under the exact same conditions mentioned above.

A binary matrix was constructed based on the bands representing fragment polymorphisms that were generated by digestion of PCR with restriction enzymes (RFLP-PCR).

This matrix was used to construct a table of genetic similarity and a phylogram with the software FREETREE version 0.9.1.50, using UPGMA grouping and the Jaccard coefficient. Its visualization was made possible with the aid of the software TREEVIEW, bootstrapped 1000 times.

RESULTS

Bacterial isolates

The procedure for isolation of *F. columnare* yielded 37 new isolates, listed in Table 1. Among these, 22 were derived from tambaqui, 7 were from tilapia and 8 were from matrinxã. These isolates were characterized as *F. columnare* based on the characteristics of the colonies and their culture and biochemical features.

When the bacteria were cultured in liquid media, we observed the development of bacilli that were thin, long and exhibiting flexing movements by gliding. When grown on a solid medium, the colonies produced by strains were flat, rather small, yellow-orange and had rhizoid edges.

By gram staining, all strains were gram-negative, filamentous bacteria showed morphology of thin bacilli. The bacteria were grouped in columns in smears.

Biochemical tests

After a preliminary diagnosis of the isolates conducted by observation of colony morphology and behavior in a hanging drop under a microscope, 22 biochemical tests

were performed and their results are given in Table 2.

The TSI test showed four distinct behaviors among the isolates. A total of 49% of the isolates showed an alkaline response and production of H₂S; 8% presented little fermentation, alkaline character and production of H₂S; 27% performed anaerobic fermentation and had alkaline as well as acid characteristics in the test tube; 16% showed total fermentation.

The fact that the samples were positive for absorption of Congo red; produced flexirrubin, H₂S showed nitrates reduction and motility indicates that the isolates could be classified as *F. columnare* (Table 2).

Of the 22 biochemical tests, which required seven days for completion, 12 resulted in the same response for 100% of the isolates. The other tests showed variations between strains and could therefore not be used as standards for the identification of the species.

PCR-RFLP technique

The PCR-RFLP technique was used to develop a faster and more conclusive test.

The restriction of the amplified products with *EcoRI* generated two fragments between 900 and 700 bp. These fragments were produced by *EcoRI* digestion of genetic material from all of the microorganisms analyzed, resulting in a single genetic profile. Digestion with *EcoRI* was not even able to differentiate between the controls strains of *Aeromonas hydrophila* and *Streptococcus agalactiae* (Figure 1).

Restriction of amplified products with *PstI* generated two fragments between 1000 and 700 bp that were only observed in isolates 7, 8, 9, 11, 19, 20, 21, 38 and 39 of *F. columnare.* The other isolates were not cleaved by this enzyme.

The restriction of amplified products with *HhaI* produced four fragments between 700 and 200 bp in isolates 1 - 37 of *F. columnare*, three fragments between 1000 and 300 bp in strains 38 and 39 of *F. columnare,* three fragments between 600 and 300 bp in *A. hydrophila* and four fragments between 700 and 200 bp in *S. agalactiae.* These fragments were of similar molecular size to those observed in *F. columnare.*

Restriction with *MboI* produced four fragments between 1000 and 100 bp in the 37 new isolates of *F. columnare*, no cleavage in strains 38 and 39 of *F. columnare*, no cleavage in *A. hydrophila* and four fragments in *S. agalactiae* that were of different molecular sizes relative to those produced in *F. columnare.*

Restriction of the amplified products with *HpaII* generated four fragments between 750 and 100 bp in isolates of *F. columnare,* and the genetic profile did not differ from the fragments produced by digestion of sequences amplified from other species. Thus, there is

Table 1. Number, name and origin of the new *Flavobacterium columnare* strains isolated from the tropical fishes tilapia, matrinxã and tambaqui naturally infected with columnaris disease in Brazil.

Indication	Isolate name	Origen
1	F3	KIDNEY MATRINXÃ
2	F5	SKIN TAMBAQUI
3	F7	SKIN TAMBAQUI
4	F8	SKIN TAMBAQUI
5	F9	SKIN TAMBAQUI
6	F10	SKIN TAMBAQUI
7	F11	SKIN TAMBAQUI
8	F15	KIDNEY MATRINXÃ
9	F16	KIDNEY MATRINXÃ
10	F17	KIDNEY TAMBAQUI
11	F18	SKIN TAMBAQUI
12	F19	SKIN TAMBAQUI
13	F20	SKIN TAMBAQUI
14	F21	SKIN TAMBAQUI
15	F22	SKIN TAMBAQUI
16	F23	SKIN TILAPIA
17	F24	KIDNEY TILAPIA
18	F25	SKIN MATRINXÃ
19	F26	SKIN MATRINXÃ
20	F27	SKIN MATRINXÃ
21	F28	SKIN TAMBAQUI
22	F29	SKIN TILAPIA
23	F32	SKIN TILAPIA
24	F33	SKIN TILAPIA
25	F35	SKIN TAMBAQUI
25	F36	SKIN TILAPIA
27	F37	SKIN TAMBAQUI
28	F38	SKIN TAMBAQUI
29	F39	SKIN TAMBAQUI
30	F40	SKIN TAMBAQUI
31	F41	SKIN TAMBAQUI
32	F42	SKIN TAMBAQUI
33	F43	SKIN TILAPIA
34	F45	SKIN TAMBAQUI
35	F46	SKIN MATRINXÃ
36	F47	SKIN TAMBAQUI
37	F48	SKIN MATRINXÃ

no possibility of discrimination between the species studied by means of restriction with this enzyme.

Analysis of the phylogram (Figure 2) and the similarity matrix revealed three main branches between the new isolates of *F. columnare* in addition to the control group of this species. The similarity between species in group I was greater than or equal to 81%; in group II, the similarity was greater than or equal to 93%; in group III, the isolates shared 94% similarity or greater, and in the control group of *F. columnare*, the similarity was 100%.

Aeromonas hydrophila differed from group I of *F. columnare* by 65.5% on average. This species differed from group II by 58%, from group III by 55% and from the control group of *F. columnare* by 73%. *Streptococcus*

Table 2. Biochemical characteristics of *Flavobacterium columnare* strains isolated from the tropical fishes tilapia, matrinxã and tambaqui naturally infected with columnaris disease in Brazil.

Characteristics	Results (%)
Starch hydrolysis	+19 / - 81
Arginine	+ 92 / - 8
Catalase	+ 100
Citrate	+43/ - 57
L-Tryptophane desamination	+63 / - 37
Lysine decarboxylation	+ 90/ - 10
Ornithine decarboxylation	+100
Esculin	+59 / - 41
Glucose fermentation	+100
Flexirrubin-type pigment	+100
Gas from glucose	- 100
Gelatin hydrolysis	+100
Gram staining	-100
Indole production	+100
Inositol	+89/ - 11
Mannitol	+89/ - 11
Gliding movements	+100
H$_2$S and gas production	+66 / - 34
Nitrate reduction	+100
Rhamnose	+10 / - 90
Sucrose	+100
Congo red test	+100

(+) positive response, (-) negative response, (+/-)part of isolates a positive response and the other part, negative.

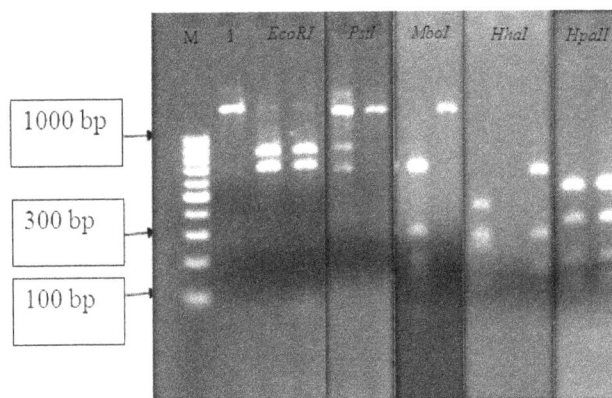

Figure 1. Gel containing the polymorphisms found among the PCR products of the 16S rDNA gene amplified from the DNA of *Flavobacterium columnare* isolates and digested with the restriction endonucleases *EcoRI, MboI, PstI, HhaI* and *HpaII*; M refers to 1 kb DNA ladder molecular marker; 1 refers to the amplified gene, without addition of restriction enzymes.

of this bacterium are carried out in fishes from temperate waters (Pilarski et al., 2008).

The lack of a specific medium for culturing *F. columnare* is another obstacle for isolation of the bacterium. Decostere et al. (1998) reported on the difficulty of obtaining pure colonies.

In this work we used the Carlson and Pacha (1968) medium modified by Pilarski et al. (2008), which enabled the development of *Flavobacterium* colonies, as well as probable strains of *Aeromonas* and *Pseudomonas,* which grew in an overlapping pattern. A large number of replications were required to purify the colonies of the desired species.

The antibiotic tobramycin was not used to eliminate contaminating species as was previously described by Decostere et al. (1998) in order not to induce the appearance of L form bacteria (Davis et al., 1973).

The pure *F. columnare* colonies were studied with respect to their morphophysiological and biochemical characteristics as described by the Decostere et al. (1998). Their molecular characteristics (PCR-RFLP of 16S rDNA) were assessed according to Arias et al. (2004) and Figueiredo et al. (2005).

In accordance with the results described by Pilarski et al. (2008), the 37 isolates found in this study showed rod-shaped, Gram-negative, long, slender and aerobic cells with optimum growth at 30 °C.

In liquid medium, the isolates showed sliding motility in a hanging drop analysis. On solid medium, the isolates produced colonies of yellow-grayish color with irregular borders. The colonies had the form of roots and were not strongly adherent to the Agar. Similar results were obtained by Shamsudin and Plumb (1996), Descostere et al. (1998) and Pilarski et al. (2008).

agalactiae differed by only 27% on average from group I of *F. columnare*, by 21% compared from group II, by 23% from group III and by 56% from the control group of *F. columnare*. The genetic difference between strains of *A. hydrophila* and *S. agalactiae* was 53%.

Regarding the preference of each *F. columnare* subgroup for the studied fish species, we found that for isolates belonging to subgroup I, 54% were from tambaqui, 31% were from tilapia and 15% were from matrinxãs. In subgroup II, 83% were from tambaqui and 17% were from tilapia. Finally, 57% of the isolates in subgroup III were derived from matrinxãs, 29% were from tambaquis and 14% were from tilapia. In the control group of *F. columnare,* a strain was isolated from Nile tilapia and the other was from tambaqui.

DISCUSSION

Information on the occurrence, pathogenicity and isolation of *F. columnare* in fish cultures in Brazil is scarce. Most of the publications on isolation and characterization

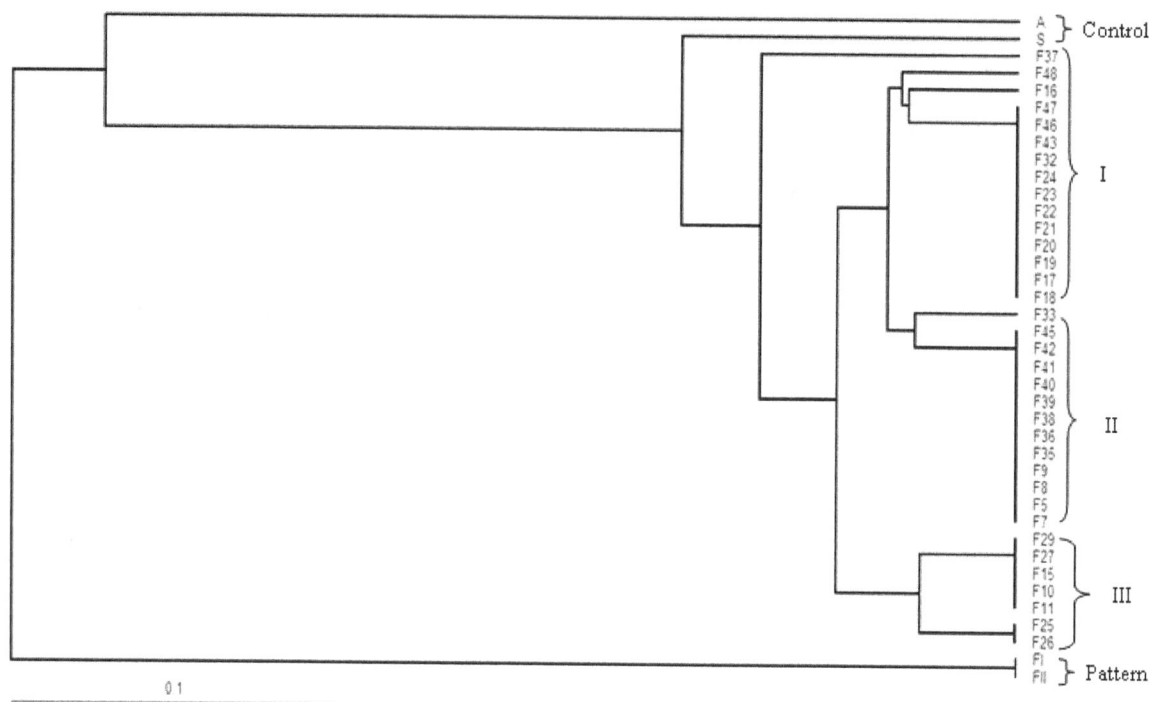

Figure 2. Phylogram obtained by analysis of PCR products amplified from DNA isolates of *Flavobacterium columnare*(F), *Aeromonas hydrophila* (A) and *Streptococcus agalactiae* (S) and digested with the restriction endonucleases *EcoRI, MboI, PstI, HhaI* and *HpaII*.

Bernadet and Grimont. (1989), Griffin (1992), Shamsudin and Plumb (1996), Descostere et al. (1998) and Pilarski et al. (2008) have reported that cultures grown on solid medium showed an organoleptic property similar to the smell of fruit. Among the isolates described in this work, this feature was not noticeable.

Regarding the biochemical characterization, a wide variety of behaviors could be observed among the various isolates. Among the tests for characteristics that are specific to *F. columnare* (motility, nitrate reduction, flexirrubin production, absorption of Congo Red, fermentation of glucose and sucrose, desamination of L-tryptophan, gelatin and catalase hydrolysis), all isolates tested uniformly positive regardless of the fish species from which they were collected. This result is consistent with results reported by Pilarski et al. (2008).

The absorption of Congo red, which occurred in all isolates, is indicative of galactosamine glucan production by *F. columnare*, and this fact is corroborated by Bernadet et al. (1989), Griffin (1992), Decostere et al (1999) and Pilarski et al. (2008).

All isolates in this study were positive for gelatin hydrolysis, and this is one of the main features of the bacterium *F. columnare*. Because the production of certain enzymes promotes the degradation of macrom-

olecules in the medium such as gelatin, there is a clearing of the medium in which there is bacterial growth (Bernadet and Grimont 1989; Griffin, 1992; Pilarski et al., 2008).

Among the molecular techniques, PCR is undoubtedly the most efficient tool for the diagnosis of columnaris. Several protocols were applied, but due to the lack of data on the genome sequence of *F. columnare* from water and tropical climate hosts, the use of specific primers for conserved regions of bacterial DNA is the most accessible technique (Bader et al., 2003; Darwish et al., 2004; Figueiredo and Leal, 2008).

Thus, using PCR-RFLP of the 16S rRNA gene for identification and differentiation of *F. columnare* isolates from existing strains such as *A. hydrophila* and *S. agalactiae*, it was possible to map genetic profiles of each microorganism in relation to the applied restriction enzyme (Figure 1).

Restriction of the amplified products with *EcoRI* and *HpaII* showed only one genetic profile for the three species in question and did not allow for differentiation. While restriction with *PstI* allowed for the detection of two genetic profiles of *F. columnare* (Figure 1). However, according to a simulation performed with the software "pdraw32" on one 16S gene of *F. columnare* that has been partially sequenced (deposited in the National

Center of Genes and Genomes Information), the enzyme *PstI* has no restriction site in this gene. The fact that the enzyme recognized a site in some isolates suggests the presence of a mutation.

After restriction with *MboI*, two genetic profiles of *F. columnare* were observed and these were distinct from the profiles of *A. hydrophila* and *S. agalactiae*. Similar results were also observed after restriction of the amplified products with *HhaI*, but there was no difference in the profile of *S. agalactiae* (Figure 1).

The phylogram (Figure 2) shows three main branches between the isolates of *F. columnare* in addition to the grouping of *F. columnare* strains that was used as a control. Several authors have found intraspecific variation between the different *F. columnare* isolates that were studied. Shoemaker et al. (2007) found two subgroups among his isolates using the ISR-SSCP technique; Arias et al. (2004) and Darwish and Ismaiel (2005) described three *F. columnare* subgroups when using PCR-RFLP; and Flemming et al. (2007) identified two *Flavobacterium johnsoniae* subgroups by PCR-RFLP with the enzyme *HaeIII*.

Amplification of the 16S rDNA gene region and the respective application of RFLP with the enzymes *EcoRI, MboI, PstI, HhaI* and *HpaII* is a simple technique to be performed for the characterization of large numbers of isolates; nevertheless, the RFLP patterns produced depend on the variability of the sequences produced in determined regions of the gene.

Data from this study show the existing genetic variation within a supposedly conserved gene, 16S rDNA, and they can explain the difficulty of establishing diagnostic protocols for molecular characterization of the studied species. Analysis of the phylogram and its ramifications also suggests the existence of subspecies of *F. columnare*.

The results are in accordance with Sebastião et al., 2007 (unpublished data) who showed by means of the RAPD-PCR technique that both strains are 78% similar and sufficient for grouping.

Dependence of the molecular technique on certain regions of the gene makes comparisons between isolates from different studies a challenge unless similar methods are applied in the analysis (Darwish and Ismaiel, 2005).

Although the number of fish samples was small (18 samples), this study represents an initial attempt to correlate the genetic subgroups of *F. columnare* with their host species, that is, tambaquis are more susceptible to subgroups I and II of tropical *F. columnare* and matrinxã is more sensitive to colonization by strains belonging to the bacteria of subgroup III. No assumptions can be made regarding the control group because it only contains two strains of different origins. Similar relationships have also been observed by Olivares-Fuster et al. (2007a), but for temperate climate species.

Conclusion

Therefore, based on 37 isolates obtained from three fish species (matrinxã, tambaqui and Nile tilapia), the existence of broad genetic diversity in tropical *F. columnare* strains was verified, despite the similarity of phenotypic and biochemical characteristics. For Brazilian aquaculture, this work signifies a first step in identifying the existing *F. columnare* strains for a faster and more precise diagnosis and the correlation of three genetic subgroups with host species.

REFERENCES

Arias CR, Welker TL, Shoemaker CA, Abernathy JW, Klesius PH (2004). Genetic fingerprinting of *Flavobacterium columnare* isolates from cultured fish. J. Appl. Microbiol., 97: 421-428.

Bader JA, Shoemaker CA, Klesius PH (2003). Rapid detection of columnaris disease in channel catfish (*Ictalurus punctatus*) with a new species-specific 16S rRNA gene based PCR primer for *Flavobacterium columnare*. J. Microbiol. Meth., 52: 209-220.

Bernardet JF, Nakagawa Y, Holmes B (2002).Proposed minimal standards for describing new taxa of the family *Flavobacteriaceae* and emended description of the family.Int.J.Syst.Evol.Micr.,52:1049-1070.

Carlson RV, Pacha RE (1968). Procedure for the isolation and enumeration of myxobacteria from aquatic habitats. J. Appl. Microbiol. 16: 795-796.

Chagas EC, Val AL (2003). Efeito da vitamina C no ganho de peso e em parâmetros hematológicos de tambaqui. Pesq. Agropec. Bras., 38: 397-402.

Darwish AM, Ismaiel AA, Newton JCE, Tang J (2004). Identification of *Flavobacterium columnare* by a species-specific polymerase chain reaction and renaming of ATCC 43622 strain to *Flavobacterium johnsoniae*. Mol. Cell. Probes., 8: 421-427.

Darwish AM, Ismaiel AA (2005). Genetic diversity of *Flavobacterium columnare* examined by restriction fragment lengh polymorphism and sequencing of the 16S ribosomal RNA gene and the 16S-23S rDNA spacer. Mol. Cell. Probes., 19: 267-274.

Davis BD, Dulbecco R, Eisen HN, Ginsberg HS, Barry WW (1973). Infecções bacterianas e micóticas. Microbiologia, São Paulo: EDART – São Paulo Ltda. pp. 310-311.

Decostere A, Haesebrouck F, Devriese LA (1998). Characterization of four *Flavobacterium columnare* (*Flexibacter columnaris*) strains isolated from tropical fish. Vet. Microbiol., 62: 35-45.

Decostere A, Haesebrouck F, Charlier G, Ducatelle R (1999). The association of *Flavobacterium columnare* strains of high and low virulence with gill tissue of black mollies (*Poecillia sphenops)*. Vet. Microbiol. 67: 287-298.

Figueiredo HCP, Klesius PH, Arias IR, Evans J, Shoemaker CA, Pereira FI, Peixoto MTD (2005). Isolation and characterization of strains of *Flavobacterium columnare* from brazil. J. Fish. Dis., 28: 199-204.

Figueiredo HR, Leal CAG (2008). tecnologias aplicadas em sanidade de peixes. R. Braz. J. Anim. Sci. 37:08-14

Flemming L, Rawlings D, Chenia H (2007). Phenotypic and molecular characterisation of fish-borne *flavobacterium johnsoniae*-like isolates from aquaculture systems in south africa. Res. Microbiol., 158: 18-30.

Géry J (1977).Characoids of the world. Neptune: Trop. Fish Hob, pp. 672.

Griffin BR (1992). A simple procedure for identification of *cytophaga columnaris*. J. Aquat. Anim. Health., 4: 63-66.

Kuske CR, Barns SM, Busch JB (1997). Diverse uncultivated bacterial groups from soils of the arid southwestern united states that are present in many geographic regions. J. Appl. Env. Microbiol., 63: 3614–3621.

Marengoni NZ (2006). Produção de tilápia do nilo *oreochromis niloticus* (linhagem chitralada), cultivada em tanques-rede, sob diferentes densidades de estocagem. Arch. Zootec., 55: 127-138.

Olivares-Fuster O, Baker JL, Terhune JS, Shoemaker CA, Klesius PH, Arias CR (2007a). Host-specific association between *Flavobacterium columnare* genomovars and fish species. Syst. Appl. Microbiol. 30: 624-633.

Olivares-Fuster O, Shoemaker CA, Klesius PH, Arias CR (2007b). Molecular typing of isolates of fish pathogen, *Flavobacterium columnare*, by single-strand conformation polymorphism analysis. FEMS Microbiol. Lett., 269: 63–69.

Pilarski F, Rossini AJ, Ceccarelli PS (2008). Isolation and characterization of *Flavobacterium columnare* (Bernardet et al. 2002) from four tropical fish species in Brazil. Braz. J. Biol. 68: 409-414.

Shamsudin MN, Plumb JA (1996). Morphological, biochemical and physiological characterization of *Flexibacter columnaris* isolates from four species of fish. J. Aquat. Anim. Health., 8: 335-339.

Shoemaker CA, Olivares-Fuster O, Arias CR, Klesius PH (2008). *Flavobacterium columnare* genomovar influences, mortality in channel catfish (*Ictalurus punctatus*). J. Vet. Microbiol., 127: 9-353.

Song YL, Fryer JL, Rohovec JS (1998). Comparison of gliding bacteria isolated from fish in north America and other areas of The Pacific rim. J. Fish Pathol., 23: 197-202.

Tavares-Dias M, Frasca-Scorvo CMD, Campos-Filho E, Moraes FR (1999). Características hematológicas de teleósteos Brasileiros. Iv. Parâmetros Eritroleucométricos, trombométricos e Glicemia do matrinxã (*Brycon cephalus* Günther, 1869) (Osteichthyes: Characidae). Ars. Vet., 15: 149-153.

Toyama T, Kita-Tsukamoto K, Wakabayashi H (1996). Identification of *Flexibacter maritimus, Flavobacterium branchiophilum* and *Citophaga columnaris* by PCR targeted 16S ribossomal DNA. J. Fish Pathol., 31: 25-31.

Wilson K (1987). Large scale CsCl preparation of genomic DNA from bacteria. In: Ausubel FMR, Brent RE, Kingston DD, Moore JG, Seidman JA, Smith K, Struhl ŽEds, Cur. Prot. in Mol. Biol. Greene Publishing Associates and Wiley-Interscience, NewYork., 1: 243–245.

Zimmermann SE, Hasper TOB (2003). Piscicultura no Brasil: o processo de intensificação da tilapicultura. In: Reunião Anual da Sociedade Brasileira de Zootecnia, 40 Santa Maria. Anais, SBZ. CD ROOM.

7

Extended spectrum beta lactamases among multi drug resistant *Escherichia coli* and *Klebsiella* species causing urinary tract infections in Khartoum

Akram Hassan Mekki[1], Abdullahi Nur Hassan[2]* and Dya Eldin M Elsayed[3]

[1]Department of Microbiology, Faculty of Health Sciences, Omdurman Ahlia University, Sudan.
[2]Department of Clinical Microbiology and Infectious Disease, Faculty of Medicine, Alzaiem Alazhari University, Sudan.
[3]Department of Community Medicine, Faculty of Medicine, Alzaiem Alazhari University, Sudan.

This is a descriptive laboratory based case study carried out in Khartoum state hospitals during the period of June, 2007 to April, 2008. The study aimed to evaluate emergence of ESBL among multi drug resistant *Escherichia coli and Klebsiella* species causing nosocomial UTI. Hundred strains of multi drug resistant (MDR) *E. coli and Klebsiella* species causing nosocomial urinary tract infections (UTIs) from two main hospitals from Khartoum (Omdurman teaching hospital and Fedail Hospital) were included in this study. Susceptibility testing was performed against antibiotics commonly used in treatment of urinary tract infections. *E. coli*, *Klebsiella pneumoniae* and *Klebsiella oxytoca* (49, 38 and 13% respectively) were among the studied isolates. β-Lactamase was produced by all isolates; high resistance level for 3rd generation cephalosporin was noticed. ESBLs were detected in high prevalence among all multi drug resistant *E. coli* and *Klebsiella* species isolates 53%. All isolates were found sensitive to Imipenem and Meropenem. In this study it's recommended that developing guidelines for the early phenotypic detection of ESBL in microbiology laboratories and seeking knowledge of antibiotic susceptibility pattern for empirical antibiotic therapy. Further studies about ESBL occurrence among UTIs are also recommended.

Key words: ESBL in Sudan, multi-drug resistant, MDR *Escherichia coli*, MDR *Klebsiella* spp, urinary tract infection, beta lactamase.

INTRODUCTION

Urinary tract infection (UTI) is a very common infection both in the community and hospital patients and ranks high amongst the most common reasons that compel a patient to seek medical attention (Gastmeier et al., 1998, Magee et al., 1999, Mobley, 2000).

Uropathogens have shown a slow but steady increase of resistance to several agents over the last decade. *Escherichia coli* and other *Enterobacteriaceae* have become less susceptible to commonly used antimicrobials such as trimethoprim/sulfamethoxazole and, in some areas, fluoroquinolones (Winstanley et al., 1997,

1997, Bishara et al., 1997). Study done by (Hassan et al., 2007) in Sudan showed multidrug resistance among uropathogens.

It is well known that the mechanism of antimicrobial resistance could happen by enzymatic inactivation, altered receptors or by altered antibiotic transport (Koneman et al., 1997). Current knowledge on antimicrobial susceptibility pattern of uropathogens is mandatory for appropriate therapy. Extended Spectrum Beta Lactamases (ESBL) hydrolyses expanded spectrum cephalosporins, which are used in the treatment of UTI. They arise by mutations in genes for common plasmid-mediated beta lactamases that alter the configuration of the enzyme near its active site to increase the affinity and hydrolytic ability of the beta lactamases for oxyimino

*Corresponding author. E-mail: Abdullahi2001@yahoo.com.

compounds while simultaneously weakening the overall enzyme efficiency. Some ESBLs confer high-level resistance to all oxyimino beta lactams, but for other ESBLs resistance is only slightly increased or increased selectively for particular beta lactams. This creates a problem for the clinical laboratory, since organisms producing less active ESBLs can fail to reach current National Committee for Clinical Laboratory Standard's (NCCLS) break points for resistance but can cause significant disease (Katsanis et al., 1994).

The aim of this study is to evaluate emergence of ESBL among multi drug resistant *E. coli and Klebsiella* species causing nosocomial UTI, since the failure of treatment of UTI both complicated and uncomplicated, which are most commonly caused by *Enterobacteriacae*, will increase the rate of morbidity and mortality among UTI patients. Also study sought alternative drugs which can be used to these MDR strains.

MATERIALS AND METHODS

A laboratory based descriptive case study was carried out in order to evaluate emergence of ESBL among multi drug resistant *E. coli and Klebsiella* species causing nosocomial UTI. One hundred bacterial strains were collected from two main (Omdurman Teaching (64) and Fedail (36)) hospital laboratories in Khartoum state during June 2007 to April 2008 as pure isolates in nutrient agar slopes. The study was conducted in the laboratories of Medical Microbiology Laboratory, Faculty of Health Sciences, Omdurman Ahlia University. Personnel data were collected from file records.

Strains of *E. coli and Klebsiella* species with a significant growth (>10^5 CFU) that showed resistance to amoxicillin and at least one cephalosporin and collected from urine of patients who fulfill the definition of nosocomial UTI regardless of age were included in this study. No hospital infection outbreaks were reports during study period.

Non-probability sampling technique was used and selected samples were taken as isolated multidrug resistant *E. coli and Klebsiella* species.

Antimicrobial susceptibility testing was performed, to confirm the multidrug resistance, by using Kirby-Bauer susceptibility testing technique with commercially available disks (Oxoid-UK). Amoxycillin (30 µg) nalidixic acid (30 µg), nitrofurantoin (50 µg), ciprofloxacin (1 µg), gentamicin (10 µg), amikacin (10 µg), cefuroxime (30 µg) and trimethoprim / sulphamethoxazole (1.2 / 23.8 µg) were disks used (National Committee for Clinical Laboratory Standards, 2001).

Beta lactamase test was performed to confirm the production of beta lactamase enzymes by using Nitrocefin disk which is recommended by National committee for Clinical Laboratory Standards (NCCLS) and the World Health Organization (WHO) (O'Callaghan et al., 1972).

Etest was used to determine the MIC of ceftazidime and cefotaxime (Cormican et al., 1996).

ESBL production was detected by double disk diffusion method. Discs containing ceftazidime (30 µg), cefotaxime (30 µg) respectively, were placed 25 mm (centre to centre of the discs) from the co-amoxiclav disc. Inoculated plates were incubated aerobically at 37°C for 18 - 24 h. After overnight incubation at 37°C, a clear extension of the edges of the inhibition zone of any of the antibiotics towards the disc containing clavulanic acid was regarded as a phenotypic confirmation of the presence of ESBL (Therrien et al.,

2000).

Antimicrobial susceptibility test for alternative drugs were performed by Kirby-Bauer susceptibility testing technique. These include Imipenem (30 µg), Meropenem (50 µg), Aztreonam (30 µg), Pipercillin / tazobactum (1 µg). The data was analyzed by using Statistical Package for Social Studies (SPSS)

Ethical consideration

This study didn't involve human subject. No personnel identifying data were used so, informed consent was not sought.

RESULTS

One hundred bacterial strains identified as *E. coli* (49%), *K. pneumoniae* (38%) or *Klebsiella oxytoca* (13%) and resistant to amoxicillin and at least one cephalosporin were studied. Strains were identified from urine sample collected from patients with defined nosocomial UTI in ages between four days to ninety-one years. Pre susceptibility patterns of 100 selected isolates were done (Table 1).

Isolates that were found to be resistant to amoxicillin were suspected of harbouring one or more β-lactamases. Nitrocefin tests were performed and they all gave positive results.

All of the isolates were resistant to amoxicillin, naldixic acid, gentamycin, trimethoprim/sulphamethoxazole. Nitrofurantoin, ciprofloxacin and cefuroxime resistance was seen among most of isolates. Amikacin showed a better susceptibility pattern, especially in *Klebsiella* species (Table 1).

MIC for cefotaxime and ceftazidime by using Etest was obtained in this study. Cefotaxime was susceptible 13.6% and ceftazidime 7.8% of all multidrug resistant *E. coli and Klebsiella* species.

Fifty three isolates (53%) indicated the production of ESBL with double disc diffusion, using cefotaxime and ceftazidime disks (Photo 1).

All isolates were 100% susceptible for imipenem and meropenem. Pipracillin/tazobactum showed a good activity (74.05%) against tested isolates, while azterionam showed a poor activity as there were only 19.23% susceptible isolates. The study revealed that there were a high resistance pattern for cefoxitin and cefipime, 100 and 97.09% respectively.

DISCUSSION

Antimicrobial resistance is now recognized as an increasingly global problem, especially Gram-negative bacteria (Slama, 2008).

Increasing resistance to broad spectrum cephalosporins amongst *E. coli* and *Klebsiella* species predominantly due to the production of ESBLs were reported from different countries (Bouchillon et al., 2002, Khanfar et al.,

Table 1. Antimicrobial resistance profiles of isolates by disk diffusion method from urine samples (n = 100).

Isolate	E. coli (%)	K. pneumoniae (%)	K. oxytoca (%)
AMX	100	100	100
NA	100	100	100
F	100	97.37	100
CIP	97.96	100	92.31
CN	100	100	100
AK	69.39	39.47	30.77
CXM	95.92	100	100
SXT	100	100	100

Key: AMX, Amoxicillin; NA, Naldixic Acid; F, Nitrofurantoin; CIP, Ciprofloxacin; CN, Gentamicin; AK, Amikacin, CXM, Cefuroxime; SXT, Trimethoprim / sulphamethoxazole.

2008).

Many other reports from different countries and regions showed different prevalence rates of ESBLs producing *Entrobacteriaceae* causing urinary tract infections. *K. pneumoniae* and *E. coli* are the most common ESBL-positive species, but all *Enterobacteriaceae* can harbor plasmid-mediated ESBL genes (Bouchillon et al., 2002, Lautenbach et al., 2001). In this study the ESBLs producing MDR uropathogen *E. coli and Klebsiella* species were 53%. This finding is a little bit higher than those obtained from the studies done by (Bouchillon et al., 2002) from Egypt and (Kadar et al., 2005) in Saudi Arabia, where ESBL were produced by 40.9 and 40.3% respectively. Another study carried out in India (Mohammed et al., 2005) showed that ESBL was positive in 42% and Supriya et al. (2003) detected ESBL production in 48.3%, as determined by the double disc synergy test, which was much higher than that obtained by Ibukun et al. (2001) from Nigeria in which ESBL production was only 20.8%. The result of this study is much less than Mohanty et al. in India 2003, where they observed ESBL production in 71.5% of the Gram-negative bacilli.

Resistance to additional classes of antibiotics rather than beta lactams has been noted among ESBLs producing *E. coli* and *Klebsiella* species. With resistance to each additional class of antibiotics, infections related with ESBL producing bacteria become a greater therapeutic challenge (Emily et al., 2005).

This study showed that all isolates were sensitive to the carbapenems which are the most common alternative drugs used for treatment of ESBL producing bacteria. Similar results were observed in the study done by Kadar et al. (2005) who revealed that, more than 89% of the ESBL producers were susceptible to imipenem and meropenem. However, use of alternative drug which is very broad spectrum and expensive drug as first line for treatment of ESBL-positive bacteria will significantly increases cost of treatment and will contribute to carbapenem resistance in other organism (Nordman et al., 2002; Wright et al., 2008).

High prevalence rate of ESBL-producing bacteria

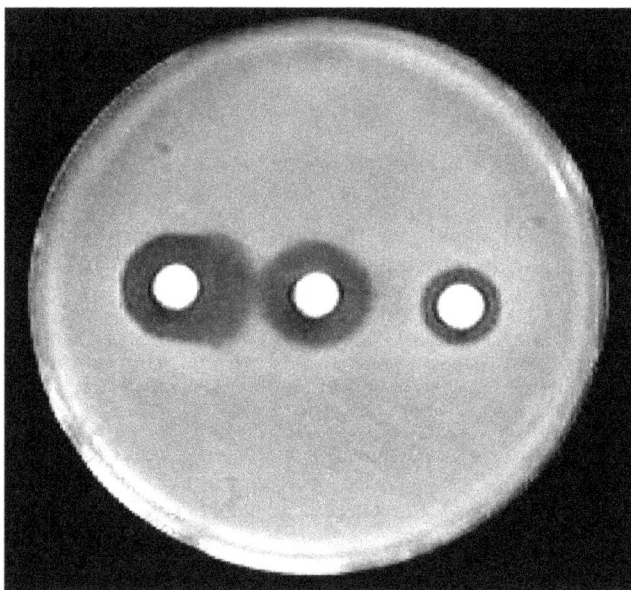

Photo 1. *Klebsiella pneumoniae* showing ESBL production by double disk diffusion method. Ceftazidime (left disc) a clear extension of the edges of the inhibition zone of the antibiotic towards the disc containing clavulanic acid (middle disc) while cefotaxime (right disc) showed negative test result.

among MDR *E. coli* and *Klebsiella* species were shown in this study. This rate is alarming and need special consideration. This study investigated only the frequency of ESBL producing among MDR *E. coli and Klebsiella* species causing nosocomial UTI and this is limitation of the study. The prevalence rate of ESBL producing non-MDR *E. coli and Klebsiella* species in the country were not studied and it was out of this study aim. Further studies about this issue are needed.

ESBL producers are associated with increased morbidity and mortality, especially amongst patients on intensive care and high-dependency units (John et al., 2002). Updated knowledge of the common antibiotic-sensitivity patterns must be sought when starting empirical antibiotic therapy in Sudanese patients with urinary tract infection. Guidelines for the early phenotypic detection of ESBL in microbiology laboratories are needed. Further studies about ESBL occurrence among UTIs are also recommended.

ACKNOWLEDGMENTS

We are thankful to our colleagues Ahmed Mohd Ibrahim, Muaz Osman Fagery, Mohamed Hussain Arbab and Sahar Mohamed Seid Ahmed for their help and given us valuable advices throughout this work. Also the staffs of the Department of Medical Microbiology and infection control committee for both hospitals.

Finally, we would like to thank all my friends for their encouragement and support during the preparation of this research.

REFERENCES

Bishara J, Leibovici L, Huminer D, Drucker M, Samra Z, Konisberger H, (1997). Five-year prospective study of bacteraemic urinary tract infection in a single institution. Eur J. Clin Microbial. Infect. Dis. 16 (7): 563-567.

Bouchillon SK, Johnson BM, Hoban DJ, Johnson JL, Dowzicky MJ, Wu DH. (2002). Interscience Conference. Antimicrob. Agents Chemother.: 42: 27-30.

Cormican MG, Marshall SA, Jones RN, (1996). Detection of extended-spectrum β-lactamase (ESBL)-producing strains by the Etest ESBL screen. J. Clin.l Microbiol. 34: 1180–4.

Emily PH, Adam DL, Theoklis EZ, Irving N, Neil OF, Warren BB, Xiangquin M, Ebbing L, (2005). Risk Factors for Increasing Multidrug Resistance among Extended Spectrum β-Lactamase Producing Escherichia coli and Klebsiella Species. Clin. Infect. Dis. 40:1317–1324.

Gastmeier P, Kampf G, Wischnewski N, Hauer T, Schulgen G, Schumacher M. (1998). Prevelance of nasocomial infections in representative German hospitals. J. Hosp. Infect. 38: 37-49.

Hassan AN, Elsayed DE, Mahjoub M, (2007). Uropathogens and their antibiotic resistance patterns. Sudan Med. Monitor. 2(2): 51-54

Ibukun A, Tolu O, Brian JM, (2003). Extended-Spectrum ß-Lactamases in isolates of Klebsiella spp and Escherichia coli from Lagos, Nigeria. Nig. J. Health Biomed. Sci., 2: 53-60.

John T, Jan B, Douglas J, Biedenbach N, Ronald J, (2002). Pathogen occurrence and antimicrobial resistance trends among urinary tract infection isolates in the Asia-Western Pacific Region: report from the SENTRY Antimicrobial Surveillance Program, 1997–2000. Int. J. Antimicrob. Agents. 20: 10-17.

Kadar AA, Angamathu K, (2005). Extended-spectrum beta-lactamases in urinary isolates of Escherichia coli, Klebsiella pneumoniae and other gram-negative bacteria in a hospital in Eastern Province, Saudi Arabia. Saudi Med. J. 26(6): 956-9.

Katsanis GP, Spargo J, Ferraro MJ, Sutton L, Jacoby GA, (1994). Detection of Klebsiella pneumoniae and Escherichia coli strains producing extended spectrum beta lactamases. J. Clin. Microbial. 32: 691-6.

Khanfar HS, Bindayna KM, Senok AC, Botta GA (2009). Extended spectrum beta-lactamases (ESBL) in Escherichia coli and Klebsiella pneumoniae: trends in the hospital and community settings. J. Infect. Dev Ctries. 1;3(4): 295-9

Koneman EW, Allen SD, Janda WM, Schreckenberger P, Winn WC, (1997). Antimicrobial Resistance, Color Atlas and Text book of Diagnostic Microbiology, Fifth Edition, USA (Philadelphia), Lippincott-Raven Publishers. 15: 798-800.

Lautenbach E, Patel JB, Bilker WB, Edelstein PH, Fishman NO, (2001). Extended-Spectrum β-Lactamase-Producing Eschericia coli and Klebsiella pneumoniae: Risk factors for infection and impact of resistance on outcomes. Clin. Infect. Dis., 32: 1162-1171.

Magee JT, Pritchard EL, Fitzgerald KA, Dunstan FDJ, Howard AJ, (1999). Antibiotic prescribing and antibiotic resistance in community practice: retrospective study 1996-1998. BMJ. 319: 1239-1240.

Mobley HL, (2000). Virulence of the two primary uropathogens. ASM News. 66: 403-410.

Mohammed A, Mohammed S, Asad K, (2007). Etiology and antibiotic resistance patterns of community-acquired urinary tract infections in J N M C Hospital Aligarh, India Ann. Clin. Microbiol. Antimicrob. 6(4):1 7.

Mohanty S, Kapil A, Das BK, Dhawan B, (2003). Antimicrobial resistance profile of nosocomial uropathogens in a tertiary care hospital. Ind. J. Med. Sci. 57(4): 148-154.

National Committee for Clinical Laboratory Standards: (2001). Performance standards for antimicrobial susceptibility testing. International Supplement. NCCLS Committee for Clinical Laboratory Standards. Wayne, Pa 11th edition.

Nordmann P, Poirel L, (2002). Emerging carbapenemases in gram-negative aerobes. Clin. Microbiol. Infect. 8: 321-31.

Callaghan CH, Morris A, Kirby SM, Shingler AH, (1972). Novel method for detection of β-lactamases by using a chromogenic cephalosporin substrate. Antimicrob. Agents Chemother. 1(4): 283-8.

Slama TG, (2008). Gram-negative antibiotic resistance: there is a price to pay. Crit. Care. 12(4): 1-7

Supriya ST, Suresh VJ, Sarfraz A, Umesh H, (2004). Evaluation of extended spectrum beta Bishara Therrien C, Levesque RC, (2000). Molecular basis of antibiotic resistance and β-Lactamase inhibition by mechanism-based inactivators: perspectives and future directions. FEMS Microbiol. Rev. 24: 251-262.

Winstanley TG, Limb DI, Eggington R, Hancock F, (1997). 10 year survey of the antimicrobial susceptibility of urinary tract isolates in the UK: the Microbe Base project. J. Antimicrob. Chemother. 40: 591–594.

Wright BM, Eiland EH, (2008). Current Perspectives on Extended-Spectrum Beta-Lactamase-Producing Gram-Negative Bacilli. J. Pharm. Pract. 21(5): 338-345

8

Asymptomatic bacteriuria in patients with type-2 diabetes mellitus

E. A. Ophori[1]*, P. Imade[2] and E. J. Johnny[3]

[1]Department of Microbiology, University of Benin, Benin City, Edo State, Nigeria
[2]Department of Medical Microbiology, University of Benin Teaching Hospital, Benin City, Edo State, Nigeria.
[3]University of Benin Teaching Hospital, Benin City, Edo State, Nigeria.

This study was to investigate the prevalence of asymptomatic bacteriuria (ASB) in patients clinically diagnosed with type 2 diabetes mellitus and to determine the antibiotic sensitivity pattern of bacterial isolates. One hundred and thirty type 2 diabetics comprising 56 males and 74 females (aged between 30 - 59 years) attending the Central Hospital, Benin- City, Nigeria were studied. Mid-stream urines were collected from patients who gave informed consent aseptically into sterile McCartney bottles and examined microscopically, culturally using standard techniques and tested for glucose, post-prandial glucose, protein and ketone using a dipstick. Samples were cultured on blood agar, McConkey agar and cysteine lactose electrolyte deficient (CLED) media and incubated at 37°C aerobically for 24 h. Isolates were tested against antibiotics which included tetracycline, chloramphenicol, ciprofloxacin and cotrimoxazole by the disc diffusion method. White blood cells (WBC) and red blood cells (RBC) were detected in 87 and 6% of samples while ketones and proteins were 6% and 96% respectively present in the samples. Significant bacteriuria ($\geq 10^5$ cfu /ml) was observed in some samples. Bacteria isolated included *Escherichia coli* with a prevalence of 56.9%, followed by *Klebsiella pneumoniae* (12.7%), *Staphylococcus aureus* (8.5%) and *Proteus* sp. (6.3%). *E. coli, K. pneumoniae, S. aureus* and *Proteus* sp. were most sensitive to cotrimazaxole, amoxicillin, nalidixic acid and ciprofloxacin but a large number of bacteria were resistant to tetracycline, chloramphenicol and ampicillin. The misuse of some antibiotics is a major factor responsible for bacterial resistance. Therefore, treatment of ASB in diabetics must be by drugs prescribed by physicians after proper laboratory analysis.

Key words: Bacteriuria, type 2 diabetes mellitus, antibiotics, Nigeria, white blood cell, red blood cell.

INTRODUCTION

Diabetes mellitus has a number of long term effects on the genitourinary system. This effect predisposes to bacterial urinary tract infection (UTI) in the patient with diabetes (Nicolle, 2003). Diabetes mellitus is a major health problem in Nigeria. The prevalence of asymptomatic bacteriuria (ASB) in women has been reported as in school children (6 - 7%), during pregnancy (6% asymptomatic) and 10 - 12% among elderly women (Meiland et al., 2006). Development of asymptomatic UTI in diabetic women has been reported to be much more common than in non-diabetic women, men, and from

diabetic out-patients with urinary tract infections (Geerling et al., 2001; Stapleton, 2002; Nicolle, 2005).

Asymptomatic bacteriuria is a major risk factor for the development of UTI in pregnancy due to physiological changes (Assel et al., 2009). Other factors include low socio-economic status, sickle cell trait, and grand multi-parity. The term bacteriuria means the presence of bacteria in urine and it is taken to be significant if 10^5 organisms per millilitre of a fresh "clean catch" urine specimen are present in any patient (Alebiosu et al., 2003; Kaas, 1956).

Most bacterial aetiologic agents in asymptomatic bacteriuria have been reported to include *Klebsiella pneumoniae, Escherichia coli, Streptococcus agalactiae, Enterococcus faecalis,* coagulase negative

*Corresponding author. E-mail: eaophori@yahoo.com.

Table 1. The age and sex distribution of diabetics with asymptomatic bacteriuria (ASB).

Age (years)	Males number with ASB / %	Females number with ASB / %	Total number with ASB /%
30 - 34	2 (15.4)	2(5.9)	4(8.5)
35 - 39	1 (7.7)	16 (47.1)	17 (36.2)
40 - 44	1 (7.7)	6 (17.6)	7 (14.9)
45 - 49	4 (30.7)	4(11.8)	8(17.0)
50 - 54	3(23.1)	4(11.8)	7(14.9)
55 - 59	2 (15.4)	2(5.9)	4(8.5)
Total	13(27.66)	34(72.34)	47(100)

Staphylococcus and *Streptococcus pyogenes* (Alebiosu et al., 2003; Olaitan, 2006; Assel et al., 2009). A number of these organisms such as *K. pneumoniae, E. coli* and *E. faecalis* have been reported to be very sensitive to nitrofurantoin, gentamycin, ciprofloxacin and ofloxacin but resistant to ampicillin, tetracycline and septrin (Alebiosu et al., 2003). However, Olaitan (2006) reported that cotrimoxazole was the most effective antimicrobial agent against *E. coli, S. aureus* and *E. faecalis,* but resistant to ampicillin and tetracycline. Also, increased resistance against amoxicillin, cotrimoxazole and nalidixic acid has been shown against *K. pneumoniae, E. coli,* and *Enterobacter* sp. (Dytan and Chua, 1999).

Untreated ASB predisposes the individual to recurrent UTI which can cause renal disease. Patients with diabetes mellitus have been reported to have increased rates of UTI infections (Baqai et al., 2008; Vazquez and Sobel, 1995). Increased UTI in pregnant women has led to high morbidity and mortality with the subsequent increase in nosocomial infections (Vazque and Sobel, 1995).

The objective of this study was to investigate the aetiologic agents of ASB in patients with type 2 diabetes mellitus and the antibiotic sensitivity pattern among isolates.

MATERIALS AND METHODS

A total of 130 urine samples were collected from patients with type 2 diabetes mellitus with UTI attending the Central Hospital, Benin - City, Nigeria. Urinary tract infection here is characterized by a urinalysis result showing more than ten pus cells per high power field and no previous history of antibiotic intake, non-steroidal anti-inflammatory drugs and immuno-suppressors in the preceeding 2 weeks. The study had the approval of the Ethics Committee of the Central Hospital Benin-City. All patients gave their informed consents.

Patients were aged 30 - 59 years (females were aged range 30 - 44 years and males 45 - 59 years). The females were all pregnant. Clinical parameters including duration of diabetes mellitus, drug therapy, clinical symptomatology especially urinary complaints, and last menstrual period were recorded. Fasting blood sugar and two-hour post prandial blood sugar were estimated as reported (Alebiosu et al., 2003).

Asymptomatic bacteriuria in patients with diabetes mellitus is the presence of a significant quantity of bacteria in a urine specimen properly collected from a person without symptoms or signs of UTI.

For women, two consecutive specimens with isolation of the same species in quantitative counts of at least 10,000 colony forming units /ml (cfu/ml) of urine while for men, a single specimen with one bacterial species isolated in a quantitative count of at least 10,000 cfu / ml is regarded as asymptomatic bacteriuria (Hajeri et al., 2008).

Midstream urine samples were collected from patients into sterile MacCarthey bottles for urinalysis. Samples were inoculated on MacConkey agar, Blood agar and cysteine lactose electrolyte deficient (CLED) media using a calibrated loop to determine colony forming unit. The plates were incubated at 37°C aerobically for 24 h. Cultures with colony counts $\geq 10^5$ cfu/ ml were considered as significant bacteriuria (Flower et al., 1991; Harding et al., 2002). The organisms were identified using standard cultural, morphological and biochemical techniques (Bucchana and Gibbon, 1974).

Urinalysis was carried out using a dipstick while antibiotic sensitivity test was done by the disc diffusion test (Kirby-Bauer method) using Mueller-Hinton agar (Lab.39)???? What is this?. The antibiotics used included tetracycline, chloramphenicol, amoxicillin, cotrimoxazole, ciprofloxacin, and gentamycin (Mast Diagnostics, Merseyside, and U.K). Resistance and sensitivity to antibiotics was measured by the method of Baker and Breach (1980). When the antibiotic agent was 16 mm or higher, it was recorded as sensitive and resistant when less than 16 mm. The sensitivity plates were incubated aerobically for 18 h, and the zone of inhibition was recorded.

RESULTS

There were 74 females and 56 males with diabetes mellitus. The mean fasting blood glucose and two hour post-prandial blood glucose levels were 6.2 ± 1.4 mmol[1]/L (range 5.1 - 7.3 mmol/L) and 9.4 ± 1.0 mm[1]/L (range 9.0 - 9.8 mmol/L), respectively. Also, urinalysis showed that ketone bodies were present in 6%, protein 96%, WBC 87% and RBC in 6% cases.

Forty-seven (36.15%) of urine samples had significant bacteriuria (34 females and 13 males). Their ages ranged from 30 - 59 years with a mean of 48 ± 3.0 years.

Table 1 shows ASB with age and sex of patients. ASB was highest in age groups 45 - 49 in males and 35 - 39 in females. The highest percentage from females was the age group of females with a high child bearing rate. Table 2 shows the bacteria isolated in ASB and the percentage prevalence of the organisms.

E. coli were found to be the most prevalent in ASB (59.6%), followed by *K. pneumoniae* (12.7%) and

Table 2. Bacteria isolated from urine samples of patients with type -2 diabetes mellitus with ASB.

Bacteria	No. of occurrence	% occurrence
Escherichia coli	28	59.6
Klebsiella pneumoniae	6	12.7
Staphylococcus aureus	4	8.5
Proteus sp.	3	6.3
Streptococcus pyogenes	2	4.3
Enterococcus faecalis	2	4.3
Strept. saprophyticus	2	4.3
n =	47	100

n = number of occurrence / percent occurrence.

Table 3. Bacterial isolates from diabetic patients that was sensitive to the different antibiotics.

Antibiotics	Number of each bacterial isolates sensitive to the different antibiotics						
	EC	Isolates KP	SA	P	SP	EF	SS
Tetracycline	4	1	2	0	0	1	0
Choramphenicol	6	2	0	0	1	2	0
Amoxycillin	8	6	3	3	1	2	1
Ampicillin	0	0	0	1	0	0	1
Cotrimoxazole	16	6	4	3	0	0	1
Nalidixic	10	6	2	2	0	1	1
Gentamycin	4	0	0	1	1	0	1
Ciprofloxacin	12	6	4	2	1	1	0
n = number of isolate	28	6	4	3	2	2	2

An isolate with a zone of inhibition ≥ 16 mm is sensitive while ≤ 15 mm is resistant to a particular antibiotic.
EC- *Escherichia coli*, KP- *Klebsiella pneumonia*,SA- *Staphylococcus aureus*, P- *Proteus sp.* EF- *Enterococcus faecalis*, SS- *Strept. Saprophyticus.*

S. aureus (8.5%). The least prevalent organisms were S*treptococcus pyogenes*, *E. faecalis* and *S. saprophyticus* (4.3% each). Of the 28 *E. coli* isolated, 22 (78.5%) were from females.

The antibiotic sensitivity test is shown in Table 3. The results showed that *E. coli, K. pneumoniae, S. aureus* and *Proteus* spp. were highly sensitive to cotrimoxazole, amoxycillin, nalidixic acid and ciprofloxacin. Very few bacteria were sensitive to tetracycline, chloramphenicol and erythromycin. Only *Proteus* and *S. saprophyticus* were sensitive to ampicillin.

DISCUSSION

The present study showed that asymptomatic bacteriuria (ASB) was present in 47 (36.15%) out of 130 patients with type 2 diabetes mellitus. This is higher than results of previous studies that showed 21% in Karachi (Baqai et al., 2003), 26% in Nigeria (Alebiosu et al., 2003), 9.3% in Ethiopia (Uncu et al., 2002), in Ghana (Turpun et al., 2007) and 19% in Bahrain (Hajeri, 2008). The population

studies in these reports are comparable to the number of patients in this study. Some studies have even reported much lower values of between 5.8 - 19% (Alebiosu et al., 2003). The variations in percentages of ASB have been attributed to factors such as geographical variations, ethnicity of the subjects and variation in the screening test (Assel et al., 2009).

E. coli was the most common pathogen isolated in this study (59.6%). This is in contrast to the report of Alebiosu et al. (2003) where *K. pneumoniae* was the most common isolates from ASB. However, the result is consistent with the majority of reports where *E. coli* had been reported to be the major pathogen in ASB (Assel et al., 2009; Olaitan 2006; Hajeri, 2008; Baqai et al., 2008). This is why in general practice most work on pathogenesis of UTI focuses on *E. coli* because of its high prevalence in UTI (Johnson, 2003). In chronic UTI, a slow growing *E. coli* with atypical colony morphology and multi-drug resistance has been reported (Triizsch et al., 2003). Most of the *E. coli* isolated in this study was from the pregnant females, the highest prevalence being among the age group of 35 - 39 years. This is in agreement with previous reports where a high incidence of *E. coli* in ASB in diabetics women have been reported (Assel et al., 2009).

Other bacteria isolated include *K. pneumoniae* (12.7%), *S. aureus* (8.3%), *Proteus* (6.3%) and *S. pyogenes*, *S. faecelis* and *S. saprophyticus* (43% each). *Klebsiella* spp. and *E. coli* are known to be important causes of both nosocomial and community acquired UTI.

Patton (1991) reported that *Klebsiella*, *Proteus* and *Enterococcus* spp. are the next most frequent isolates after *E. coli* in ASB. This is in line with the results of the present study where *Klebsiella* and *Proteus* had occurrences of 12.7% and 6.3% respectively, next to *E. coli* of 59.6%.

Results of antibiotic sensitivity showed that cotrimoxazole and ciprofloxacin where very effective against most of the isolates. However, *S. saprophyticus* was resistant to both antibiotics. The sensitivity of *E. coli* in this study is in agreement with previous reports (Olusanya and Olutiola, 1984). The high sensitivity to these two antibiotics may be due to their broad spectra on bacteria. Very few isolates were sensitive to tetracycline, ampicillin, chloramphenicol and erythromycin. Olaitan (2006) reported that the high prevalence of resistance to some of the commonly used antibiotics such as ampicillin and tetracycline may be due to their abuse and low cost of purchase. These factors are common in the study environment where some patients buy drugs without a physician's prescription.

In conclusion, a high prevalence of ASB was established in both males and females. The high prevalence of ASB of 36.15 % in this study is of major public health importance. The predominant pathogen was *E. coli* and this organism is beginning to acquire resistance to some of the clinically used antibiotics. The level of resistance to antibiotics recorded in this study is of great concern. The authors recommend improved personnel hygiene which is likely to reduce ASB that may be complicated in UTI. The use of unprescribed antibiotics and their abuse is a problem and appropriate public health programmes would help resolve this issue.

ACKNOWLEDGEMENTS

The authors wish is thank the patients who gave their informed consent. We also thank Staff and Management of Central Hospital, Benin City, Nigeria for their support and assistance during this study. We are grateful to you all.

REFERENCES

Alebiosu CO, Osinupebi OA, Olajubu FA (2003). Significant asymptomatic bacteriuria among Nigeria type 2 diabetics. J. Nat. Med. Assoc. 95(5): 344-348

Assel MT, Al-Meer FM, Al-Kuwari MG, Ismail MF (2009). Prevalence and predictor of asymptomatic bacteriuria among pregnant women attending Primary health care in Qatar Middle East. J. Fam. Med. 4: 14-17

Baker FJ, Breach MR (1980). Medical Microbiological Techniques (1st ed). Butterworths, London.

Baqai R, Aziz M, Rasool G (2008). Urinary tract infection in diabetics patients and biofilm formation of uropathogens. Infect. Dis. J. Pak. 17(1): 21-24.

Bauer, AW, Kirby WN, Sheis JC, Tuck M (1966). Antibiotic susceptibility by standard simple disc method. Am. J. Clin. Pathol. 36: 493-496.

Bucchanan RE, Gibbons ME (1974). Bergey's Mannual for Determinative Bacteriology (8th) Williams and Wilkins Baltimaore, USA.

Dytan AT, Chua JA (1999). Surveillance of pathogens and resistance patterns in urinary tract infections. Phil. J. Microbiol. Infect. Dis. 28(1): 11-14.

Flower CH, Levett PN, Singh S, Fraser HS (1991). Prevalence of asymptomatic bacteriuria in diabetic females in Barbados 40(1): 34.

Harding GK, Zhanel GG, Nicolle LE, Cheang M (2002). Antimicrobial treatment in Diabetic women with asymptomatic bacteriuria N. Engl. J. Med. 347: 1576-1583.

Hajeri A (2008). When to treat asymptomatic bacteriuria. Bahrain. Med. Bull. 30(2): 1-4.

Johnson JR (2003). Microbial virulence determinants and the pathogenesis of Urinary tract infection. Infect. Dis. Clin. North. Am. 17: 261-265.

Kaas EH (1956). Asymptomatic infections of urinary tract. Trans. Assoc. Am. Phys. 69: 56-64.

Meiland R, Geerlings SE, Stolk RP, Netten PM, Schechberfor PM, Hoepelman AI (2006). Asymptomatic bacteriuria in women with diabetes mellitus: effect on renal function after 6 years of follow- up. Archives Int. Med. 1666(20): 2222-2227.

Nicolle LE (2003). Asymptomatic bacteriuria: when to ereen and when to treat. Infect. Dis. Clin. North Am. 17: 367-394.

Nicolle LE (2005). Urinary tract infection in diabetes. Curr. Opin. Infect. Dis. 18(1): 49-53.

Olaitan JO (2006). Asymptomatic bacteriuria in female students population of a Nigeria University. Int. J. Microbiol. 2(2): 4-9.

Olusanya N, Olutiola PO (1984). Studies of bacteriuria in patients and students, in Ile-Ife, Nigeria. West Afr. J. Med. 3(3): 177-183.

Stapleton A (2002). Urinary tract infections in patients with diabetes. Am. J. Med. 113(1): 805-845.

Triilzsch K, Hoffman H, Christian K, Schub SN, Luts BR (2003). Highly resistant metabolically of chromic urinary tract in fraction. J. Clin. Microbiol. 141: 5689-5694.

Turnpin-Cam-Minkah B, Danso KA, Frimpon EH (2007). Asymptomatic bacteriuria in pregnant women attending antenatal clinic at Komfo Anokye Teaching Hospital Kumasi Ghana. Ghana Med. J. 41(1): 26-29.

Uncu Y, Uncu G, Esmer A, Bilgel N (2002). Should asymptomatic bacteriuria be screened in pregnancy? Clin. Exp. Obt. Gyneco. 29(4): 281-285.

Vazquesz JA, Sobel JD (1995). Fungal infections in diabetes. Infect. Dis. Clin. North Am. 9(1): 97-116.

9

Bacterial contamination: A comparison between rural and urban areas of Panipat District in Haryana (India)

Tyagi Shruti[1], Tyagi Pankaj K.[1*], Panday Chandra Shekhar[1] and Kumar Ruchica[2]

[1]Department of Biotechnology, Meerut Institute of Engineering and Technology, Meerut UP, India.
[2]Department of Biotechnology, N. C. College of Engineering, Israna, Panipat Haryana, India.

A randomized sampling from open air of the kitchens in rural vs urban households to determine bacterial contamination of Haryana (India) were carried out by taking 80 samples between July to September 2009. 40 samples of each in rural and urban area were collected in culture plates. The inoculation procedures were varied from direct inoculation of the kitchen air into the nutrient agar medium. Identification by bacterial taxonomy key, different morphological and biochemical tests in rural households, numbers of bacteria revealed *Salmonella* spp., *Acinetobacter* spp., *Pseudomonas* spp. and *Paenibacillus* spp. with 9 different strains and in urban households, numbers of bacteria revealed *Bacillus* spp., *Pseudomonas* spp., *Micrococcus* spp., *Paenibacillus* spp. and *Acinetobacter* spp. with 27 strains. Among the isolates, *Salmonella* spp. (80%) followed by *Acinetobacter* (63%), *Pseudomonas putida* (38%) and *Paenibacillus polymyxa* (30%) were observed in rural areas. In urban areas *Bacillus* spp. (88%), *Pseudomonas* spp. (75%), *Micrococcus* spp. (70%), *Paenibacillus* spp. (38%) and *Acinetobacter* spp. (30%) were observed. The bacteriological quality of air of kitchens in rural households was found to be more pathogenic and virulent as compared to that of kitchen in urban households. These opportunistic pathogens may be harmful, especially in immunocompromised host. In this setting, there is a constant risk of contamination and transfer to willing host. Hence, better quality of air can be achieved by manipulating sanitation and hygiene within houses, kitchens and surrounding areas.

Key words: Air of kitchens, households, bacteriological quality, sanitation and hygiene.

INTRODUCTION

Bacteria are naturally everywhere: in the water we drink; air we breathe; in our skin; and inside our body. However, they are not visible to the naked eyes. Tiny organisms like bacteria can only be examined under a compound light microscope. The structure or morphology of bacterial cells includes the shape, size and morphological arrangement. The best way to utilize this study guide is to understand the characteristics by description and be familiar with the visual representations (Kristina, 2010). Bacteria are one-celled microorganisms or unicellular microscopic living organisms. They are the smallest organisms which have all the needed protoplasmic equipment for self-multiplication and growth. They are classified as prokaryotes because the cell does not contain a nucleus. Microorganisms such as bacteria that can cause disease are known as pathogenic and those that are harmless are considered as non-pathogenic microorganisms. Many living rooms in urban areas of Meerut district favor carpeting over vinyl flooring, table top cover, carpeting and curtains improves aesthetics, reduces noise, and helps prevents slips and falls. But the possible spread of infectious diseases and odors caused by micro organisms, and the treat of allergies resulting from inhibited growth of microorganisms is a concern. (Jaakkola et al., 2006).

*Corresponding author. E-mail: pktgenetics@gmail.com.

Airborne microbes, allergens and chemicals cause respiratory disease, inflammation in the nose, throat, sinuses, upper airway and the lung. Many infections are acquired by inhalation of pathogens that may remain in the respiratory system but also invade the rest of the body through lymphatic and blood circulations. Under air-borne diseases, by a somewhat loose usage, we may include a long list of diseases in which the channels of entrance and exit are the air passages. The medium of transfer is the air, in which the bacteria, usually breathed, coughed, or sneezed out in droplets, pass from person to person. This group of infections includes some of the most important diseases which affect mankind. In some of these diseases we do not know the actual germ, but it has been shown that they are carried from person to person in the air such as:

Typhoid fever: This is an infectious disease which is caused by the *Salmonella typhimurium*. It is also spread by air (By Roger I. Lee, The Air-Borne Diseases, Part I). Tuberculosis: This is an infectious disease which is caused by the tubercle Bacillus. The disease may infect any tissue of the body and may assume a wide variety of manifestations, but the most common form of tuberculosis is that of the lungs, consumption or phthisis (By Roger I. Lee, The Air-Borne Diseases, Part I). Influenza: True influenza is a distinct disease and is not an ordinary cold or "grippe." The disease is due to a specific Bacillus which can be isolated readily and which, of course, is entirely distinct from the microorganism which causes the cold or "grippe." (By Roger I. Lee, The Air-Borne Diseases, Part I).

Pathogenic organisms continuously enter the home with foods (foodborne) or through water (waterborne), through foods prepared in the home by an infected person (person to person spread), through the air (airborne), by the insects or via pets (Beumer et al., 1999). These are considered as a primary source of potential harmful microorganisms in the home. In the domestic environment, the kitchen is particularly important in spreading infectious diseases. Bryan (1988) indicated that a colonized person handling the implicated food was the most frequently identified factor that contributed to staphylococcal food poisoning, shigellosis and typhoid fever. Several studies on bacterial contamination in the kitchen were carried out in the past decades (Finch et al., 1978; Speirs et al., 1995). Bacterial load of hand towels, dishcloths, tea towels, steel sinks and working surfaces were implicated to be the frequent sites (Finch et al., 1978; Borneff et al., 1988; Josephon et al., 1997; Ikawa and Rossen, 1999; Kusumaningrum et al., 2002). Foodborne diseases associated with foods prepared in contaminated kitchen include salmonella as the most common culprit (Holah and Thorpe, 1990; Dufrenne et al., 2001; Kusumaningrum et al., 2004). Some other bacterial infections associated with contaminated kitchen environment are caused by Campylobacter, Listeria,

Staphylococcus aureus, Bacillus cereus and Escherichia coli (Dufrenne et al., 2001; Regnath et al., 2004). Infectious diseases are known as a serious health risk for many centuries. The mortality rate due to these was a great concern even as in the late 18th and early 19th centuries. About 80% of the diseases prevalent in India can be attributed to rate of safe drinking water, poor sanitation and unhygienic practices followed by inhabitants.

The first well known bacterial transmission in the kitchen was documented in the early part of the 20th century, when Mary Mallon, who worked as a cook in private New York households, was identified as a healthy chronic carrier of the typhoid fever bacterium. She had been spreading typhoid fever through the food she prepared. Due to poor personal sanitary habits, she caused more then 30 cases of typhoid fever with three deaths, while Mallon herself had never been sick with typhoid fever (Olsen et al., 2000; Lerner et al., 1996; Porter et al., 1996). This through epidemiological discovery work and the finding of typhoid bacterium in Mallon's stool, proved the significant role of household environment in transmission of food borne diseases and had a great impact on the science of microbial hygiene.

Historically, salmonella has caused the largest proportion of reported food borne diseases outbreak associated with private homes in Europe and other bacterial infections associated with this environment are caused by campylobacter, *S. aureus, B. cereus and E. coli* (Olsen et al., 2000). Salmonella and campylobacter contamination rates of poultry product are found to be up to 60 and 80% respectively (Dufrenne et al., 2001; Scott et al., 1982). Indeed both clean air and water are the basic necessities of human life. In many Indian villages supply of safe drinking water is a serious problem and the same holds true for air in urban areas. Any carelessness in adoption of routine hygienic measures can lead to transmission of infection through air droplets. The present research work revealed that the status of air of kitchens and living rooms have been found repeatedly contaminated with a variety of bacterial contaminations. In India and specially Haryana states of India very few literatures have been identified of bacterial contamination in the air of kitchens and living rooms in rural and urban areas and their comparison. This study therefore aimed to investigate, identify, and comparison of the bacterial contamination in air of kitchens and living rooms of 160 different samples from 80 homes in rural and urban areas of Panipat district.

METHODOLOGY

One hundred and sixty samples of open air of kitchens and living rooms were collected from Panipat district and its surrounding villages between July to September 2009. In this   study, we randomly selected four villages from rural areas namely Israna, Balana, Naultha, Mehrana and four urban sites namely Modal Town, HUDA Sector, NFL Colony and Sukhdev Nagar. 160 air samples from 80 homes were collected for bacteriological analysis from 40 homes (kitchens and living rooms) in each of the rural and urban areas. All these samples were analyzed by conventional techniques as described by Buchanan and Gibbons (1974) and Carter and Cole (1995). Some samples were identified by IMTECH (Institute of Microbial Technology) Chandigarh, India (CSIR Institute, Govt. of India).

Study area

Randomly selected 80 different homes (kitchens and living rooms) in Panipat district of Haryana were surveyed for potentially harmful pathogens in the domestic kitchens and living rooms of rural and urban areas between July to September 2009.

Sample collection

The 160 samples were collected from kitchens and living rooms of rural and urban areas in Panipat district. The samples were aseptically collected in already clean prepared culture plates of Nutrient Agar. The samples were taking from the open air from kitchens and living rooms. Nutrient agar plate (NA) were using for the collection of samples. For each site two replicates were place on the kitchens and living rooms, left open for 1 to 2 h, and then incubated for 24 h at 30 to 34 °C.

Sample analysis

All samples were analyzed by conventional techniques as described by Buchanan and Gibbons (1974); Carter and Cole (1995). After collection of samples, culture plates were incubated in BOD incubator at 30 to 34 °C for 24 h. After incubation samples were analyzed by morphological or biochemical methods. Microbiological direct analysis of air requires quantitative determination, that is, total population of microorganisms. The densities of cells, spores/conidia of microorganisms were measured in the laboratory through several methods of direct microscopic or colonies counter. In the direct microscopic counts, a known volume of liquid is added to the slide and the numbers of microorganism are counted by examining the slide with the bright field microscope. For colony counter Neubauer or Petroff-Haussér counting chamber, breed smears or electric cell counter (or Coulter counter) were used. The samples were again analyzed by 13 different biochemical tests for kitchens sample and 12 biochemical test for living rooms sample such as catalase test, oxidase test, hydrogen sulphide production test, nitrate reduction test, indole production, MR reaction, VP reaction, citrate use test, urease test, litmus milk test, lactose fermentation, sucrose fermentation, dextrose fermentation.

Identification of isolates

After 24 h of incubation, the colonies that appeared morphologically dissimilar were chosen, counted, subcultured to fresh appropriate culture media and incubated at 30 to 34 °C for 24 h. Identification of microorganisms did not commence, due to the fact that inhibition was evident that a pure culture had been obtained. Colonies identifiable as discrete on the different agar medium (EMB, Blood agar, MacConkey agar, XLD etc) will carefully examined macroscopically for culture characteristics such as the shape, color, size texture and hemolytic reactions. Colonies are gram stained and individual bacterial cell were observed. The bacteria were speciated using there isolated colonies (Beumer et al., 1996). Further identification of enteric organisms was done using different taxonomical methods given by Aneja (2003). Anaerobes and many

traditional morphological and biochemical test were selected for this study.

RESULTS AND DISCUSSION

A total of 160 samples from 80 homes (40 kitchens and 40 living rooms of each rural urban area) were collected and analyzed for bacterial contamination and their comparisons. Samples obtained from rural and urban kitchens from near dustbins, sink, washing-up areas, food shelf, cutlery and crockery, refrigerator, vegetables racks, floor, back side of door and near gas cylinder. Samples obtained from in rural and urban living rooms near carpets, tabletop, curtains, dressing tables and ceiling fans etc. The higher positive bacterial growth was observed 98% in kitchens of rural areas and 95% in kitchens of urban areas. On the other hand, in living rooms 97.5% of rural areas and 92.5% in urban areas are observed in Panipat district of Haryana.

Bacterial contamination in kitchens of rural and urban area

After bacterial isolation from kitchens, they were subjected to various tests and the results were obtained and summarized in Table 1. On the basis of primary characterization, the samples were subjected to morphological and biochemical analysis to confirm the identify bacteria. The presence of bacteria was discerned in 77 samples of air of kitchens in rural and urban households out of 80 samples. Only three samples (one in rural and two in urban) of air of kitchens were found to be bereft of bacteria. In urban areas isolated bacteria on nutrient agar from open air reveled growth of the *Bacillus* spp. (10), contributed the major fraction of bacteria in kitchen air followed by *Pseudomonas* spp.(7), *Micrococcus* spp. (6), *Paenibacillus* spp. (3) and *Acinetobacter* spp. (1). However, in rural areas the *Salmonella* spp. (4) contributed the major fraction of bacteria in the kitchen air followed by *Acinetobacter* spp. (3) and (1) in both *Pseudomonas putida* and *Paenibacillus polymyxa*. The total numbers of bacterial isolates from the air of kitchens in rural and urban areas were 52 and 55 respectively with the notable fact that *S. typhimurium* was found only in the kitchens in rural households. On the other hand *Micrococcus luteus and Bacillus flavus* were seen only in the kitchens of urban households (Table 2).

In the air of kitchens in rural areas, the presence of *Salmonella* spp. which is virulent and pathogenic in nature was recorded, whereas in urban areas it was not seen in any of the samples. Notably, a reverse trend was observed with respect to *Bacillus* spp., which is generally harmless and causes food spoilage only. The pathogenic and non pathogenic status of all the isolated Bacteria are shown in Table 2, which indicates that the bacteria

Table 1. Bacterial analysis in the air of 80 domestic kitchens in rural and urban areas in Panipat district of India.

Type of samples	Source of samples	Total no. of samples processed	No. of samples devoid of bacteria	Total no. of bacteria isolated	Number of genus isolated	Bacteria identified
Kitchens of rural households	Israna	10	Nil	11	2	[1]
	Balana	10	1	15	2	[2]
	Naultha	10	Nil	17	1	[3]
	Mehrana	10	Nil	09	2	[4]
Kitchens of urban households	Model town	10	Nil	10	3	[5]
	Huda colony	10	1	19	4	[6]
	NFL colony	10	1	12	4	[7]
	Sukhdev Nagar	10	Nil	14	3	[8]

Kitchens of rural households: [1] *Salmonella* spp.- 2 strains, *Pseudomonas* spp. [2] *Salmonella typhimurium*, *Acinetobacter* spp.- 2 strains, [3] *Paenibacillus polymyxa*, [4] Acinetobacter, *Salmonella typhimurium*. Kitchens of urban households: [5] *Bacillus* spp.- 5 strains, *Micrococcus* spp.-3 strains, *Pseudomonas* spp.-4 strains [6] *Bacillus* spp.-3 strains, *Paenibacillus polymyxa*, *Micrococcus luteus*, *Pseudomonas putida* [7] *Bacillus flavus*, *Paenibacillus* spp. -2 strains, *Micrococcus luteus*, *Pseudomonas* spp. -2 strains [8] *Bacillus flavus*, *Micrococcus luteus*, Acinetobacter.

Table 2. The pathogenic status of bacteria and number of colonies isolated from the air of kitchens of rural and urban households.

S. No.	Culture medium	Sampling	Bacteria identified (Genus)	Sources	{No. of colonies (%)} / 40 plates in each (rural and urban areas)	Pathogenic/non Pathogenic
1	NA	Open air	*Bacillus* spp.	Urban	35 (88)	Generally harmless, causes food spoilage
2	NA	Open air	*Paenibacillus* spp.	Rural Urban	12 (30) 15 (38)	Plant growth promoter
3	NA	Open air	*Micrococcus* spp.	Urban	28 (70)	Harmless
4	NA	Open air	*Pseudomonas* spp.	Rural Urban	15 (38) 30 (75)	Saprophytic soil bacteria
5	NA	Open air	*Acinetobacter* spp.	Rural Urban	25 (63) 12 (30)	Causes nosocomial infections
6	NA	Open air	*Salmonella* spp.	Rural	32 (80)	Human pathogen cause typhoid fever

NA: Nutrient agar.

isolated from air of kitchens in rural areas are more virulent and show higher pathogenic activity as compared to the bacteria isolated from air of the kitchens in urban areas. The total number of 6 genus isolates from kitchens in rural and urban areas. Table 2 shown in among these isolates, in rural areas *Salmonella* spp. for 80% of isolates (32 colony types from 40 plates), followed by *Acinetobacter* spp. (63%), *P. putida* (38%) and *P. polymyxa* (30%). In urban areas, *Bacillus* spp. accounted for 88% of isolates (35 colony types from 40 plates) and

allowed by *Pseudomonas* spp. (75%), *Micrococcus* spp. (70%), *Paenibacillus* spp. (38%) and *Acinetobacter* (30%). The morphological identification of the bacteria based on agar slant culture characteristic and preliminary characterization of bacteria is given in Tables 3 and 4 respectively, in which the morphological characteristics such as shape, size, colour, texture, and hemolytic growth are help to identify the bacteria. Some preliminary characters such as motility, gram positive, gram negative and growth in broth are again help to identify the bacteria.

Table 3. Morphological identification of the bacteria based on agar slant culture characteristics.

S.No	Agar slant culture character	Probable bacteria
1	Abundant, opaque, White waxy growth	*Bacillus* spp.
2	Whitish, grayish, slightly transparent glistening appearance	*Paenibacillus* spp.
3	Soft, smooth, yellow growth	*Micrococcus* spp.
4	Abundant thin, white growth, media turning green	*Pseudomonas* spp.
5	Rough surface growth, paper like	*Acinetobacter* spp.
6	Thin, even, grayish growth	*Salmonella* spp.

Table 4. Preliminary characterization of based on following parameters.

S.No.	Characters	No. of strains
1	Motility	25
2	Gram(+)	22
3	Gram(-)	14
4	MacConky agar	14
5	Shape	7 round, 29 rod
6	Anaerobic growth	4
7	Oxidative fermentation	19
8	Acid production by glucose	22

Table 5. Biochemical characterizations based on 13 different tests for a total 36 strains of gram positive and gram negative bacteria.

S.No.	Test	No. of strains (+)	No. of strains (-)
1	Catalase test	32	4
2	Oxidase test	6	30
3	Hydrogen sulphide production test	4	32
4	Nitrate reduction test	29	7
5	Indole production	-	36
6	MR reaction	5	31
7	VP reaction	14	22
8	Citrate Use test	12	24
9	Urease test	8	28
10	Litmus milk test	36	-
11	Lactose fermentation	4	32
12	Sucrose fermentation	19	17
13	Dextrose fermentation	19	17

The biochemical characterization based on 13 different tests for both gram positive and gram negative strains in the total number of 36 strains is given in Table 5.

Bacterial contamination in living rooms of rural and urban area

After bacterial isolation from living rooms, they were subjected to various tests and the results were obtained and summarized in Table 6. On the basis of primary characterization, the samples were subjected to morphological and biochemical analysis to confirm the identified bacteria. The presence of bacteria was discerned in 76 samples of air of living rooms in rural and urban areas out of 80 samples. Only 04 samples (one in rural and three in urban areas) of air of living rooms were found to be bereft of bacteria. In rural living rooms the

Table 6. Bacterial analysis in the air of 80 living rooms in rural and urban areas in Panipat district of Haryana, India.

Type of samples	Source of samples	Total no. of samples processed	No. of samples devoid of bacteria	Total no. of bacteria isolated	Number of genus isolated	Bacteria identified
Living rooms of rural households	Israna	10	Nil	16	7	[1]
	Balana	10	Nil	13	7	[2]
	Naultha	10	1	12	5	[3]
	Mehrana	10	Nil	14	7	[4]
Living rooms of urban households	Model Town	10	Nil	12	4	[5]
	Huda colony	10	1	12	4	[6]
	NFL colony	10	Nil	10	3	[7]
	Sukhdev Nagar	10	2	07	4	[8]

Living rooms of rural households: [1] *Staphylococcus* spp., *Proteus* spp., *E. coli*, *Shigella* spp., *Klebsiella* spp., *Alcaligenes* spp., *Bacillus* spp. [2] *Bacillu* spp., *Klebsiella* spp., *Salmonella* spp., *Micrococcus* spp., *Proteus* spp., *E. coli*, *Staphylococcus* spp. [3] *E. coli*, *Micrococcus* spp., *Salmonella* spp *Proteus* spp., *Shigella* spp. [4] *Klebsiella* spp., *Micrococcus* spp., *Shigella* spp., *E. coli*, *Bacillus* spp., *Salmonella* spp., *Proteus* spp. Living rooms of urba households: [5] *Streptococcus* spp., *Staphylococcus* spp.-2 strains, *Acinetobacter* spp.-2 strains, *Bacillus* spp.- 2 strains. [6]*Micrococcus* spp.-2 strain *Paenibacillus* spp.-2 strains, *E. coli*, *Lactobacillus* spp.-2 strains. [7] *Bacillus* spp-2 strains, *Streptococcus* spp.-2 strains, *Pseudomonas* spp. [8 *Staphylococcus* spp.-2 strains, *Micrococcus* spp.-2 strains, *E. coli*, *Paenibacillus* spp.-2 strains.

Proteus spp. and *E. coli* (04) contributed the major fraction of bacteria in the living rooms air followed by *Salmonella* spp., *Shigella* spp., *Klebsiella* spp., *Bacillus* spp. and *Micrococcus* spp. (03), *Staphylococcus* spp. (02), *Alcaligenes* spp. (01). However, in urban living rooms, isolated bacteria on nutrient agar from open air revealed growth of the *Staphylococcus* spp., *Bacillus* spp., *Paenibacillus* spp., *Micrococcus* spp. (04) and contributed the major fraction of bacteria in living rooms air followed by *Streptococcus* spp.(03), *Acinetobacter* spp., *E. coli*, *Lactobacillus* spp.(02) and *Pseudomonas* spp.(01),. The total number of bacterial isolates from the air of living rooms in rural and urban areas was 54 and 41 respectively. The total numbers of genus identified 14 in both living rooms of rural areas and living rooms of urban areas. It is notable fact that the more pathogenic bacterial genus were found only in the living rooms of rural areas such as *Proteus* spp., *Salmonella* spp., *Shigella* spp., *Klebsiella* spp. and non-pathogenic genuses such as *Staphylococcus* spp., *Bacillus* spp., *E. coli*, *Micrococcus* spp and *Alcaligenes* spp. On the other hand, the pathogenic bacterial genuses were found in the living rooms of urban areas such as *Acinetobacter* spp. and *Streptococcus* spp. and non-pathogenic bacterial genuses *Pseudomonas* spp. *and Paenibacillus* spp., *Klebsiella* spp., *Bacillus* spp., *Micrococcus* spp., *E. coli*, *Lactobacillus* spp. *and Staphylococcus* spp. (Table 7).

In the air of living rooms of rural areas, the maximum fraction of *Proteus* spp. (04), *Salmonella* spp., *Shigella* spp., *Klebsiella* spp (03). Which is virulent and pathogenic in nature was recorded whereas; in urban areas it was not seen in any of the samples. In urban areas living rooms generally harmless and causes food spoilage bacterial genus are observed. The pathogenic and non-pathogenic statuses of all the isolated bacteria from living rooms are shown in Table 7. The present

result shows that, the bacteria isolated from air of living rooms in rural areas are more virulent and show higher pathogenic activity as compared to the bacteria isolated from air of the living rooms in urban areas with lower pathogenic activity. Table 7 shows among these isolates, in rural areas the bacterial growth on the basis of colonies forming /plate are observed in *E. coli* for 80% of isolates (32 colony types from 40 plates), followed by *Proteus* spp. and *Salmonella* spp. (60%), *Micrococcus* spp. (55%), *Klebsiella* spp. (50%), *Bacillus* spp. (50%), *Staphylococcus* spp. (30%), *Shigella* spp. (22.5%), *Alcaligenes* spp. (15%). In urban areas, *Staphylococcus* spp. accounted for 60% of isolates (24 colony types from 40 plates), followed by *Bacillus spp.* (45%), *Paenibacillus* spp. and *Streptococcus* spp. (40%), *Micrococcus* spp. (37.5%), *E. coli* (30%), *Acinetobacter* (20%), *Lactobacillus* spp. (17.5%), *Pseudomonas* spp. (10%).

The maximum number of bacterial growth with pathogenic bacterial growth is 60% in *Proteus* spp. and *Salmonella* spp. are observed in rural areas. On the other hand, the maximum number of bacterial growth with non-pathogenic 60% is observed in *Staphylococcus* spp. in urban areas. The present results show that, the rural living rooms are highly contaminated as compared to urban living rooms. The morphological identification of the bacteria isolated from living rooms based on agar slant culture characteristic and preliminary characterrization of bacteria is given in Tables 8 and 9 respectively, in which the morphological characteristics such as shape, size, colour, texture, and hemolytic growth help to identify the bacteria. Some preliminary characters such as motility, gram positive, gram negative and growth in broth, again help to identify the bacteria. The biochemical characterization based on 12 different tests for both gram positive and gram negative strains in the total number of 52 strains is given in Table 10.

Table 7. The pathogenic status of bacteria and number of colonies isolated from the air of living rooms of rural and urban areas.

S.No.	Culture medium	Sampling	Bacteria identified (Genus)	Sources	{No. of colonies (%)}/ 40 plates	(Pathogenic / non-pathogenic) effect
1	N.A	Open air	*Staphylococcus* spp.	Rural	12 (30)	Harmless
				Urban	24 (60)	
2	N.A	Open air	*Proteus* spp.	Rural	24 (60)	Cause human Urinary Tract Infections
				Urban	16 (40)	
3	N.A	Open air	*Bacillus* spp.	Rural	20 (50)	Human pathogen causes Typhoid fever.
				Urban	18 (45)	
4	N.A	Open air	*E. coli*	Rural	32 (80)	Cause Dysentery and Hemolytic Uremic Syndrome.
				Urban	12 (30)	
5	N.A	Open air	*Micrococcus* spp.	Rural	22 (55)	Generally harmful, cause food spoilage
				Urban	15 (32.5)	
6	N.A	Open air	*Klebsiella* spp.	Rural	20 (50)	Cause Pneumonia, UTI, Septicemia.
7	N.A	Open air	*Shigella* spp.	Rural	09 (22.5)	Harmless
8	N.A	Open air	*Salmonella* spp.	Rural	24 (60)	Harmless
9	N.A	Open air	*Alcaligenes* spp.	Rural	06 (15)	Harmless
10	N.A	Open air	*Acinetobacter* spp.	Urban	08 (20)	Cause nosocomial infections
11	N.A	Open air	*Lactobacillus* spp.	Urban	07 (17.5)	Harmless
12	N.A	Open air	*Pseudomonas* spp.	Urban	04 (10)	Saprophytic soil bacteria
13	N.A	Open air	*Paenibacillus* spp.	Urban	16 (40)	Plant growth promoter
14	N.A	Open air	*Streptococcus* spp.	Urban	04 (10)	Pain on swallowing, tonsillitis, high fever, headache.

NA: Nutrient agar.

A comparison between kitchen vs. kitchen of both rural and urban areas, more contaminated kitchens of rural areas with pathogenic virulent bacteria as compared to urban kitchens. In concern of living rooms the same trends are observed. The rural living rooms are more contaminated with pathogenic virulent bacteria as compare to urban living rooms. The status of kitchen vs. living rooms of both rural and urban areas, the kitchen are more contaminated with pathogenic virulent bacteria compared to living rooms. The pathogenic and non-pathogenic bacterial contamination spread out through air and by other factors into kitchen to living rooms because the same bacterial contamination are observed in both kitchen and living rooms in both rural and urban areas.

Domestic kitchen environment are potential places for harboring and spreading pathogenic bacteria including *pseudomonas* spp., *Bacillus* spp., *Paenibacillus* spp., *Micrococcus* spp., *Acinetobacter* spp., *Salmonella* spp. according to Kusumaningrum et al. (2002); Tumwine et al. (2003); Borneff et al. (1985,1989) these pathogen survive on the surface for hours or days, depending on the species. They also stated that wiping of surfaces (physical removal) tends to transfer and spread microorganisms from one surface to the other (Ojima et al., 2002a). Bacteria are readily spread from cloths and sponges during wiping (Cogan et al., 2002; Ojima et al., 2002b; Gorman et al., 2002). *Pseudomonas* spp., an opportunistic pathogen causes UTI, respiratory tract infection, dermatitis, soft tissue infection, bacteremia, bone and joint infection, gastrointestinal infections and a variety of systemic infections. Ragnath et al. showed that, *Pseudomonas* spp. can also be found in households drains of showers and kitchens (Regnath et al., 2004). Its prediction to moist environment makes it more possible to exist in kitchen surfaces, dustbins and used sponges. Once infection with Pseudomonas is established, it is hard to control since this organism is frequently resistant to many commonly used antibiotics (Qarah et al., 2006; Humphrey, 2001).

Generally, *Bacillus* species are neither morphologically nor phylogenetically indistinguishable from each other. Though most of the members of this genus is considered

Table 8. Morphological identification of the bacteria based on agar slant culture characteristics of kitchens and living rooms of rural and urban households samples.

S.No.	Agar slant culture characteristics	Probable bacteria
1	Abundant, opaque, white waxy growth	*Bacillus spp.*
2	Whitish, grayish, slightly transparent, glistening appearance	*Paenibacillus spp.*
3	Soft, smooth, yellow growth	*Micrococcus spp.*
4	Abundant thin, white growth, media turning green	*Pseudomonas spp.*
5	Rough surface growth, paper like	*Acinetobacter spp.*
6	Thin, even, grayish growth	*Salmonella spp.*
7	Abundant, opaque, golden growth	*Staphylococcus spp.*
8	White, moist, glistening	*Escherichia coli*
9	Slimy, white, translucent, raised growth	*Klebsiella spp.*
10	Thin, even, grayish growth	*Shigella spp.*
11	Thin, blue-gray, spreading growth	*Proteus spp.*
12	White, irregular, big circular	*Lactobacillus spp.*
13	Irregular, white, rough surface	*Alcaligenes spp.*
14	Thin, even growth, white	*Streptococcus spp.*

Table 9. Preliminary characterization of based on following parameters of living rooms.

S.No.	Characters	No. of strains of living rooms
1	Motility	22
2	Gram(+)	29
3	Gram (-)	23
4	MacConkey agar	23
5	Shape	16 round, 36 rod
6	Anaerobic growth	41
7	Oxidative fermentation	48
8	Acid production by glucose	37

Table 10. Biochemical characterization based on 12 different biochemical tests of living rooms samples.

S.No.	Biochemical tests	No. of strains(+) of living rooms	No. of strains(-) of living rooms
1	Catalase test	29	23
2	Oxidase test	09	43
3	Hydrogen sulfide production test	07	45
4	Nitrate reduction test	44	08
5	Indole production test	10	42
6	MR reaction	29	23
7	VP reaction	16	36
8	Citrate use test	15	37
9	Urease test	14	38
10	Lactose fermentation	18	34
11	Sucrose fermentation	31	21
12	Dextrose fermentation	39	13

contaminants, there are 2 members which are of significant medical importance, *Bacillus anthracis* and *B. cereus*. *B. anthracis cause* anthrax *and B. cereus* causes food poisoning (Cunha, 2006). *Salmonella* infections are zoonotic and can be transferred between human and non human animals. Many infections are due to ingestion of contaminated food. A distinction is made between enteritis Salmonella and typhoid/paratyphoid

Salmonella, due to special virulence factor and a capsule protein (virulence antigen) can cause serious illness, such as *Salmonella enterica* subspecies or typhi, *Salmonella typhi* is adapted to human and does not occur on animals (www.wikkipedia.com). Oberoi et al. (1994) observed that bacterial quality of household's air was found to be better in large farm size category. It might be due to better cleanliness maintained in such houses as well as their kitchens. The households' air was mostly loaded with non-pathogenic bacteria in these cases. A survey conducted by Dhillon et al. (1990) in five colonies of Ludhiana city and five adjoining villages of Ludhiana district revealed that, a lack of family finance was the root cause of households pollution, followed by other factors such as places of dwelling and defection, poor drainage systems, improper disposal of refuse and traditional style of living.

Staphylococcus epidermidis has become the most important cause of nosocomial infections in recent years. Its pathogenicity is mainly due to the ability to form biofilm on indwelling medical devices. In a biofilm, *S. epidermidis* is protected against attacks from the immune system and against antibiotic treatment, making *S. epidermidis* infections difficult to eradicate (Vuong and Otto, 2002). *S. aureus is* ubiquitous and may be a part of human flora, however, the organism may cause disease through invasion and toxin production such as abscess, pneumonia, diarrhea and the most feared toxic shock syndrome (Tolan, 2007). *Klebsiella pneumonia* can cause pneumonia, septicemia, wound infection, burn infection, UTI and ankylosing spondylitis. Like Pseudomonas, it is an opportunistic pathogen. Pneumonia caused by Klebsiella has around 50% mortality, due to the underlying disease but may reach 90% in untreated cases (Umeh and Berkowitz, 2006).

The bacteria isolated from the air of kitchens samples in rural areas were found to be virulent with pathogenic activity. On the other hand, the bacteria isolated from the air of kitchens in urban areas were generally harmless and caused food spoilage only except *Acinetobacter* spp. Our study showed that, potentially harmful pathogens are easily accessible to every individual through contaminated sources present in kitchens. Similar results with those of Joeshson and Rubino (1997), Scott et al. (1982), and De Boer and Hahne (1990). Bacterial contamination in kitchen spread out through sponges and washcloths use in normally in kitchens were similarly reported (Suaad, 2007). Bacterial contamination spread out into kitchens to living rooms and it surrounded areas through air and other factors such as dustbin, dusting cloth, utensils etc. were similarly reported (Tyagi et al., 2011).

The results of our study have several implications on the preference for floor, carpet, tabletop in living rooms and unwashed hands, damaged vegetables, dust beans, sink, washing-up areas, food shelf, cutlery and crockery, refrigerator, vegetables racks, floor, back side of door and near gas cylinder in kitchens. The primary sources of these

bacteria are kitchens in which the food spoilage and stored dustbin contain for many days and directly entered vegetables (some infected with higher pathogens). After sometime, bacteria spread out to its surrounding areas which is more suitable for growth. In living rooms such as carpet, curtains, toilet doors, table top, dressing tables and ceiling fans etc. are the best places in which the bacterial growth are more conditionable and when the favorable conditions start (seasonal variation) these bacteria infected the individuals. This explains why most people experience a lot of respiratory symptoms from acute allergic rhinitis to pneumonia during climate changes. Avoiding these infections, we have made some arrangement in our kitchens, living rooms and its surrounding areas. When possible, floor carpeting in homes should be minimized or avoided, since this serves as habitat for opportunistic infection agents that pose harm to one's health.

Proper ventilation and sanitation in both kitchens and living rooms should be provided. Renewable kitchen sponges should be dried after use or immersed in boiling water for 5 to 10 min. Furthermore, hygienic measures and precaution in the kitchen should be well maintained to reduce harmful bacteria levels. Vegetables are entered in the kitchen after proper washing and checking. Hence, the better quality of air can be achieved by manipulating sanitation and hygiene within houses, kitchens and its surrounding.

ACKNOWLEDGMENT

Authors are highly thankful to the Director of IMTECH (Institute of Microbial Technology) Chandigarh, India for providing facilities to help identify some bacteria.

REFERENCES

Aneja KR (2003). Experiments in Microbiology Plant Pathology, Tissue culture and mushroom production technology, pp. 245-282.

Beumer RR, Te Giffel MC, Spooranberg E, Rombouts FM (1996). Listeria species in domestic environments. Epid. Infect., 117: 437-442.

Borneff J (1989). Efficient hygiene precautions in the household today. Zentralbl Bacterial Mikrobial Hyg., 187: 404-413.

Borneff J, Hassinger R, Wittig J, Edenharder R (1988). The distribution of microorganism in household kitchens I. Problem, experiment and results. Zenttralbl Bakteriol. Mikrobiol. Hyg., 186: 1-29.

Borneff J, Witting JR, Borneff M, Hartmetz G. (1985). Occurrence of enteritis- causing agents in private households- A pilot study. Zentralbl Bacterial Mikrobial, 180: 319-334.

Bryan FL (1988). Risk of practices, procedures and processes that lead to outbreak of foodborne diseases. J. Food Prot., 57: 663-673.

Buchanan RE, Gibbons NE (1974). Berge's manual of determinative bacteriology (8th Ed.) The Williams and Wilkins Co, Baltimor, p. 1246.

Carter GR, Cole JR (1995). Diagnostic procedures in veterinary bacteriology and mycology (5th Edn.) Academic press inc., California,

Cogan TA, Slader J, Bloomfield SF, Humphrey TJ (2002). Achieving hygiene in the domestic kitchen: the effectiveness of commonly used cleaning procedures. J. Appl. Microbiol., 92: 885-892.

Cunha BA (2006). *Bacillus* infections. www.emedicine.com.

De Boer E, Hahne M (1990). Cross-contamination with *Campylobacter jejuni* and *Salmonella* spp. from raw chicken product during food preparation. J. Food Prot., 53: 1067-1068.

Dhillion MK, Cheema H, Dhaliwal GS (1990). Source of environmental pollution in rural and urban habitats awareness among housewives. Indian J. Ecol., 17(1): 13-16.

Dufrene J, Ritmeester WE, Asch ED, van Leusden F, de Jonge R (2001). Quantification of the contamination of chicken and chicken products in the Netherlands with *salmonella* and *campylobacter*. J. Food Prot., 64: 538-541.

Finch JE, Prince J, Hawksworth M (1978). A bacteriological survey of the domestic environments. J. Appl. Bacteriol., 45: 357-364.

Gorman R, Bloomfield S, Adley CC (2002). A study of cross contamination of food borne pathogens in the domestic kitchen in the Republic of Ireland. Int. J. Food Microbiol., 76(1-2): 143-150.

Holah JT, Thorpe RH (1990). Cleanability in relation to bacterial retention on unused and abraded domestic sink materials. J. Appl. Bacteriol., pp. 599-608.

Humphrey T (2001). The spread and persistances of Campylobacter and Salmonella in the domestic kitchen. J. Infect., 43: 50-53.

Ikawa JY, Rossen JS (1999). Reducing bacteria in household sponges. J. Environ. Health, 62: 18-22.

Josephson KL, Rubino JR (1997). Paper IL. Characterization and quantification of bacterial pathogens and indicator organisms in household kitchens with and without the use of a disinfectant cleaner. J. Appl. Microbiol., 83: 737-750.

Jaakkola JJ, Leromnimon A, Jaakkola MS (2006). 'Interior surface material and asthma in adults: A population-based incident case-control study.' Am. J. Epidemiol., 164(8): 742-749.

Josephson KL, Rubino JR (1997). Paper IL. Characterization and quantification of bacterial pathogens and indicator organisms in household kitchens with and without the use of a disinfectant cleaner. J. Appl. Microbiol., 83: 737-750.

Kristina   LS (2010). 'Microbiology Study Guide: Understanding the Bacterial Cell Structure.' Bright hub.

Kusumaningrum HD, Van Asselt ED, Beumer RR, Zwietering MH (2004). A quantitative analysis of cross-contamination of *Salmonella* and *Campylobacter* spp. via domestic kitchen surfaces. J. Food Prot., 67(9): 1892-1903.

Kusumaningrum HD, Van Putten MM, Rombouts FM, Beumer RR (2002). Effect of antibacterial dishwashing liquid on foodborne pathogens and competitive microorganisms in kitchen sponges. J. Food Prot., 65: 61-65.

Lerner   BH (1996). Typhoid Marry: Captive to the public's health. Am. J. Public Health, p. 86.

Oberoi K, Poonam, Sharma S (1994). An investigation of the bacteriological quality of air and water of rural households in the village of Ludhiana District. J. Res.-Punjab Agric. Univ., 31(1): 113-123.

Ojima M (2002a). Bacterial contamination of Japanese households and related concern about sanitation. Environ. Health Res., 12(1): 41-52.

Ojima   M (2002b). hygiene measures considering actual distributions of microorganisms in Japanese households. J. Appl. Microbiol., 93(5): 800-809.

Olsen SJ, Mackinon LC, Goulding JS, Bean NH, Slutsker   L (2000). Surveillance for foodborne-disease outbreaks –United States, 1993-97. MMWR CDC Surveillance Summaries, 49: 1-62.

Porter R (1996). Femmae fatale? (Review of the book Typhoid Mary: Captive to the Public's Health by Leavitt JW). Nature (London, U.K.), 383: 781-782.

Qarah S (2006). *Pseudomonas* infections. www.emedicine.com.

Regnath   T, Kreutzberger M, Illing S, Oehme R, Liesenfeld O (2004). Prevalance of *Pseudomonas aeruginosa* in households of patient with cystic fibrosis. Int. J. Hyg. Environ. Health, 207(6): 585-588.

Roger   IL,   The   Air-Borne   Diseases,   Part   I. http://www.cliffsnotes.com/study_guide/Bacterial-Diseases-of-the-Respiratory   System.topicArticleld-8524,articleld-8496.html#ixzzyXPRWZRA.

Scott E, Bloomfield SF, Baelow CG (1982). An investigation of microbial contamination in the home. J. Hyg. (Lond), 89(2): 279-293.

Speirs JP, Anderson JG (1995). A study of microbial content of domestic kitchen. Int. J. Environ. Health, 5(13): 109-122.

Suaad   SA (2007). Bacterial and *Aspergillus* spp. contamination of domestic kitchens in Riyadh, Saudi Arabia. Saudi J. Biol. Sci., 14(1): 1-6.

Tolan Jr RW (2007). *Staphylococcus aureus* infection. www. emedicine.com.

Tumwine J, Thompson J, Katui-katua M, Mujwahuzi M, Johnstone N, Porras I (2003). Sanitation and hygiene in urban and rural households in East Africa. Int. J. Environ. Health Res., 13(2): 107-115.

Tyagi PK, Tyagi S, Kumar R, Panday CS (2011). Bacteriological analysis of air of kitchens in rural and urban areas of Panipat district in Haryana (India). Int. J. Pharm. Biol. Sci., 2(1): B247-256.

Umeh   O,   Berkowitz   LB   (2006).   *Klebsiella*   infections, www.emedicine.com.

Vuong C, Otto M (2002). *Staphylococcus epidermidis* infections. Microb. Infect., 4(4): 481-489.

10

Genetic variability and correlation analysis for quantitative traits in chickpea genotypes (*Cicer arietinum* L.)

Qurban Ali*, Muhammad Hammad Nadeem Tahir[1], Hafeez Ahmad Sadaqat[1], Saeed Arshad[1], Jahenzeb Farooq[1], Muhammad Ahsan[1], Muhammad Waseem[2] and Amjad Iqbal[3]

[1]Department of Plant Breeding and Genetics, University of Agriculture, Faisalabad, Pakistan.
[2]Department of Entomology, University of Agriculture, Faisalabad, Pakistan.
[3]Department of Agronomy, University of Agriculture, Faisalabad, Pakistan.

The present studies were conducted to the estimation of correlation for quantitative traits in chickpea (*Cicer arietinum* L.) in the field of the department of Plant Breeding and Genetics, University of Agriculture, Faisalabad, during the crop season 2009 to 2010. Correlation studies showed that biomass per plant, number of pods per plant, number of secondary branches per plant, number of seeds per pod and 100-seed weight were positive and significant at genotypic level but positive and highly significant at phenotypic level. Whereas number of days taken to flowering, number of days taken to maturity, primary branches per plant, secondary branches per plant were positively correlated with the grain yield per plant at genotypic and phenotypic levels. Plant height was negative and non-significantly correlated with grain yield per plant at both genotypic and phenotypic levels.

Key words: *Cicer arietinum,* correlation, genotypic, phenotypic, chickpea, Pakistan.

INTRODUCTION

Among the pulses, chickpea (*Cicer arietinum* L.) is the third leading grain legume in the world and first in the South Asia. Ninety two percent of the area and eighty nine percent of the production of grain are concentrated in semi-arid tropical countries (Anonymous, 1995). Its range of cultivation extends from the Mediterranean basin to the Indian sub-continent and south ward of Ethiopia and the East African highlands. Two types of chickpea, one namely Kabuli is grown in temperate regions while the desi type chickpea is grown in the semi-arid tropics (Muehlbauer and Singh, 1987). Chickpea is the principal rabi pulse crop and important source of calories in Pakistan which is predominantly grown in the vast rainfed areas of the country. Pakistan ranks second to India in terms of acreage under chickpea which is 1050 thousand hectares with an annual production of 571 thousand tones (Anonymous, 2009, 2010). It is rich and readily available source of protein both for human and animals. The average yield of chickpea is low as compared to other chickpea growing countries. In Punjab about 90% gram is cultivated in rainfed areas; the major chickpea production belt is Thal including the districts of Bhakhar, Mianwali, Layyah, Khushab and parts of Jhang. Chickpea is the cheapest and readily available source of protein (19.5%), fats (1.4%), carbohydrates (57 to 60), ash (4.8%) and (4.9 to 15.59%) moisture (Huisman and Van der Poel, 1994). It makes up the deficiency of cereal diets. It also helps in replenishment of soil fertility by

_____
*Corresponding author. E-mail: qurbanalisaim@yahoo.com

Table 1. Estimates of genetic components.

Quantitative traits	G V	PV	EV	GCV (%)	PCV (%)	ECV (%)	Broad-sense heritability(h²)%	Genetic advance (%)
Days taken to flowering	50.6521	36.3710	2.5410	1.9466	1.6495	0.4359	56.6	1.5219
Days taken to maturity	12.5410	9.3915	1.1564	0.9686	0.8382	0.71	51.6	1.1181
Plant height	16.8033	19.6403	2.8369	1.1212	9.28	3.53	85.6	5.3028
Primary branches per plant	5.0021	0.0127	0.0058	0.6117	4.45	3.01	54.1	0.0854
Secondary branches per plant	30.0141	21.283	0.1553	1.4985	1.2618	4.84	47.9	0.3665
Biomass per plant	9.0809	17.3115	8.2708	1.1380	4.92	3.40	52.2	3.0420
Pods per plant	58.1481	65.2329	17.0445	2.0857	16.52	8.45	73.9	8.3529
Seeds per pod	16.0021	12.014	9.2141	1.0941	0.9480	0.8305	47.0	0.0438
100-seed weight	20.886	19.1431	12.8043	3.6912	0.9787	4.13	48.8	0.7151
Grain yield per plant	32.2914	5.3477	1.1554	0.6325	5.69	3.53	61.5	1.9926
Grains per plant	21.7217	34.5260	5.8043	5.38	6.78	0.6589	62.9	5.1755

fixing of atmospheric nitrogen through symbiosis coupled with deep root system.

MATERIALS AND METHODS

The present studies were conducted in the field of the Department of Plant Breeding and Genetics, University of Agriculture, Faisalabad, during the crop season 2009 to 2010. The experimental material comprised twenty chickpea genotypes namely: Noor 91, Bittle 98, 210, 1288, 9605, 220, 1276, 1017, 2006, 848, 214, 405, 880, 2008, 219, 4009, 846, 1154, 290 and 868. Analysis of variance for all characters were carried out using the method of Steel and Torrie (1997) and individual comparison of varietals mean was accomplished by Duncan's new multiple range test. Genotypic and phenotypic correlations were calculated to observe the association between different traits (Kwon and Torrie, 1964).

RESULTS AND DISCUSSION

Genetic parameters of yield and their components are given in Table 1. In the present study, the highest genotypic variances were found for NPP (58.1481), NDF (50.6521), GYP (32.2914), NSB (30.0141), and GPP (21.7217) while lowest genotypic variance was found for BM (9.0809) and NPB (5.0021). The highest phenotypic variances were found for NPP (65.2329), NDF (36.3710), GPP (34.526), NSB (21.283) and PH (19.6403) while lowest for NPB (0.0127). The highest environmental variance was found for NPP (17.0445) while lowest for NPB (0.0058). The highest genotypic coefficient of variances was found for GPP (5.38%) while lowest genotypic coefficient of variance was found for NBP (0.6117%). The highest phenotypic coefficient of variances was found for NPP (16.20%) while lowest phenotypic coefficient of variance was found for NDM (0.8382%). The highest environmental coefficient of variances was found for NPP (8.45%) while lowest

environmental coefficient of variance was found for NDF (0.4359%). Similar findings were reported by Adhikari and Pandey (1982). The higher values of genetic advance were found for NPP (8.3529%), PH (5.3028%), and GPP (5.1755%) while lower for SPP (0.0438%) and NPB (0.0854%). The greater values of genetic advance indicated that NPP, PH and GPP can be used for selecting higher yielding genotypes (Raval and Dobariya, 2003). The highest heritability (85.60%) was found for plant height. The range of heritability from 47.0 to 85.60%. The greater values of heritabilities were found for PH, GPP, GYP and NPP while lowest values were for 100-seed weight, SPP and NPB and others have moderate type of heritability. The higher value of heritability for grain yield per plant, number of grains per plant and pods per plant indicates that these characters can be used as the genetic parameters for the improvement and selection of high yielding genotypes. These results were in accordance with the findings of Dasgupta et al. (1992). The BMP and GY per plant indicated high heritability coupled with genotypic variation by using Mather and Jinks (1982) model of heritability. Crop improvement could be possible by simple selection because high heritability coupled with high genotypic variation revealed the presence of an additive gene effect (Noor et al., 2003). On the other hand, low heritability coupled with low genotypic variation was observed for 100-seed weight, NPB and NSP. The results indicated that these traits were greatly influenced by the environment (Arshad et al., 2002).

A study of Table 2 shows that the genotypic and phenotypic correlations coefficients of number of days taken to flowering with biomass, primary branches per plant and secondary branches per plant were negative and non-significant. A positive but non- significant association was recorded between days taken to flowering and 100-seed weight, days taken to maturity,

Table 2. Genotypic and phenotypic correlation of various quantitative traits.

Traits	r	NDM	PH	NPP	NPB	NSB	DW	100-SW	NSPP	NGP	GY
NDF	G	0.175863	0.106041	0.081193	0.054416	-0.17061	-0.03977	0.179816	-0.97764**	0.547283*	0.739361*
	P	0.204305	0.113903	0.04101*	-0.02984	-0.21656	-0.04951	0.100843	-0.41997*	0.283169	0.317849*
NDM	G		-0.03723	0.245055	-0.25962*	0.227331	-0.40894	-0.42416*	-0.26711	-0.25621	0.918132**
	P	.	0.020765	0.006464	-0.25827	0.101509	-0.21399	-0.19709	-0.1254	-0.2153	0.451*
PH	G			-0.01889	0.062383	-0.60747**	-0.12268	-0.31721*	-0.46792*	-0.12127	-0.04263
	P			0.043174	0.030369	-0.38764*	-0.06821*	-0.13808	-0.27691*	-0.10022	-0.05191
NPP	G				-0.23156	0.550029**	0.293602*	-0.20052	-0.30691*	-0.18041	0.080245
	P				-0.0664	0.351766	0.191958	-0.15202	-0.05795	-0.09649	0.092244
NPB	G					0.141073	-0.12771	-0.22406	0.475881*	0.31826**	-0.77384**
	P					0.07444	0.023474	-0.13853	0.217878*	0.296044*	-0.50716*
NSB	G						0.220983	-0.47098**	0.365295*	-0.59944*	0.479671*
	P						0.114553	-0.3505	0.285029*	-0.21083*	0.265961*
DW	G							-0.40878	0.776128*	0.352163**	0.3406**
	P							-0.1603	0.29634	0.409597**	0.082493
100-SW	G								0.468291**	-0.08899	-0.44762**
	P								0.125769*	-0.14323*	-0.22423**
NSPP	G									0.400528*	-0.34051*
	P									0.151283	-0.20138*
NGP	G										-0.13967
	P										-0.23277*

* = Significant at 5% probability level, ** = Highly significant at 1% probability level, NDF = Number of days taken to flowering, NPP= Number of pod per plant, NDM = Number of days taken to maturity, NSPP = Number of seeds per pod, PH = Plant height, HSW = 100-seed weight, NPB = Number of primary branches per plant, GYP = Grain yield per plant , NSB = Number of secondary branches per plant , NGP =Number of grains per plant.

plant height and number of pods per plant. Days to flowering were significant and positively correlated with number of grains per plant and grain yield per plant. Similar results have been obtained by Yadav et al. (2001). Positive and significant correlation coefficient of number of days taken to maturity with number of pod per plant and grain yield at genotypic level but for grain yield highly significant at phenotypic level. Similar results have been obtained by Raval and Dobariya (2003) and Obaidullah et al. (2006). Genotypic correlation between plant height and number of primary branches per plant was positive and non-significant as well as non-significant at phenotypic level. The significant correlation between plant height and grain yield per plant could be attributed to the disruption in pod filling and grain development. Similar results have already been reported by Obaidullah et al. (2006). Genotypic and phenotypic correlation coefficients of number of primary branches per plant with number of secondary branches per plant were not significant. Genotypic and phenotypic correlation coefficients of number of primary branches per plant with of number of pods per plant and seeds per pod were positively significant. The grain yield per plant was negatively and significantly correlated with number of primary branches per plant. Similar results have been obtained by Wadud and Yaqoob (1989) and Bhaduoria et al. (2003) observed positive correlation between primary branches per plant and seeds per pod. Genotypic

correlation between secondary branches per plant and seeds per pod was negatively significant at phenotypic level. Since secondary branches per plant seemed to be an important yield component and in present studies this character exhibit an association with grain yield per plant. The secondary branches per plant were positively and significantly correlated with grain yield per plant. Similar findings have also been reported by Singh et al. (1997) and Jeena and Arora (2001).

Correlation coefficients of biomass per plant with number of pods per plant, number of branches per plant, number of secondary branches and plant height were positive and significant at phenotypic level and genotypic levels. The correlation of biomass per plant was positive and significant with the seeds per plant, number of grains per plant and grain yield per plant. Strong positive genotypic and phenotypic correlation of biomass per plant was with number of grain per plant but not significant association was with 100-seed weight. Almost similar results have already been reported by Jeena and Arora (2001) and Arshad et al. (2002).

A positive and significant genotypic and phenotypic correlation of number of pods per plant with biomass per plant and number of secondary branches per plant but highly significant genotypic correlation with secondary branches per plant. So number of pods per plant should be used as selection for yield improvement in chickpea (Chavan et al., 1994. A positive but significant genotypic correlation of number of seeds per pod with secondary branches per plant and biomass per plant but highly significant at phenotypic level for biomass per plant. Seeds per pod were positive and non-significant with number of grain per plant at genotypic and phenotypic level. Similar results were found by Ozcelik et al. (2004) and Bicer (2005). A positive but significant correlation of 100-seed weight with seeds per pod at genotypic but not significant at phenotypic level. Genotypic and phenotypic correlation coefficient between 100-seed weight and grain yield per plant and secondary branches per plant was negative and significant. A positive and significant genotypic and phenotypic correlation of number of pods per plant with biomass per plant, days to flowering, and number of primary branches per plant but highly significant genotypic correlation with biomass per plant. So number of grains per plant should be used as selection for yield improvement in chickpea (Chavan et al., 1994).

REFERENCES

Adhikari G, Pandey MP, (1982). Genetic variability in some quantitative characters on scope for improvement in chickpea (Cicer arietinum L.). Chickpea Newslett., June Icn., 7: 4-5.

Anonymous (1995). Agricultural Statistics of Pakistan, Ministry of Food, Agriculture and Cooperatives, Islamabad.

Anonymous (2009). Economic Survey. Government of Pakistan, Finance Division,Economic Advisor's Wing Islamabad (2009-2010).

Arshad M, Bakhsh A, Bashir M, Haqqani AM (2002). Determining the heritability and relationship between yield and yield components in chickpea (Cicer arietinum L.). Pak. J. Bot., 34: 237-245.

Bhaduoria P, Chaturvedi SK, Awasthi NNC (2003). Character association and path coefficient analysis in chickpea (Cicer arietinum L.). Ann. Agric. Res., 24: 684-685.

Bicer BT, (2005). Evaluation of chickpea (Cicer arietinum L.) landraces. Pak. J. Biol. Sci., 8: 510-511.

Chavan VW, Path HS, Rasal PN (1994).Genetic variability, correlation studies and their implications in selection of high yielding genotypes of chickpea (Cicer arietinum L.). Madras Agric. J., 81: 463-465.

Dasgupta T, Islam MO, Gayen (1992). Genetic variability and analysis of yield components in chickpea (Cicer arietinum L.). Ann. Agric. Res., 13: 157-160.

Huisman J, Van der Poel AFB (1994). Aspects of the nutritional quality and use of cool season food legumes in animal feed, pp. 53-76.

Jeena AS, Arora PP (2001). Correlation between yield and its components in chickpea (Cicer arietinum L.). Legume Res., 24: 63-64.

Kwon SH, Torrie JH (1964). Heritability and interrelationship of two soybean (Glycine max L.) populations. Crop Sci., 4: 196-198.

Mather K, Jinks JL (1982). The study of continuous variation. 3rd ed. London: Chapman and Hall; Biometrical genetics.

Muehlbauer FJ, Singh KB (1987). Genetics of chickpea (Cicer arietinum L.). In: M.C. Sexana and K.B. Singh (eds:), The chickpea CAB International, Wallingford, Oxon, OX10 8DE UK, p. 99-126.

Noor F, Ashaf M, Ghafoor A (2003). Path analysis and relationship among quantitative traits in chickpea (Cicer arietinum L.). Pak. J. Biol. Sci., 6: 551-555.

Obaidullah S, Munawar K, Iqbal A, Hamayun K (2006). Regression and correlation analysis in various cultivars of chickpea (Cicer arietinum L.). Ind. J. Pl.Sci., 5: 551-555.

Ozcelik H, Bozoglu H (2004). The determination of correlations between seed yield and some characters of chickpea (Cicer arietinum L.). Ondokuz Mays University, J. Fac. Zirrat, 19: 8-13 [CABB Abst.].

Raval LJ, Dobariya KL (2003). Yield components in improvement of chickpea (Cicer arietinum L.). Ann. Agric. Res., 24: 789-794.

Singh D, Sharma PC, Kumar R (1997). Correlation and path coefficient analysis in chickpea (Cicer arietinum L). Crop Res., 13: 625-629.

Steel RGD, Torrie JH (1997). Principles and procedures of statistics. McGraw Hill Book Co., NY. USA.

Wadud A, Yaqoob M (1989). Regression and correlation analysis in different cultivars of chickpea (Cicer arietinum L.). Sarhad J. Agric., 5:171-176.

Yadav NP, Sharma CH Haque MF (2001). Correlation and regression study of seed yield and its components in chickpea (Cicer arietinum L.). J. Res. Birsa Agric. Univ., 13: 149-151.

11

Some new results on affinity hemodialysis and T cell recovery

R. O. Ayeni[1]*, A. O. Popoola[1] and J. K. Ogunmoyela[2]

[1]Department of Pure and Applied Mathematics, Ladoke Akintola University of Technology Ogbomoso, Nigeria.
[2]Department of Mathematical Sciences, Federal University of Technology, Akure Nigeria.

We provide criteria under which affinity hemodialysis could provide a stable infected equilibrium.

Key words: Affinity hemodialysis, HIV/AIDS envelope protein, stability criteria.

INTRODUCTION

Important studies on HIV/ AIDS include papers of Hazenberge (2000), Duffin and Tullis (2002), Tullis (2004), Tullis et al (2002ab, 2003), Oluyo (2007), Oluyo and Ayeni (2007), Stafford et al (200) and Wang and Li (2006). The hallmark of HIV disease is the gradual loss of CD4+ T cells, which ultimately leaves the immune system unable to defend opportunistic infections. Recent studies Hazenberge (2000) and Tullis (2004) suggest that CD4 + T cells are lost through infection and binding of gp 120 to uninfected CD4 + T cells. The envelope ultimately leads to the death of healthy cells. On the other hand, hemodialysis assists infected T cells to recover. The present study investigates the criteria under which the rate of recovery of infected cells through hemodialysis could lead to the stability of the equilibrium point.

MATHEMATICAL MODEL

We modify the model of Duffin and Tulis (2002) by incorporating recovery through affinity hemodialysis:

Production

$S \xrightarrow{\pi} T$ rate of T-cell production from stem cells.

$T + V \xrightarrow{k_1} T_i$ infection of T cells

$P + T \xleftrightarrow{k_3} PT$ reversible gp 24 binding to normal T-Cells

T_i + hemodialysis $\xrightarrow{\mu} T$ recovery of some T infected cells as a result of hemodialysis

Clearance

$T \xrightarrow{d_1}$ death of normal T cell

$T_i \xrightarrow{d_2}$ death of infected T cells

$V \xrightarrow{c}$ viral clearance rate,

Where T_i is infected T cell, V is virus and P is concentration of gp120.

Arising from above, the relevant mathematical equations are:

$$\frac{dT}{dt} = \pi - d_1 T - k_1 TV + \mu T_i \quad , \qquad T(0) = T_0$$

$$\frac{dT_i}{dt} = k_1 TV - d_2 T_i - \mu T_i \quad , \qquad T_i(0) = 0$$

$$\frac{dV}{dt} = k_2 T_i - cV \quad , \qquad V(0) = 0$$

STABILITY OF THE CRITICAL POINTS

To obtain the critical points, we set In infected free equilibrium

---

*Corresponding author. E-mail: ayeni_ro@yahoo.com.

$$\frac{dT}{dt} = \frac{dT_i}{dt} = \frac{dV}{dt} = 0$$

$$V = T_i = 0$$

So $\pi - d_1 T = 0$ and $T = \frac{\pi}{d_1}$

To obtain the infected equilibrium, we obtain

$$0 = k_2 T_i - cV \Rightarrow V = \frac{k_2 T_i}{c}$$

Also $0 = k_1 TV - d_2 T_i - \mu T_i$ and Substituting for V, we obtain

$$k_1 T \frac{k_2 T_i}{c} - (d_2 + \mu) T_1 = 0$$

i.e $\left[\frac{k_1 k_2 T}{c} - (d_2 + \mu) \right] T_1 = 0$

$$T_i \neq 0 \Rightarrow T = \frac{(d_2 + \mu)c}{k_1 k_2}$$

Substituting for T and V in

$$0 = \pi - d_1 T - k_2 TV + \mu T_i$$

gives T_i and subsequently V.

So, the un-infected equilibrium is $\left(\frac{\pi}{d_1}, 0, 0 \right)$ and the infected equilibrium is

$$\left(\frac{\mu + d_2}{k_1 k_2} c, \frac{\pi}{d_2} - \frac{d_1(d_2 + \mu)c}{d_2 k_1 k_2}, \frac{k_2}{c} \left(\frac{\pi}{d_2} - \frac{d_1(d_2 + \mu)c}{d_2 k_1 k_2} \right) \right)$$

Theorem 1

If $\frac{\pi}{d_1} \neq (\mu + d_2)c/k_1 k_2$ there exist two equilibria.

Let us denote this infected equilibrium by $\left(T_*, T_{i*}, V_* \right)$ where each component corresponds to an earlier specified value. We let

$$x = T - T_*, \quad y = T_i - T_{i*}, \quad z = V - V_*$$

Then

$$\frac{dx}{dt} = -(d_1 + k_1 V_*)x + \mu y - k_1 T_* z$$

$$\frac{dy}{dt} = k_1 V_* x - (d_2 + \mu) y + k_1 T_* z$$

$$\frac{dz}{dt} = k_2 y - cz$$

Thus

$$\begin{matrix} x' \\ y' \\ z' \end{matrix} = A \begin{matrix} x \\ y \\ z \end{matrix}$$

Where;

$$A = \begin{pmatrix} -(d_1 + k_1 V_{*1}) & \mu & -k_1 T_* \\ k_1 V_* & -(d_2 + \mu) & k_1 T_* \\ 0 & k_2 & -c \end{pmatrix}$$

Thus

$$|A - \lambda I| = 0$$

Implies

$$p(\lambda) = \lambda^3 + \lambda^2 (\mu + d_1 + k_1 V_* + c + d_2)$$

$$+ \lambda (ck_1 V_* + d_1 d_2 + d_2 k_1 V_* + \mu k_1 V_*)$$

$$+ d_1 d_2 c + k_1 V_* d_2 c + c \mu k_1 V_* = 0$$

3.1

Theorem 2

Equation (3.1) has no positive root.

Table 1. Description of variables and constants.

Terms	Description	Values
π	Rate of production of T cells	dT_0/day
d_1	Natural death rate of healthy T cells	0.01/day
k_1	Viral infection rate (CD4+Tcells)	0.00027/virus day
μ	Infected T cell recovery rate	μ/day
d_2	Death rate of infected T cells	0.39/day
k_2	Viral production for T cell	850/day
c	Clearance rate of the virus	3/day
S	Stem cell	
T	Uninfected activated CD4+T cell	
T_i	Infected CD4+T cells	
V	Virus produced by T cells and macrophages	
P	Concentration of pg 120	

Theorem 3

Equation (3.1) has three negative roots or one negative root and two complex roots.

Theorem 4

The infected equilibrium is globally asymptotically stable.
   Routh – Hurwitz criteria (Wang and Li, 2006). All zeros of $\lambda^3 + \alpha\lambda^2 + \beta\lambda + \gamma = 0$ have negative real parts if and only if $\alpha\beta - \gamma > 0$.
   Therefore, all zeros of (3.1) have negative real parts if and only if

$$\left(\mu + d_1 + k_1 V_* + c + d_2\right)\left(c k_1 V_* + d_1 d_2 + d_2 k_1 V_* + \mu k_1 V_*\right) - \left(d_1 d_2 c + k_1 V_* d_2 c + c\mu k_1 V_*\right) > 0$$

That is,

$$\left(d_1 + k_1 V_* + d_2\right)\left(c k_1 V_* + d_1 d_2 + d_2 k_1 V_* + \mu k_1 V_*\right) + \mu\left(d_1 d_2 + d_2 k_1 V_* + \mu k_1 V_*\right) + c\left(c k_1 V_* + \mu k_1 V_*\right) > 0$$

Proof of theorems

Clearly all coefficients and the constants of $p(\lambda)$ are positive. The proof of the theorems 2 to 4 involved using the Descartes rule of signs: The number of positive zeros of a polynomial with real coefficients is either equal to the number of variations in sign of the polynomial or is less than this by an even number. Table 1

Proof of theorem 1

The infection – free equilibrium is given by

$$\left(\frac{\pi}{d_1}, 0, 0\right), if\ T_i \neq 0,\ V \neq 0,$$

then $T = \left(\mu + d_2\right)c / k_1 k_2$. Hence the other equilibrium is

$$\left(\frac{\left(\mu + d_2\right)^c}{k_1 k_2}, \frac{\pi}{d_2} - d_1 \frac{\left(d_2 + \mu\right)^c}{d_2 k_1 k_2}, \frac{k_2}{c}\left(\frac{\pi}{d_2} - \frac{d_1\left(d_2 + \mu\right)^c}{d_2 k_1 k_2}\right)\right)$$

Proof of theorem 2

The number of variations in sign is zero. Hence by Descartes' rule of signs the polynomial equation has no positive root.

Proof of theorem 3

From p(λ) in (3.1), we obtain

$$p(-\lambda) = \lambda^3 - \lambda^2\left(\mu + d_1 + k_1 V_* + c + d_2\right)$$

$$+ \lambda\left(c k_1 V_* + d_1 d_2 + d_2 k_1 V_* + \mu k_1 V_*\right)$$

$$- \left(d_1 d_2 c + k_1 V_* d_2 c\right) + c\mu k_1 V_*\right) = 0$$

So the number of changes in sign is 3. Hence by Descartes' rule of signs, $p(\lambda)$ has either three negative roots or one negative root and two complex roots.

Proof of theorem 4

Since all the parameters are positive, the inequality holds.
By theorem 3 and Routh – Hurwitz criteria (3.1) has;

(i) Either three negative roots or
(ii) One negative root and two complex roots whose real parts are equal and negative. So in either case the equilibrium is globally asymptotically stable.

RESULT AND DISCUSSION

The main purpose of this paper is to verify, beyond earlier papers, the effect of affinity hemodialysis on HIV/AIDS as a potential treatment option for HIV patients resistant to drugs. A key factor in this analysis is μ. When μ is zero, the possibility of a quasi-steady infected equilibrium does not exist. Thus a stable infected equilibrium does not arise. This paper shows, further, that affinity hemodialysis is a potentially useful adjunctive therapy which can be employed to treat HIV-infected patients either directly or in conjunction with drug therapy (Tullis, 2004)

ACKNOWLEDGMENT

This revised version (medical and mathematical aspect) has benefited from thoughtful comments of the referees.

REFERENCES

Duffin RP, Tullis RH (2002). Mathematical model of the complete course of HIV infected and AIDS, J. Theor. Med. Pp. 1-7.
Oluyo TO (2007). A mathematical model of HIV epidemic/verification using data obtained under contact tracing of Nigeria, Ph.D Thesis, LAUTECH Ogbomoso, Nigeria.
Oluyo TO, Ayeni RO (2007). A mathematical model of virus neutralizing antibody response Res J. Appl. Sci.2: 889 – 891.
Tullis RH, Scammura D, Ambrus J (2002). Affinity hemodialysis for anvirus therapy with specific application to HIV J. Theor Med. 3: 157 – 166.
Tullis RH, Duffin RP, Zech M, Ambrus JL (2002). Affinity hemodialysis for antiviral Therapy 1. Removal of HIV – 1 from cell culture supernatants plasma and blood. Ther. Apher. 6: 213-220.
Stafford MA, Covey L, Cao Y, Daar ES, Ho DD, Perelson AS (2000). Modelling Plasma virus concentration during primary HIV infection, J. Theor. Biol. 203: 285 – 301.
Tullis RH (2004). Mathematical model of the effect of affinity hemodialysis on the T-Cell depletion leading to AIDS. Blood purification 22: 84 – 91.
Tullis RH, Duffin RP, Zech M, Ambrus JL (2003). Affinity hemodialysis for antiviral therapy II. Removal of HIV – 1 viral proteins from cell culture supernatants and whole blood; Blood Purif. 21: 58-63.
Wang L, Li MY (2006).Mathematical analysis of the global dynamics of a model for HIV infection of CD4+T cells, Math Biosccnces.
Hazenberg MD, Hamann D, Schuitemaker H, Miedema FT (2000). Cell depletion in HIV – 1 infection. How CD4+T cells go out of stock. Nat. Immunol. 1: 285 – 289.

12

Prevalence of bacteremia in patients with diabetes mellitus in Karbala, Iraq

Mohammed A. K. Al-Saadi[1], Alaa H. Al-Charrakh[1]* and Salim H. H. Al-Greti[2]

[1]Department of Microbiology, College of Medicine, Babylon University, Babylon Province, Iraq.
[2]Medical Institute of Karbala, Karbala Province, Iraq.

The present study is designed to study bacteremia and to measure some immunological parameters of diabetic patients in Kerbala City, Iraq during the period from November 2006 until May 2007. This study included a total of 125 patients with diabetes mellitus (30 type I and 95 type II), and 55 healthy persons as Control subjects. Blood samples were collected from both patients and Controls, blood culture was done for bacterial isolation and identification, virulence factors as well as antibiotic susceptibility tests were assessed for each isolate. This study also included the estimation of T-cells count, interferon-gamma (IFN-γ) concentration, interleukin-4 concentration, IgG, and IgM concentration. The obtained results showed that bacteremia was observed in 24% of the diabetic patients. Gram-positive bacterial isolates were more predominant; 21:30 (70%); than Gram-negative isolates; 9:30 (30%). Cefotaxime, tetracycline and trimethoprime-sulphamethazole antibiotics were the most effective drugs on both Gram-positive and -negative bacteria. Immunological tests showed decrease in T-cells count significantly ($p < 0.05$) in type I and II diabetic patients (9.1%, 10.63%, respectively). Concentration of IFN-γ also decreased significantly ($p < 0.05$) in same patients (0.285 and 0.313 I.U/ml, respectively) as compared with control subjects (0.860 I.U/ml). Levels of IL-4 decreased non-significantly ($p > 0.05$) in patients with type I and II diabetes (7.050 and 7.703 pg/ml, respectively). The levels of IgG were increased significantly ($p < 0.05$) in both types I and II (1674.45 and 2095.86 mg/dL) respectively, but the levels of IgM were increased significantly ($p < 0.05$) in type II (177.64 mg/dL) and non-significantly ($p > 0.05$) in type I.

Key words: Bacteremia, diabetes mellitus, antibiotics, IL-4, IFN-γ, Iraq.

INTRODUCTION

Diabetes mellitus (D.M.) is defined as an abnormal metabolic state in which there is glucose intolerance due to inadequate insulin action as well as the late development of many complications (Boon et al., 2006). The World Health Organization recognizes two main forms of D.M: Type I and II. Type I is usually due to autoimmune destruction of the pancreatic beta cells (β-cells) which produce insulin (WHO, 1999). Type II is a heterogeneous group of disorders characterized by tissue-wide insulin resistance and varies widely (Kasper, 2005). Diabetic subjects probably have a higher risk of

many infections. In addition, immune-suppression which occurs in those patients because of increased sugar levels in the blood stream and as a result of dysfunction of the immune system, make diabetic patients more prone to microbial infections, especially bacteremia (Joshi et al., 1999). Several factors could predispose diabetic patients to infections. These factors include genetic susceptibility to infection; altered cellular and humoral immune defense mechanisms; local factors include poor blood supply and nerve damage and alterations in metabolism associated with diabetes (James, 2000). Chronically ill and immunocompromized patients with D.M have an increased risk of bacteremia. They may also develop bacteremia and fungimia (Vincent et al., 2004). On the other hand, their antimicrobial function is inhibited by hyperglycemia due to inhibition of

---
*Corresponding author. E-mail: aalcharrakh@yahoo.com.

glucose -6- phosphate dehydrogenase (G6PD) (Thomson et al., 2004). Another mechanism which can lead to increased prevalence of infections in diabetic patients is an increased adherence of microorganisms and they growth better in glucose, in addition to their expression of different virulence factors (Geerlings and Hoepelman, 1999). Impaired host defense mechanisms such as impaired granulocyte function, decreased cellular immunity, impaired complement function and decreased lymphokine response may be influenced by glycemic control (James, 2000).

The objective of this study was to investigate the etiologic agents of bacteremia in patients with diabetes mellitus and study the virulence factors and antibiotic sensitivity pattern among recovered bacterial isolates, in addition to studying some humoral and cellular immunological parameters in diabetic patients with and without bacteremia. Also a comparison was made between diabetic patients and the healthy control group.

MATERIALS AND METHODS

Patients and samples collection

This study included a total of 125 diabetic patients as diagnosed by clinical physicians (35 Type I and 90 Type II) with age ranging between17 to 65 years who attended Al-Hussein Hospital in Kerbala, Iraq during the period from November 2006 until May 2007. Case history of patients involving patient's name, age, type of therapy and accompanying disease were recorded. Only diabetic patients with no malignancy, asthma, cardiovascular accident, heart failure and with no previous administration of antibiotics were considered in the present study. Also, a total of 55 healthy persons (volunteers) with age matching to the patients group were enrolled in the present study as control subjects.

Blood samples (withdrawn by disposable syringe under aseptic techniques) were collected from each patient and control subject. The blood samples were divided into three parts; 1 ml was immediately inoculated into sterilized blood culture bottles for bacteriological investigation. 2 ml were put in EDTA anti-coagulant tubes for testing with the E-rosette technique. The remaining 3 ml were put in sterilized plane tubes and allowed to clot and then the serum was separated (by centrifugation at 3000 rpm for 15 min) and stored (refrigerated) until use for further immunological tests.

For bacteriological investigation, the blood samples were introduced directly into Brain Heart Infusion (BHI) broth bottles with the ratio 1:10 of blood to medium used. Blood culture bottles were incubated at 35°C for 7 days and shaken for the first 48 h. Bottles were tested for bacterial growth once on the day of receipt, twice on day 2, and once daily on day 3 through 7 (Forbes et al., 2007). Blood culture suspected of being positive were sub-cultured onto MacConkey agar (Mast, U.K.) plates of 5% human blood agar, and chocolate agar (supplemented with 10% CO_2). All agar plates were incubated for 48 h at 35°C.

Bacterial isolates

The isolation and identification of bacteria recovered from patient's blood culture were performed through colony morphology of bacterial isolates, microscopic Gram stain investigation and biochemical tests (Forbes et al., 2007). Isolates identification was also confirmed using API system strips (Biomerieux/France).

Detection of virulence factors

Virulence factors of bacterial isolates were detected according to standard procedures. The following virulence factors; coagulase, capsule production, lipase, gelatinase, hemolysin and colonization factor antigens; were detected according to Forbes et al., (2007). Bacteriocin production was assayed using cup assay method in Brain Heart infusion medium supplemented with 5% glycerol according to (Al-Qassab and Al-Khafaji, 1992).

Antimicrobial susceptibility testing

The susceptibility of the bacterial isolates to antimicrobial agents was determined using disk diffusion method as recommended and interpreted accordingly by the Clinical and Laboratory Standards Institute (CLSI, 2006; 2007). The following antimicrobial agents were obtained (from Oxoid, U.K.) as standard reference disks for known potency for laboratory use: Amoxicillin (Ax, 10 µg), cefotaxime (CTX, 30 µg), oxacillin (OX, 1 µg), cefoxitin (Fox, 30 µg), amikacin (Ak, 30 µg), tetracycline (Te, 30 µg), doxycycline (Do, 30 µg), ciprofloxacin (CIP, 5 µg), nalidic acid (NA, 30 µg) and trimethoprim-sulphamethazole (SXT, 1.25/23.75 µg).

All tests were performed on plates of Muller- Hinton agar (Oxoid, U.K.). A 0.5 MacFarland suspension (provided by Biomérieux/ France) of tested bacterial isolates was applied to the plates which were dried in an incubator at 35°C for 15 min. Antimicrobial disks were placed on the agar with sterile forceps. The agar plates were incubated and inverted at 35°C for 18 h. Results were recorded by measuring the inhibition zone (ml) and interpreted according to Clinical and Laboratory Standards Institute documents (CLSI, 2007).

Estimation of immunological parameters

Erythrocyte-rosette formation test (E-rosette) was carried out according to Gengozian et al. (2002) for estimation of T-lymphocytes.

The concentration of IFN-γ was estimated according to the procedure provided by the manufacturer (Biosource, Eumpe, S.A). The biosource IFN-γ is a solid phase enzyme amplified sensitivity Immune assay (EASIA) performed on a microtiter plate.

The concentration of IL-4 in patient's sera was carried out according to the procedure provided by the manufacturer (Biosource, Eumpe, S.A). The assay is based on an oligoclonal system in which a blend of monoclonal antibodies (MAbs) directed against distinct epitopes of IL-4 are used. IL-4 that are present in the sample reacts with captured monoclonal antibodies (MAb1) coated on the microtiter well and with a monoclonal antibody (MAb2) labeled with horseradish peroxidase (HRP).

The concentrations of immunoglobulin G and M (IgG and IgM) were estimated using the technique of single radial immunodiffusion test (SRID) (Biomaghrab, Tunisia) according to Lewis et al. (2001).

Statistical analysis

The mean, standard deviation (SD), and analysis of variance (ANOVA) tests were calculated according to Cochran (1974).

RESULTS AND DISCUSSION

Results of blood culture revealed that positive bacterial blood cultures (bacteremia) were observed in 10 diabetic patients (28.57%) with type I, and 20 patients (22.2%) with type II (Table 1). These results indicated the

Table 1. Distribution of bacteremia among diabetic patients and controls.

Patient group	Total number	Number of bacteremia (%)	Number of non-bacteremia (%)
Diabetes mellitus type I	35	10 (28.57)	25 (71.43)
Diabetes mellitus type II	90	20 (22.2)	70 (77.8)
Total	125	30 (24)	95 (76)
Control subject	55	0	55 (100)

Table 2. Gram-positive and Gram-negative bacteria isolated from diabetic patients.

Bacterium species	Number of isolates (%)
Gram positive	
Staphylococcus epidermidis	13 (61.90)
Streptococcus mitis	4 (19.05)
Staphylococcus aureus	3 (14.29)
Bacillus cereus	1 (4.76)
Total	21 (100)
Gram negative	
Klebseilla pneumoniae	4 (44.44)
Escherichia coli	2 (22.22)
Citrobacter freundii	1 (11.11
Proteus mirabilis	1 (11.11
Pseudomonas aeruginosa	1 (11.11
Total	9 (100)

presence of bacteremia in both types of diabetic patients which were also obtained by other investigators (Akbar, 2000; Cisterna et al., 2001).

Results of distribution of bacterial isolates according to type of Gram staining, revealed that out of 30 bacterial isolates recovered from patients with diabetes, 21 (70%) were Gram-positive bacteria while 9 (30%) were Gram-negative bacteria. The predominance of Gram-positive bacterial isolates in blood cultures of diabetic patients in the present study was in accordance with findings of several authors who mentioned that Gram-positive bacteria were predominant agents of bacteremia in diabetic patients (Thomson et al., 2004; Moutschen, 2005).

Results of identification of Gram-positive and -negative bacterial isolates based on cultural characteristics and biochemical properties of these isolates revealed that these isolates were of different bacterial species (Table 2). *Staphylococcus epidermidis* was the major Gram-positive species (61.90%), while *Bacillus cereus* (4.76%) was the less commonly isolated Gram-positive species in diabetic bacteremia. These results agreed with VonEiff et al. (2001) who reported that *S. epidermidis* was the predominant agent of bacteremia in immunocompromized patients.

Klebsiella pneumoniae was the predominant species among Gram-negative bacterial species (44.4%), while the other bacterial species such as *E. coli*, *Citrobacter*

freundii, *Proteus mirabilis* and *Pseudomonas aeruginosa* were isolated in low percentages. These findings agreed with the results of Oni et al. (2000) who mentioned that *Klebsiella* spp. represented 43% of all Gram-negative bacterial isolates recovered from bacteremia in diabetic patients. At the same manner, Al-Muslemawi (2007) showed that *Citrobacter freundii* may cause bacteremia in immunocompromized patients including patients with D.M.

All Gram-positive bacteria isolated in the present study were non-capsule producers. Gram-negative bacterial isolates were also non-capsule producers except *K. pneumoniae* isolates (Table 3) which exhibited a large and clear capsule when examined by negative staining method.

All *S. aureus* isolates were able to produce coagulase which is considered a virulence factor for pathogenicity of these bacteria by clumping the fibrin around the bacteria (Forbes et al., 2007). Possibly, coagulase could provide an antigenic disguise if it clotted fibrin on the cell surface or could make the bacterial cells resistant to phagocytes or tissue bacterial target (Humphreys, 2004).

Table 3 showed that *S. aureus*, *E. coli*, and *P. aeruginosa* isolates expressed β –hemolysis on blood agar medium while all *Streptococcus mitis* (4:4) isolates exhibited α-hemolytic pattern, showing a greenish zone around the bacterial colonies. On the other hand,

Table 3. Virulence factors of bacterial isolates recovered from diabetic patients.

Bacterial isolate	Coagulase	Lipase	Haemolysis	Gelatinase	Capsule formation	*CF-I	*CF-III	Bacteriocin production
Staphylococcus aureus 1	+	+	β	+	–	+	+	+
S. aureus 2	+	+	β	+	–	+	+	+
S. aureus 3	+	+	β	+	–	+	+	+
Staphylococcus epidermidis 1	–	–	γ	–	–	+	+	–
S. epidermidis 2	–	–	γ	–	–	+	+	–
S. epidermidis 3	–	–	γ	–	–	+	+	+
S. epidermidis 4	–	–	γ	–	–	+	+	+
S. epidermidis 5	–	+	γ	–	–	+	+	+
S. epidermidis 6	–	+	γ	–	–	+	+	–
S. epidermidis 7	–	–	γ	–	–	+	+	–
S. epidermidis 8	–	–	γ	–	–	+	+	–
S. epidermidis 9	–	–	γ	–	–	+	+	–
S. epidermidis 10	–	+	γ	–	–	+	+	+
S. epidermidis 11	–	–	γ	–	–	+	+	–
S. epidermidis 12	–	–	γ	–	–	+	+	+
S. epidermidis 13	–	–	γ	–	–	+	+	+
Streptococcus mitis 1	–	–	α	–	–	+	+	–
S. mitis 2	–	–	α	+	–	+	+	+
S. mitis 3	–	–	α	–	–	+	+	+
S. mitis 4	–	–	α	–	–	+	+	+
Bacillus cereus	–	–	γ	+	–	+	+	+
Klebsiella pnuemoniae 1	–	+	γ	–	+	+	+	+
K. pnuemoniae 2	–	+	γ	–	+	+	+	+
K. pnuemoniae 3	–	+	γ	–	+	+	+	+
K. pnuemoniae 4	–	–	γ	–	+	+	+	+
Escherichia coli 1	–	+	β	–	+	+	+	+
E. coli 1	–	+	β	–	–	+	+	+
Citrobacter freundii	–	+	γ	–	–	+	+	+
Proteus mirabilis	–	+	γ	+	–	+	+	–
Psuedomonas auroginosa	–	+	β	+	–	+	+	+

*CF: colonization factor antigens.

Staphylococcus epidermidis (13:13) isolates and each of *B. cereus*, *C. freundii* and *Proteus mirabilis* isolates were γ-hemolytic (non hemolytic).

The production of hemolysin by *S. aureus* is well known and considered as a main virulence factor for these bacteria and it is associated with increased severity of infections (Humphreys, 2004).

In this study, gelatin was used as a protein for

Table 4. Antibiotic resistance of gram-positive isolates recovered from diabetic patients.

Antibiotic	Bacterial isolate (Number)							
	S. epidermidis (13)		*S. aureus* (3)		*S. mitis* (4)		*B. cereus* (1)	
	R	S	R	S	R	S	R**	S*
Amoxcillin	0	1	3	1	2	1	11	2
Oxacillin	0	1	3	1	2	1	11	2
Cefoxitin	0	1	1	3	1	2	3	10
Cefotaxime	0	1	0	4	0	3	0	13
Amikacin	0	1	2	2	1	2	0	13
Tetracycline	0	1	0	4	1	2	3	10
Doxycycline	0	1	1	3	1	2	4	9
Ciprofloxacin	1	0	2	2	0	3	0	13
Nalidixic acid	1	0	2	2	3	0	5	8
Trimethroprim-sulphamethozole	0	1	1	3	0	3	0	13

* Sensitive; ***Resistant.

detection of protease activity in bacterial isolates. In Gram-positive bacteria, *S. epidermidis* (13:13) isolates were negative for this factor while *S. aureus* isolates (3:3) were positive for this test. In addition to that, a single isolate of *B. cereus* was able to produce gelatinase. One isolate of *Streptococcus mitis* was also positive for this test.

A single isolate each of *P. mirabilis* and *P. aeruginosa* were positive for gelatinase enzyme, while all isolates of *K. pneumoniae* and *E. coli* were negative for this factor (Table 3). Gelatinase is a proteases enzyme considered as a potential virulence factor of many microorganisms because of its ability to breakdown immunoglobulin and complement components that make up the host defenses against microbial infections and therefore, enable the pathogens to invade the host tissues (Travis et al., 1995).

All *S. aureus* (3:3) isolates were positive for lipase enzyme production, while only (4:13) isolates of *S. epidermidis* were positive for this factor. A single isolate of *B. cereus* and all isolates of *S. mitis* were also negative for this enzyme however, most Gram-negative isolates (8:9) were positive for this test. Host cell membranes have lipids in their components and lipase enzyme will destroy this element and aid the pathogen to penetrate the host tissue to develop the infections (Lisa et al., 1994).

Twenty (66.67%) of the total bacterial isolates recovered from diabetic patients were able to produce bacteriocin by cup assay method and form a clear inhibition zone (12 to 22 mm) on solid medium. These findings are in agreement with the results obtained by many researchers (Al-Qassab and Al-Khafaji, 1992; Al-Charrakh, 2005) who found that cup assay method was the best method used for detection of bacteriocin-production by isolates of *Lactobacilli*, and *Klebsiella pneumoniae* respectively. The importance of bacteriocin for virulence and pathogenicity of bacteria is controversial. Bacteriocin Is essential for virulence and pathogenicity of

Enterococcus in septicemia (Hancock and Gilmore, 2000) because it was found that cytolysin of *Enterococcus faecalis* (possess both hemolysin and bacteriocin activities) promotes the appearance of this bacteria in the blood thus indicating that the bacteriocin is essential for virulence of these bacteria in blood stream infections.

Antibiotics resistance of bacterial isolates

Results in (Table 4) show the antibiotic resistance of Gram-positive bacterial isolates recovered from diabetic patients. Most isolates of *S. epidermidis* (11:13) were highly resistant to amoxicillin and oxacillin, but they were highly sensitive to cefotaxime, amikacin, ciprofloxacin and trimethoprime-sulphamethazole. They also showed low levels of resistance to nalidixic acid, tetracycline and cefoxitin. Humphreys (2004) reported that the main mechanism of resistance of *S. epidermidis* to β-lactams is mediated by β-lactamase production. Results revealed that *S. epidermidis* had the highest number of multi-drug resistant isolates, and these findings are in agreement with those obtained by other investigators (Mahmood, 2001; VonEiff et al., 2001). *S. epidermidis* may act as a reservoir for resistance which can be transferred to *S. aureus*. All isolates of *S. aureus* (3:3) were resistant to nalidixic acid. Two of these isolates were resistant to amoxicillin and oxacillin, and only one isolate was resistant to cefoxitin, amikacin and tetracycline. These results are similar to the findings obtained by Brook et al. (1981) who showed that *S. aureus* isolates were resistant to amoxicillin and less resistant to cefotaxime. A further significant increase in the rate of oxacillin resistance, as was observed in the early 1990s, was not registered.

However, one may speculate that these rates would have been even higher if we had also included isolates from body sites other than blood, as was done in former studies (Kresken and Hafner, 1999). Three isolates of *S.*

Table 5. Antibiotic resistance of gram-negative isolates recovered from diabetic patients.

Antibiotic	Bacterial isolate (Number)									
	K. pnuemoniae (4)		E. coli (2)		C. freundii (1		P. mirabilis (1)		P. aeuriginosa (1)	
	S*	R**	S	R	S	R	S	R	R	S
Amoxicillin	1	0	1	0	0	1	1	1	4	0
Oxacillin	1	0	1	0	0	1	2	0	3	1
Cefoxitin	1	0	0	1	0	1	1	1	2	2
Cefotaxime	0	1	0	1	0	1	0	2	1	3
Amikacin	1	0	1	0	1	0	2	0	4	0
Tetracycline	0	1	0	1	0	1	1	1	1	3
Doxycycline	1	0	0	1	0	1	1	1	3	1
Ciprofloxacin	1	0	1	0	1	0	2	0	3	1
Nalidixic acid	1	0	0	1	1	0	1	1	3	1
Trimethroprim-sulphamethozole	1	0	0	1	0	1	0	2	1	3

* Sensitive; **Resistant.

mitis were resistant to amoxicillin and oxacillin, with two resistant to amikacin, ciproflaxin and nalidixic acid, and only one isolate was resistant to cefoxitin, doxycycline and trimethoprime-sulphamethazole (Table 4).

The results agreed with Carratala et al. (1995) who mentioned that bacteremia due to viridans streptococci were highly resistant (77%) to penicillin products. Among the streptococcal species, the viridans streptococci may be the most important pathogens causing bacteremia and sepsis in immunocompronized patients (Patrick, 1999). This problem is exacerbated by the emerging resistance of streptococci to antimicrobial agents commonly used for empirical and prophylactic treatments in those patients. The results also showed that the single isolate of B. cereus was sensitive to most antibiotics tested in the current study, and this result was confirmed by many researchers who found that Bacillus spp. isolates were highly sensitive to most antibiotics used in their studies (Weber et al., 1988).

Table 5 shows the results of antibiotic resistance of Gram-negative bacterial isolates. All isolates of K. pneumoniae were resistant to amoxicillin and amikacin with three being resistant to oxacillin, doxycycline, ciproflaxin and nalidixic acid, and two were resistant to cefoxitin and only one isolate was resistant to cefotaxime, tetracycline and trimethroprim-sulphamethazol. These results were in agreement with the reports published by some workers (Al-Charrakh et al., 2011) from Iraq and another researcher (Jarlier et al., 1996) in France who mentioned that K. pneumoniae was highly resistant to amoxicillin and ampicillin by the production of β-lactamases that render these isolates unsusceptible to most β-lactam antibiotics.

Two isolates of E. coli recovered in the present study were resistant to oxacillin, amikacin and ciproflaxin but they were sensitive to cefotaxime and trimethoprime-sulphamethazole (Table 5). This resistance could be interpreted depending on the fact that many strains of E. coli have acquired plasmids conferring resistance to one or more than one type of antibiotics, therefore antimicrobial therapy should be guided by laboratory result test of sensitivity. However, the sensitivity of E. coli isolates agreed with local reports in Iraq by several researchers (Al-Muhanna, 2001; Al–Hamawandi, 2005) who showed that all isolates of E. coli were sensitive to cefotaxime and trimethoprime-sulphamethazole. The single isolate of Citrobacter freundii was resistant to amikacin although this antibiotic is not used widely in the treatment in Iraq while the single isolate of P. mirabilis was resistant to amoxicillin, oxacillin, amikacin and ciproflaxin (Table 5). The single isolate of P. aeruginosa was resistant to most antibiotics tested and this result agreed with results reported by several investigators (Mahmoud, 2001; Kiska and Gilligan, 2003) who showed that P. aeruginosa strains isolated from blood-stream infections were resistant to 3rd generation cephalosporins and quinolones.

Immunological parameters

Table 6 shows the results of E-rosette test. The mean value of E. rosette that was positive for T-lymphocytes in both type I and II D.M was less than control subjects. T-cell counts decreased significantly ($p < 0.05$) in both types of diabetic patients, whereas the mean difference between the two types of diabetic patients was non-significant ($P > 0.05$). These results indicate the effect of hyperglycemia on the proliferation and function of T-cells that leads to immunosuppression of cellular immunity in diabetic patients. These results agreed with many reports which pointed that the main causes responsible for subnormal T-cell levels are acquired immune deficiency disorders including D.M., acute viral infections, and congenital immunodeficiency disease (Kotton, 2004).

Table 6. The T-cells count estimated and IFN-γ concentration for diabetic patients and control subjects.

Test group	T-cells count		IFN-γ concentration	
	Mean value %	Standard error	Mean I.U/ml	Standard error
Diabetes mellitus type I	9.1	1.16	0.285	0.085
Diabetes mellitus type II	10.63	0.817	0.313	0.060
Control subjects	22.55	1.16	0.860	0.085

Table 7. IL-4 concentration and Mean values of the IgG and IgM concentrations in the serum of diabetic patients and control subjects.

Test group	IL-4 concentration		IgG and IgM concentrations	
	Mean (pg/ml)	Standard error	Mean value of IgG (mg/dL)	Mean value of IgM (mg/dL)
Diabetes mellitus type I	7.050	2.523	1674.45	145.78
Diabetes mellitus type II	7.703	1.784	2095.86	177.64
Control subjects	13.650	2.523	1269.51	112

Decreasing T-cells count greatly contributes to impaired cell mediated immunity (Moretti, 1992). This impairment is due to the supporting role of T-cells activation in both arms of immune responses; cellular and humoral (T_{H1} and T_{H2} cells); in addition to its direct roles by directly killing foreign invaders by cytotoxic T-lymphocytes (Abbas et al., 2000).

Several immunological abnormalities related to D.M. such as depletion of T-cells, defective NK cells and insulinopenia-induced enzymatic defect have often been proposed to inhibit energy-requiring functions of phagocytes and lymphocytes (Moutschen et al., 1992).

Results also revealed that IFN-γ decreased significantly ($p<0.05$) in both types of diabetic patients as compared with control subjects (Table 6). These results indicate the presence of reduced cellular immunity in diabetic patients which agreed with the results obtained by several authors (Kukreja et al., 2002; Tsiavon et al., 2005) who stated that IFN-γ decreased significantly in diabetic patients. However, in the present study there was no significant ($p>0.05$) differences between the two types of D.M.

IFN-γ is secreted from T_{H1} lymphocyte, NK cells and macrophage, therefore any defect in these cells leads to decreasing IFN-γ production that reflects impaired host resistance in diabetic patients due to a dysregulation of the cytokine network. IFN-γ enhances the microbicidal function of macrophages by stimulating the synthesis of reactive oxygen intermediates and nitric oxide (NO). It promotes the differentiation of naïve CD4[+] T-cells to the T_{H1} subset and inhibits the proliferation of T_{H2} cells beside activating nuetrophils and stimulates the cytolytic activity of NK cells (Doan et al., 2005).

Between-group comparisons of IL-4, IgG and IgM are presented in Table 7. IL-4 showed significant between group differences ($p<0.05$), with no difference detected between type I and type II DM (mean of 7.05 and 7.70 pg/ml respectively), but both types of DM were signifycantly different from controls (mean of 13.65). These results agreed with Mayer et al. (1999) and Kukreja et al. (2002) who mentioned that there was no significant reduction in the levels of IL-4 among diabetic patients as compared with controls. These results expressed that T_{H2} response is slightly affected in DM as compared with T_{H1} response.

Regarding levels of IgG and IgM in diabetic patients and control subjects, the results revealed that the levels of IgG increased significantly ($p<0.05$) in both types of diabetic patients, while IgM levels increased significantly ($P<0.05$) in type II and non-significantly ($p>0.05$) in type I as compared with control subjects (Table 7). These findings agreed with those obtained by Gorus et al. (1998) who mentioned that the mean levels of IgM and IgG were higher in both types of diabetes mellitus patients.

Changes in total immunoglobulin concentrations were largely reversed under insulin therapy which may reflect exposure to environmental triggers, such as viral infections or to insulinopenia prior to clinical disease onset (Gorus et al., 1998). Alteration in humoral immune response was less in cellular immunity during immunosuppression process. Therefore levels of IgG and IgM in diabetic patients are higher than in normal persons but not like levels of IgG and IgM in persons suffering from microbial infections without diabetes. Some reports mentioned that diabetes might exhibit hypergammaglobulinemia (Al-Ardawi et al., 1994).

The results expressed in Table 8 show the relationship between bacteremia and immunological parameters studied in type I and II diabetic patients. E-rosette value, which is used as a marker for T-cells count, decreased significantly ($p<0.05$) in bacterimic diabetic patients as

Table 8. Relationship between bacteremia and immunological parameters in diabetic patients.

Test group	Total no. (%)	Immunological parameter				
		E-rosette* (%)	IFN-γ* (I.U/ml)	IL-4* (pg/ml)	IgG* (mg/dL)	IgM* (mg/dL)
Bacteremia	30 (24)	8.217**	0.109 **	5.478	1865.24	168.5
Non-bacteremia	95 (76)	15.246 **	0.577 **	10.458	1626.996	138.47
Total No.	125 (100)					

*The mean value is calculated for both types of D.M. with bacteremia..**The mean difference is significant at 0.05 level.

compared with non bacteremic diabetic patients. Mean value of IFN-γ concentration in diabetic patients with bacteremia was 0.109 I.U./ml, while in diabetes mellitus patients without bacteremia, it was 0.577 I.U/ml (Table 8). These results show significant decrease (p<0.05) in IFN-γ concentration in diabetic patients with bacteremia as compared with non bacterimic diabetic patients. Therefore, decreasing IFN-γ concentration leads to increase likelihood of bacteremia in diabetic patients (Abbas et al., 2000). Microbial products stimulate macrophages to produce tumor necrosis factor alpha (TNF-α) and IL-12. Together, these two cytokines then stimulate NK cells to produce IFN-γ (Bancroft et al., 1991). This cytokine plays an important role in the growth and differentiation of cytotoxic T-cells, activates NK cells and acts as a B-cell maturation factor. It regulates immunobulines (Ig) isotype production (Snapper and Paul, 1987; Murray, 1988). Concerning levels of IL-4 concentration in bacteremic and non-bacteremic diabetic patients results, the study revealed that levels of IL-4 decreased non- significantly (p>0.05) in bacteremic diabetic patients as compared with non bacteremic diabetic patients. IL-4 is produced by TH$_2$ type of CD^{+4} T-lymphocytes, following activation by the antigen binding to the T-cell receptor and also produced by activated mast cells and basophiles (Abbas and Lichtman, 2007). IL-4 deregulates the production of IFN-γ by TH$_1$ CD^4 T-lymphocytes, on the B-cells (Banchereau, 1990).

The levels of IgG and IgM in bacteremic diabetic patients were 1865.24 and 168.5 mg/dL respectively, while in non bacteremic patients they were 1626.996 and 138.47 mg/dL, respectively. These results indicated non-significant increase (p>0.05) in immuno-globulin levels in bacteremic diabetic patients as compared with non bacteremic diabetic patients.

Immunoglobulin increased during courses of bacteremia by 10 to 15 fold than in normal patients. Thus, the non significant increased levels of immunoglobulin in bacteremia diabetic patients is not high enough for the level of protection from infections. Therefore, there are non-significant differences between patients with bacteremia and controls without bacteremia. This may be due to shortage in the signaling of TH$_2$ required for B-cell proliferation and differentiation into immunoglobulin-secreting cells (plasma cells) (Rich et al., 2003).

REFERENCES

Abbas AK, Lichman AH (2007). Basic immunology: Function and disorders of the immune system. 2nd ed. Saunders, Elsevier. USA.

Abbas AK, Lichman AH, Pober JS (2000). Cellular and molecular immunology (4th ed). W.B. Saunders Co.

Akbar DH (2000).Adult bacteremia, comparative study between diabetic and non-diabetic patients. Saudi Med. J. 21(1): 40-44.

Al-Ardawi MS, Nasrat HA, Bahnassy AA (1994). Serum immunoglobulin concentration in diabetic patients. Diabetic Med. 11: 384-387.

Al-Charrakh AH (2005). Bacteriological and genetic study on extended-spectrum beta-lactamases and bacteriocins of *Klebsiella* isolated from Hilla city. Ph.D. Thesis. College of Science. Baghdad University, Iraq.

Al-Charrakh AH, Yousif SY, Al-Janabi HS (2011). Occurrence and detection of extended-spectrum β- lactamases in *Klebsiella* isolates in Hilla, Iraq. Afr. J. Biotechnol., 10 (4): 657-665.

Al-Hamawandi JAA (2005).Bacteriological and immunological study on infants pneumonia at Babylon Governorate. M.Sc. Thesis. College of Science. Al-Mustansiriya University, Iraq.

Al-Muhanna ASJ (2001). Identification of bacteria causing pneumonia in three medical centers in Najaf. M.Sc. thesis. College of Medicine, Kufa University, Iraq.

Al-Muslemawi TAJ (2007). Study of some biochemical, biological and pathological properties of lipopolysaccharides extracted from *Citrobacter freundii* Ph.D. Thesis. College of Science. Baghdad University, Iraq.

Al-Qassab AO, Al-Khafaji ZM (1992). Effect of different conditions on inhibition activity of enteric lactobacilli against diarrhea-causing enteric bacteria. J. Agric. Sci., 3(1): 18-26.

Banchereau J (1990). Interleukin-4. Sci. J., 6: 946-953.

Bancroft GJ, Schreiber RD, Unanue ER.(1991). Natural immunity: a T-cell-independent pathway of macrophage activation, defined in the scid mouse. Immunol. Rev., 124: 5-15.

Boon NA, Colledge NR, Walker BR, Hunter JAA (2006). Principles and practice Davidson's medicine. 20th ed. Churchill Livingstone, London.

Brook I, Yocum P, Friedman EM (1981). Aerobic and anaerobic bacteria in tonsils of children with recurrent tonsillitis. Ann. Otol., 90: 261-263.

Carratala J, Alcaide F, Fernandez S, Corbella X (1995). Bacteremia due to viridans streptococci that are highly resistant to penicillin. Clin. Infect. Dis., 20(5): 1169-1173.

Cisterna R, Cabezas V, Gomez E, Busto C, Atutxa I (2001). Community-acquired bacteremia. Rev. Esp. Quimioter Decis., 14 (4): 369-382.

Clinical and Laboratory Standards Institute (CLSI) (2006). PerformanceStandards for Antimicrobial Disk Susceptibility Tests; Approved Standard. Ninth Edition M2-A9. Clinical and Laboratory Standards Institute, Wayne, Pennsylvania, USA, 26(1).

Clinical and Laboratory Standards Institute (CLSI) (2007). Performance standards for antimicrobial susceptibility testing. Approved standard M100-S17. Clinical and Laboratory Standards Institute, Wayne, Pennsylvania, USA, 27(1).

Cochran WG (1974). Sampling techniques (2nd ed). Wiley Eastern private limited, New Delhi.

Doan T, Melvold R, Waltenbaugh C (2005). Concise Medical immunology. Lippincott Williams and Wilkins Philadelphia, USA.

Forbes BA, Daniel FS, Alice SW (2007). Baily and Scott's Diagnostic microbiology. 12th ed., Mosby Elsevier Company, USA.

Geerlings SE, Hoepelman AI (1999). Immune dysfunction in patients with diabetes mellitus. Immunol. Med. Microbiol., 26 (3-4): 259-265.

Gengozian N, Hall RE, Whitehurst CE (2002). Erythrocyte-rosetting properties of feline blood lymphocyte and their relationship to monoclonal antibodies to T lymphocytes. J. Expert Biol. Med., 9: 771-778.

Gorus FK, Vandewalle CL, Winnok F, Lebleu F, Keymenlen B, Vander A (1998). Increased prevalence of abnormal immunoglobulin M, G and A concentration at clinical onset of insulin dependent diabetes mellitus: a registry-based study. Pancreas, 16(1): 50-59.

Hancock IE, Gilmore MS (2000). Pathogencity of enterococci. In: Fischetti, V.; Novick, R.; Ferretti, J.; Poruoy, D.; and Rood, J. Gram positive pathogens. ASM publications, Washington, USA, pp. 751-758.

Humphreys H (2004). Staphylococcus. In Greenwood D, Slack R, Penethrer J. Medical Microbiology, a guide to microbial infections: pathogenesis, immunity, laboratory diagnosis and control (6thed). Churchill Livingstone, London. pp 174-188.

James ST (2000). Infectious complications in patients with diabetes mellitus. Int. Diabetes Monit. 12 (2): 33-39.

Jarlier V, Fosse T, Philipon A (1996). Antibiotic susceptibility in aerobic Gram-negative bacilli isolated in intensive care units in 39 French teaching hospitals (ICU study). Intensive Care Med., 22: 1057-1065.

Joshi N, Caputo GM, Weitekamp MR, Karchmer AW (1999). Infections in patients with diabetes mellitus. N. Engl. J. Med., 341: 1906-1912.

Kasper DL, Braunwald E, Fauci SA, Hanser SL, Longo DL, Jameson JL (2005). Harrison's Principle of internal medicine. McGraw-Hill Medical Publishing Division, London, 2(16): 2067

Kiska DL, Gilligan PH (2003). Pseudomonas. In Murry PR, Baron EJ, Jorgensen JH, Pfallor MA, Yolken RH. Manual Clin. Microbiol., 1(8): 719-728.

Kotton C (2004). Encyclopedia: T (thymus derived) lymphocyte count. The University of Pennsylvania Health System.

Kresken M, Hafner D (1999). Drug resistance among clinical isolates of frequently encountered bacterial species in Central Europe during 1975-1995. Infection, 27: S2-S8.

Kukreja A, Cost G, Marker J, Zhang C, Sun Z (2002). Multiple immuno-regulatory defects in type I diabetes. J. Clin. Invest., 109 (1): 131-140.

Lewis SM, Bain BJ, Bates I (2001). Dacie and Lewis, Practical haematology (9th ed). Churchill Livingstone, London.

Lisa TA, Lucchesi GI, Domenech CE (1994). Pathogenicity of Pseudomonas aeruginosa and its relationship to the cholin metabolism through the action of cholinesterase, acid phosphatase and phospholipase. Curr. Microbiol., 29: 193-199.

Mahmood A (2001). Blood stream infections in a medical intensive care unit: Spectrum and antibiotic susceptibility patterns. J. Pak. Med. Assoc., 51(6): 213-215.

Mayer A, Rharbaui C, Thivolet J, Madec AM (1999). The relationship between peripheral T-cells reactivity to insulin, clinical remissions and cytokine production in type I diabetes mellitus. Clin. Endocrinol. Metab., 84 (7): 2419-2424.

Moretti ML (1992). Immune impairment and the hypothesis of the acquired immune deficiency cycle. Inter. American Medical and Health Assoc. J., 1(1): 1-33.

Moutschen MP (2005). Alterations in natural immunity and risk of infection in patients with diabetes mellitus. Rev. Med. Liege, 60(5-6): 541-554.

Moutschen MP, Scheen AJ, Lefebvre PJ (1992). Impaired immune responses in diabetes mellitus: analysis of the factors and mechanisms involved. Relevance to the increased susceptibility of diabetic patients to specific infections. Diabet. Metab., 18(3): 187-201.

Murray HW (1988). Interferon gamma, the activated macrophage and host defense against microbial challenge. Ann. Int. Med., 108: 595-608.

Oni AA, Ogunkunle MO, Oke AA, Brake RA (2000). Pattern of Gram-negative rods bacteremia in diabetic patients in Ibadan, Nigeria. Afr. J. Med. Sci., 29(3-4): 207-210.

Patrick CC (1999). Viridans streptococcal infections in patients with neutropenia podiatry. Infect. Dis. 18: 280-283.

Rich R, Fleisher T, Shearer W, Kotzin B, Schroeder H (2003). Clinical immunology principles and practice (2nd ed). Mosby, Harcourt publishers, London, Vol. 1.

Snapper CM, Paul WE (1987). IFN-γ and BSF-1 reciprocally regulate immunoglobulins isotype production. Science, 236: 944-946.

Thomson RW, Hunborg HH, Lervang HH (2004). Risk of community acquired pneumococcal bacteremia in patients with diabetes. Diabetes Care, 27: 1143-1147.

Travis J, Potempa J, Maeda H (1995). Are bacterial proteinases pathogenic factors? Trends Microbiol., 3: 405-407.

Tsiavon A, Hatziagelak E, Chaidaroglon A, Raptis SA (2005). Correlation between intracellular IFN-γ production by CD+4 and CD+8 lymphocytes and IFN-γ gene polymorphism in patients with type-2 diabetes mellitus and latent autoimmune diabetes of adult. Cytokine, 31: 135-141.

Vincent JW, Russell P, Felman EL (2004). Oxidative stress in the pathogenesis of diabetic neuropathy. Endocrine Rev., 25(4): 612-628.

VonEiff C, Proctor RA, Peters G (2001). Coagulase negative staphylococci: Pathogens have major role in nosocomial infections. Postgrad. Med., 110(4): 63-76.

Weber DJ, Saviteer SM, Rutala WA, Thomann CA (1988). In vitro susceptibility of Bacillus spp. to selected antimicrobial agents. Antimicrob. Agents Chemother., 32: 642-645.

WHO (1999). Definition, diagnosis and classification of diabetes mellitus and its complications. Geneva report-1.

13

Antibacterial activity: A comparison of ripe and unripe fruit extracts of *Cissus multistriata* (*Vitaceae*) plant

Enemuor, Simeon Chukwuemeka[1], Omale, James[2]* and Joseph, Ekpa Matthew[1]

[1]Department of Microbiology, Kogi State University, Anyigba, Kogi State, Nigeria.
[2]Department of Biochemistry, Kogi State University, Anyigba, Kogi State, Nigeria.

The antibacterial activity of methanol extracts of the ripe and unripe fruit of *Cissus multistriata* against *Escherichia coli* (Swine) ISB492, *E. coli* (Swine) ISB440, *Serratia marcescens* FD5/64, *S. marcescens* FD1/62, *Staphylococcus aureus* FD1/62 and *Bacillus cereus* ISB517 were determined using agar ditch diffusion and tube dilution methods. The crude methanol extracts exhibited antibacterial activity against some of the tested bacterial isolates. Both ripe and unripe fruit extracts were inhibitory to *S. marcescens* (FD5/64). The unripe fruit extract also was inhibitory to *S. marcescens* FD1/62 and *S. aureus* FD1/62. The unripe fruit extract exhibited more antibacterial activity than the ripe fruit extract with minimum inhibitory concentration (MIC) of 50 mg/ml. The present findings have added to the fact that *C. multistriata* has some medicinal values which the traditional medical practitioners have been tapping in their treatment of ailments in their localities. Further studies are required to identify the phytochemicals involved and to know the component that is lost during ripening that contributed to loss of some antibacterial activity of the ripe fruit extract of the plant. When these facts are harnessed, it will surely be useful in the development of some new drugs with broad spectrum of antimicrobial activity.

Key words: *Cissus multistriata*, antibacterial activity, ripe and unripe fruit, bacterial isolates, minimum inhibitory concentration (MIC), Kogi State, Nigeria.

INTRODUCTION

The search for agents to cure infectious diseases began long before people were aware of the existence of microbes (Sofowora, 1982). These early attempts used natural substances, usually native plants or their extracts and many of these herbal remedies proved successful (Sofowora, 1982). According to WHO reports, over 80% of the world population depends on traditional medicine for their primary healthcare needs (Duraipandiyan et al., 2006). There is alarming increase in the incidence of new and re-emerging infectious diseases. Hence, there is a continuous and urgent need to discover new antimicrobial compounds with diverse chemical structures and novel mechanisms, especially due to development of resistance to the antibiotics in current clinical use (Bauer et al., 2003). The screening of plant extracts has been of great interest to scientists in the search for new drugs for greater effective treatment of several diseases (Dimayuga and Garcia, 1991).

Therefore, plant extracts and phytochemicals with known antimicrobial properties can be of great significance in therapeutic treatments (Diallo et al., 1999; Rojas et al., 2006; Erdogrul, 2002). The medicinal value of plants lie in some chemical substances that produce a definite physiological action on the human body. The most important of these bioactive compounds of plant

*Corresponding author. E-mail: jamesomale123@yahoo.com

Figure 1. Photograph of *Cissus multistriata*.

are alkaloids, flavonoids, tannins and phenolic compounds (Edeoga et al., 2005). Many plant parts have antimicrobial properties such as tannins, essential oils and other aromatic compounds (Kumar and Sigh, 1984). In addition, many biological activities and antibacterial effects have been reported for plant tannins and flavonoids (Scalbert, 1991; Chung et al., 1991). *Cissus* is a genus of about 350 species of tropical and subtropical, chiefly woody vines of the grape family – *Vitaceae* (Burkil, 1985). *Cissus multistriata* (Figure 1) is a well known plant to the traditional medicine practitioners in Nigeria (Omale et al., 2009a). It is used as a medicinal plant for the management of diverse ailments in different locations in Nigeria. The Ebiras use the stem prepared in form of decoction, as internal cleanser for new born babies while the Yorubas use the leaves for the treatment of infertility in women and stomach ailment in children.

It is commonly used by the Ibajis in the eastern part of Kogi State for the treatment of malnutrition –kwashiorkor. Other ethno medicinal uses of this plant include its use as

cough remedy, fracture healing and management of arthritis (Omale et al., 2009a). The leaf extract has a broad spectrum antimicrobial activity (Omale et al., 2009b). It contains some phytochemical and possesses antioxidant properties (Omale and Okafor, 2008).To the best of our knowledge, there is no record of work on the antibacterial activity of the fruit of *C. multistriata* (Vitaceae) plant. Therefore, the present study was carried out to evaluate the antibacterial potential of the ripe and unripe crude methanol extract of *C. multistriata*.

MATERIALS AND METHODS

Plant material

The plant material used was collected from a local farm located at Idah, Kogi State, Nigeria. It was fully characterized and identified in the Department of Botany, University of Ibadan, Nigeria as *C. multistriata* (Figure 1). The fruit was harvested, rinsed in clean water, and air dried for three weeks until constant weight was

Table 1. Antibacterial activity of methanolic extract of unripe fruit of *C. multistrata* on some clinical isolates.

Bacterial isolates	Mean zone diameter of inhibition (mm)			
	100 mg/ml	50 mg/ml	25 mg/ml	Control (distilled water)
Escherichia coli (ISB 492)	0	0	0	0
Escherichia coli (ISB 440)	0	0	0	0
Serratia marcescens (FDS/ 64)	10	8	2	0
Serratia marcescens (FDI/ 62)	9	5	3	0
Staphylococcus aureus (FDI/62)	7	11	6	0
Bacillus cereus (ISB 517)	0	0	0	0

obtained. It was grounded into powder and stored in a sterile glass bottle in the refrigerator.

Preparation of extracts

A portion (100 g) of the powdered ripe and unripe fruits was weighed into a 1000 ml beaker separately and 500 ml of methanol was added to each. This was stirred every 30 min for 3 h and allowed to stand for 72 h as described by the Association of Analytical Chemist, AOAC (1980). The crude extract was prepared by decanting, followed by filtration through muslin cloth, and further filtered with Whatman No. 1 filter paper (150 mm) to obtain a clear filtrate. The crude filtrates (ripe and unripe extracts) were concentrated by evaporation to semi-solid state on a water bath at 100°C. The green semi-solid extracts obtained were further concentrated using a rotary evaporator .The extracts were reconstituted in their extracting solvent to obtain a stock solution of 200 mg/ml. The stock solutions obtained were then filter-sterilized using Millipore filter (0.45 µm pore size). The sterile extracts were stored in sterile capped bottles.

Test organisms

The test organisms used were kindly supplied by Dr Odugbo from the Bacterial Research Division, National Veterinary Research Institute, Vom, Plateau State, Nigeria. They are *Escherichia coli* (Swine) ISB492, *E. coli* (Swine) ISB440, *Serratia marcescens* FD5/64, *S. marcescens* FD1/62, *Staphylococcus aureus* FD1/62, *Bacillus cereus* ISB517.

Revival of the freeze-dried bacteria cultures

Revival of the freeze-dried cultures of the test organisms were performed following the instructions from the Bacterial Research Division, National Veterinary Research Institute, Vom, Plateau State, Nigeria. The ampoules were opened just above the cotton plug by scoring the glass cutter and broken at the scored mark. Aseptic working conditions and sterility were strictly observed. Each bacterial material was suspended into a test tube containing 3 ml nutrient broth, shaken and then left in the refrigerator for 6 h. The suspension was inoculated on nutrient agar plates. Duplicate plates were made for each organism and then incubated at 37°C for 24 h. Subcultures were made on nutrient agar.

Determination of antibacterial activity of the crude methanol extracts

The antibacterial activity of the crude methanol extracts was tested by first inoculating nutrient agar plates with the test organisms. Duplicate plates were made for each test organism by flooding each plate with 5 to 6 h old revived freeze-dried culture. Thereafter, a sterile cork borer (6 mm diameter) was used to make four ditches on each plate. 0.1 ml of the filter-sterilized extract was dropped into each ditch and was labeled appropriately. Into the fourth ditch was dropped distilled water only to serve as a control. The inoculated plates were left on the lamina flow bench for 1 h to allow the extracts to diffuse into the agar (National Committee for Clinical Laboratory Standard, 1990). Thereafter, the plates were incubated aerobically at 37°C. Zones of inhibition produced after incubation was measured in millimeter.

Determination of minimum inhibitory concentration (MIC) of the extracts on the isolates

A broth dilution method as described by National Committee for Clinical Laboratory Standard (1990) was employed in the determination of MIC. Varying concentrations of the extracts (100, 50 and 25 mg/ml) were prepared. A portion (0.1 ml) of each concentration was added to each 9 ml of nutrient broth containing 0.1 m of standardized test organism of bacteria cells. The bottles were incubated aerobically at 37°C for 24 h. The tube with least concentration of extract without growth after incubation was taken and recorded as the minimum inhibitory concentration (MIC).

RESULTS AND DISCUSSION

The antibacterial activities of unripe and ripe fruit extracts on the tested isolates are as presented in Tables 1 and 2, respectively. The *E. coli* strains and *B. cereus* were not susceptible to both the unripe and ripe extracts, of *C. multistriata*. The two strains of *S. marcescens* and *S. aureus* were inhibited by the unripe fruit extract (Table 1) with zones of inhibition ranging from 2 to 10 mm. Only *S. marcescens* (FD5/64) was susceptible to the ripe fruit extract (Table 2). The largest zone of inhibition was observed for unripe fruit extract against *S. marcescens* (FDS/ 64).This antibacterial activity could be attributable to the phytochemicals in the extracts. The polyphenol

Table 2. Antibacterial activity of methanolic extract of ripe fruit of *C. multistriata* on some clinical isolates.

Bacterial isolates	Mean zone diameter of inhibition (mm)			
	100 mg/ml	50 mg/ml	25 mg/ml	Control (distilled water)
Escherichia coli (ISB 492)	0	0	0	0
Escherichia coli (ISB 440)	0	0	0	0
Serratia marcescens (FDS/ 64)	4	2	0	0
Serratia marcescens (FDI/ 62)	0	0	0	0
Staphylococcus aureus (FDI/62)	0	0	0	0
Bacillus cereus (ISB 517)	0	0	0	0

Table 3. Minimum inhibitory concentration (MIC) of the unripe and ripe fruits of *C. multistriata*.

Bacterial isolates	Minimum inhibitory concentration (MIC) (mg/ml)	
	Unripe fruit	Ripe fruit
Escherichia coli (ISB 492)	Nil	Nil
Escherichia coli (ISB 440)	Nil	Nil
Serratia marcescens (FDS/ 64)	50	100
Serratia marcescens (FDI/ 62)	100	Nil
Staphylococcus aureus (FDI/62)	100	Nil
Bacillus cereus (ISB 517)	Nil	Nil

composition of the ripe and unripe fruit of *C. multistriata* (Omale, 2010) and other chemical compositions of the plant (Omale et al., 2009a) have been determined. Both the ripe unripe fruit have been found to contain tannins, flavonoid, phenols and anthocyanins which have been found to exhibit various degrees antimicrobial activities (Rath et al., 2009; Anibijuwon et al., 2010; Pareckh and Chanda, 2007). Similar results have been reported for the leaf of *C. multistriata* when tested on some clinical isolates (Omale et al., 2009b).

In this study, the unripe fruit extract showed more antibacterial activity than the ripe one. This is suggestive that ripening may have transformed certain bioactive components which are responsible for the antibacterial activity of the fruit. The MIC results indicated that unripe fruit extracts of *C. multistriata* appeared more potent with 50, and 100 mg/ml of MIC against *S. marcescens* FD5/64, FD1/62 and *S. aureus* FD1/62 respectively (Table 3). The present findings have added to the fact that *C. multistriata* has some medicinal values which the traditional medical practitioners have been tapping in their treatment of ailments in their localities. From the results obtained in this study further research needs to be carried out to determine the exact phytochemicals (and their nature) involved in the antibacterial activity of the fruit of *C. multistriata*. It will be necessary also to know the component that is lost during ripening, that

contributed to loss of some antibacterial activity of the ripe fruit extract of the plant. When these facts are harnessed, it will surely be useful in the development of some new drugs with broad spectrum of antimicrobial activity.

REFERENCES

Anibijuwon II, Duyilemi OP, Onifade AK (2010). Antimicrobial activity of leaf of *Aspila africana* on some pathogenic organisms of clinical origin. Nig. J. Microbiol., 24 (1): 2048-2055.
AOAC (1980). Association of Analytical Chemist; Official methods of Analysis. 13th edition, Washington, DC.
Bauer J, Rojas R, Bustamante B (2003). Antimicrobial activity of selected Peruvian medicinal plants. J. Ethnopharmcol., 88: 199-204.
Burkil NM (1985). The useful plants of West Tropical Africa. The Whiferfries press Limited, London, p. 850.
Chung KT, Wong TY, Wei Y, Yuang YW, Lin Y (1998). Tannins and Human Health. A review. Crit. Rev. Food. Sci. Nutr., 8: 421-464.
Diallo D, Hveem B, Mahmoud MA, Betge G, Paulsen BS, Maiga A, (1999). An ethno botanical survey of herbal drugs of Gourm district, Mali. Pharm. Biol., 37: 80-91.
Dimayuga RE, Garcia SK (1991). Antimicrobial screening of medicinal plants from Baja California Sur, Mexico. J. Ethnopharmacol., 31: 181-192.
Duraipandiyan V, Ayyanar M, Ignacinmuthu S. (2006). Antimicrobial activity of some ethnomedicinal plants used by Paliyar tribe from Tamil Nadu, India. BMC Complementary Altern. Med., 6: 35-51.
Edeoga HO, Okwu DE, Mbaebie BO (2005). Phytochemical constituents of some Nigerian medicinal plants. Afr. J. Biotechnol., 4: 685-688.

Erdogrul OT (2002). Antibacterial activities of some plant extracts used in folk medicine. Pharm. Biol., 40: 269-273.

Kumar R, Singh NP (1984). Effect of tannins in Pala leaves (*Zizyphus nummularia*) on ruminal proteolysis. Indian J. Anim. Sci., 54: 881-884.

National Committee for Clinical Laboratory Standard (1990). Methods for antimicrobial susceptibility testing. Manual of clinical Microbiology, 5th edition, American Society for Microbiology, Washington DC, pp. 1105-1125.

Omale J, Okafor PN, Irene II (2009a). Chemical composition and effects of aqueous extract of *Cissus multistriata* on some biochemical parameters in albino rats. Int. J. Pharm. Tech. Res., 1(3): 509-513.

Omale J, Okafor PN, Irene II (2009b). Studies on the glyconutrients and antimicrobial efficacy of *Cissus multistriata* leaf extracts. Asian J. Pharm. Clin. Res., 2(3): 66-73.

Omale J, Okafor PN (2008). Comparative antioxidant capacity, membrane stabilization, polyphenol composition and cytotoxicity of the leaf and stem of *Cissus multistriata*. Afr. J. Biotchnol., 7(17): 3129-3133.

Omale J (2010). Polyphenol compositions and antioxidant capacity: A comparison of ripe and unripe fruit extract of *Cissus multistriata* (Vitaceae) plant. Afr. J. Biotechnol., 2 (5): 133-137.

Parekh J, Chanda S (2007). *In vitro* antibacterial activity of the crude methanol extract of *Woodfordia fructicosa Kurz*. Flower (Lythraceae). Braz. J. Microbiol., 38: 204-207.

Rath SK, Mohapatra N, Dubey D, Panda SK, Thatoi HN, Dutta SK (2009). Antibacterial activity of *Diospyros melanoxylon* barks from Similipal Biosphere Reserve, Orissa, India. Afr. J. Biotchnol., 8(9): 1924-1928. *Woodfordia fructicosa* kurz. Flower (Lythraceae).

Rojas JJ, Ochoa VJ, Ocampo SA, Munoz JF (2006). Screening for antimicrobial activity of ten medicinal plants used in Colombian folkloric medicine: A possible alternative in the treatment of non-nosocomial infections. BMC Complementary Altern. Med., 6: 2.

Scalbert A (1991). Antimicrobial properties of tannins. Phytochemistry, 30: 3875-3883.

Sofowora A (1982). Medicinal plants and traditional medicine in Africa. John Wily, Chichester, p. 179.

14

A potential new isolate for the production of a thermostable extracellular α- amylase

Ahmed, A. Alkando, Hanan Moawia Ibrahim*

Central Laboratory, Ministry of Science and Technology, Sudan.

Two hundred and seventy *Bacillus* species were isolated from soil samplse in Khartoum State and tested for α- amylase production. 20 potential isolates of α- amylase producer were obtained during primary screening. Secondary screening of these isolates yielded a high thermostable alkaline α-amylase producing isolate. Of all the species tested, *Bacillus licheniformis* gave maximum α-amylase activity of 0.7947 U/mg/ml at pH of 8 in iodine method, and (0.024 U/mg/ml) in 3.5.dinitrosalicylic acid (DNS) method. Characterization of the extra-cellular crude α- amylase was further evaluated for its biochemical properties as an enzyme for industrial use. The production of α- amylase following growth of the microorganism was found to be at optimum temperature and pH of 70°C and 9.0, respectively.

Key words: Thermostable α- amylase, alkaline α- amylase, *Bacillus licheniformis*.

INTRODUCTION

Amylases (1-4,α D-glucan glucano hydrolase EC 3.2.1.1) hydrolyses internal α–1-4- glycosidic linkage in starch and related substrate in an endo-fashion producing oligosaccharides, glucose and α–limit dextrin (Fogarty et al., 1999). α-amylases constitute a class of industrial enzyme having approximately 30% of the world enzyme production (Van der Maarel et al., 2002) and represent one of the three largest groups of industrial enzymes and account for approximately 60% of total enzymes sales in the world (Rao et al., 1998) and is an important enzyme, particularly in the process of starch hydrolysis. Though they originate from different sources (plants, animals and microorganism) and show varying pattern of action depending on the source and origin (Hagenimana et al., 1992), in industry, they are mainly produced from microbes. Enzymes from microbial sources generally meet industrial demands due to their higher yield and thermostability (Burhan et al., 2003). Microbial enzymes present a wide spectrum of characteristics that make them useful for specific application. Microbial α- amylase are among the most important hydrolytic enzymes and have been studied extensively. Each application of α-amylase from microbial source requires unique properties with respect to specificity, stability, temperature and pH dependence (McTigue et al., 1995). It can be produced in amounts meeting all the demands of the market because the diversity of enzymes available from microorganisms is very great. They find potential applications in industry such as food, fermentation, textile (in desizing fabric), paper and detergent (Pandy et al., 2000; Hag et al., 2003). The spectrum of amylase applications has expanded in several fields such as clinical, medicinal and analytical chemistry (Reddy et al., 2003). Despite the fact that many different α- amylases have been purified and charac-terized so far and some of them have been used in biotechnological and industrial applications, the presently known α- amylases are not sufficient to meet most industrial demands (Karbalaei-Heidari et al., 2007). High value of microbial extracellular enzyme is placed on extreme thermostability and thermostability of amylase for use in the bioprocess of starch (Adeyanju et al., 2007). Thermo stability is a feature of most of the enzymes sold for bulk industrial usage and thermophilic organisms are therefore of special interest as a source of novel thermo-stable enzymes. Recent research with thermostable α-amylase has concentrated on the enzymes of thermo-piles and extreme thermopiles (Sadhukhan et al., 1992; Arnesen et al., 1998; Lin et al., 1998), and little is known about the properties of the enzymes produced by these organisms (Eduardo et al., 2000). Screening of micro-organisms with higher α- amylase could therefore,

---

*Corresponding author. E-mail: hara111@yahoo.com.

facilitate the discovery of novel α- amylase suitable for new industrial applications such as bread and baking industries (Gupta et al., 2003).

The objective of the present study is: to isolate and carry out a taxonomic characterization of a bacterial strain isolated from Sudanese soil which was found to be a potential source of thermostable α- amylase.

MATERIALS AND METHODS

All the chemicals used in this study are of analytical grade unless otherwise stated. This research was conducted between 2006 and 2009 at the Central Laboratory, Ministry of Science and Technology, Sudan.

Average diameter index of microorganism - average diameter of colony
$$\frac{\text{Average diameter of colony}}{\text{Average diameter of colony}} \quad (1)$$

Isolation of organism

Soil samples were taken from 3 to 5 cm depth after removing 5 cm from the earth surface. These samples were collected into sterilized plastic bags and then transferred to labeled screw-caped bottles. Isolation of the microorganisms was performed by the soil dilution plate technique (Clark et al., 1958). The nutrient agar base containing soluble starch (1%) was used for bacteria isolation. The stock of the bacterial isolate was mixed with sterilized glycerol, and kept in 1 ml aliquots and preserved at -20°C for further use.

Screening and identification of bacteria strain

The isolated *Bacillus* strains were primarily screened for α- amylase synthesis after incubation at 60°C for 48 hr, the plates were flooded with a solution of 0.5% (w/v) I_2 and 5.0% (w/v) KI (Thippeswamy et al., 2006). Colonies exhibiting halo starch hydrolysis were picked up. The clear zone surrounding the colony was measured in (cm) from the edge of the colony to the limit of clearing and also the diameter of colony was recorded. The relative amylolytic activity is expressed as an index of activity calculated according to the follow equation: Identification of the selected *Bacillus* strain was identified on the basis of standard morphological and biochemical tests according to the method described in Bergey's Manual of Determinative Bacteriology (Buchanan and Gibbons, 1974). Subculture was prepared by a full loop of the suspension of vegetative cells streaked aseptically onto a nutrient agar (NA) several times until pure colonies were obtained. The pure isolates were maintained on NA slants and routinely sub-cultured every four weeks.

Inoculum preparation

50 ml of vegetative inoculum was transferred to 250 ml sterilized cotton-plugged Erlenmeyer flasks and then rotated at 200 rpm at 35°C in a rotary shaking incubator for 24 h.

Fermentation technique

α-amylase fermentation was carried out by submerge fermentation. 10% from the bacterial inoculums was inoculated to 500 ml shake-flasks containing 100 ml of a defined medium (Horikoshi medium II) (Horikoshi, 1999). Soluble starch 1%(w/v), peptone 0.5%(w/v), yeast extract 0.5%(w/v), $MgSO_4.7H_2O$ 0.02%(w/v), K_2HPO_4

0.1%(w/v), and $Na_2CO_3$1%. The flasks were incubated at 37°C in an orbital shaker at 150 rpm for 24 to 48 hr with an initial pH of 7.0. The culture broth was then centrifuged (Hettich Mikro 20) at 30 rpm for 30 min at 4°C, the free-cell supernatant was used as an extracellular crude enzyme (Hanan et al., 2004).

Determination of amylolytic potentially isolate activity (Iodine method)

α- amylase activity was determined by Iodine method to assay the amylolytic activity (Fuwa, 1954). One unit of the activity is defined as the amount of enzyme that could produce a change of 0.01 absorbency at 700 nm in the standard assay.

Saccharification activity: Dinitrosalsylic acid (DNS method)

The DNS method used involved estimating the amount of reducing sugar produced (DeMoraes et al., 1999), using 1% soluble starch as substrate. Glucose was used as standard. One unit of enzyme activity was defined as the amount of enzyme that formed 1 mg of reducing sugar in 1 min.

Optimisation of α-amylase activity assay conditions (DNS)

The influence of pH, time of incubation, temperature and soluble starch concentration were measured to optimize the enzyme assay conditions. Optimum pH was determined at different pH range (5 to 10). Reaction time was determined at different time intervals ranging from 5 to 60 min with 2% soluble starch in phosphate buffer at pH 6.0 as earlier obtained at 60°C. The optimum temperature was determined by incubation of 2% (w/v) soluble starch in phosphate buffer pH 6.0 with the crude enzyme at various temperatures ranging from 40 to 100°C for 10 min. Substrate concentration was determined at various ranges from 0.5 to 3% (w/v).

Biomass estimation

The pellets were washed with acetate buffer three times in order to remove the starchy material attached to the pellet. The suspension were then filtered using a pre-weighted filter paper and dried in an oven at 60°C for 1 to 2 days to a constant weight for the measurement of cell concentration (mg/ml) (Herbert, 1961).

High Performance Liquid Chromatography (HPLC)

For all analysis that involved HPLC, the conditions were set as follows: the column employed was of analytical column Supelcosyl LC-NH2 (5 µm) with length of 250 mm and 4.6 mm internal diameter. The mobile phase consisted of acetonitrite and water in ratio 75:25 (v/v). The flow rate was set at 1 ml/min. (Kouame et al., 2004). The internal column temperature was maintained at 30°C and the evaporation temperature was 60°C. The eluent from the column was monitored by refractive index detector, and the data werte recorded by the integrated computer system attached to the HPLC (Waters corp.).

Reaction conditions

Extracted crude α- amylase was incubated with 1% soluble starch, while glucose and maltose were incubated at 70°C without the addition of crude enzyme. Then 3.5 ml of 0.3 mM NaOH was added to stop the reaction. Samples were filtered through millipore filter

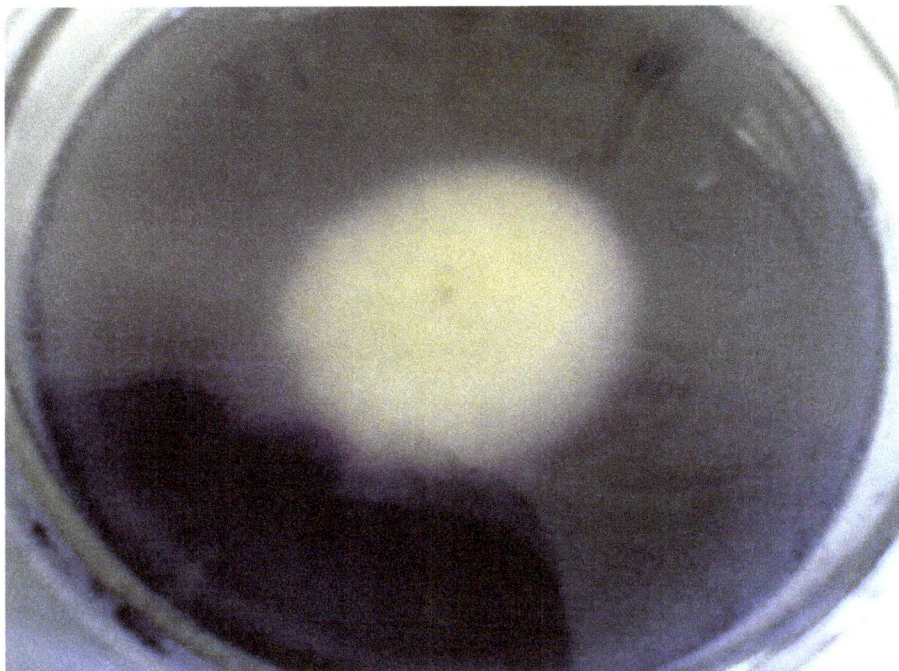

Figure 1. Zone of clearance of Strain (2) which further identified as *B. licheniformis.*

units of 0.2 μm pore size prior to injection to the HPLC.

RESULTS

The first step was to isolate the desired microorganism that produces α- amylase. Among the four selected *Bacillus* strains (1, 2, 3, and 4) that showed the highest amylase synthesis, Strain 2 was selected because it gave a larger diameter zone of clearance (6.5 cm) (Figure 1) and the highest relative amylolytic activity (calculated with Equation 1) of 1.5 compared to the other species. Therefore Strain 2 was chosen to fulfill the aim of this research which is amylase production. These culture were further screened for amylase production by submerge fermentation using the hydrolysis method.

Characterization and identification of the isolated amylolytic bacterial strain

The morphological and biochemical characteristics of the isolated amylase producing bacteria strain (2) are shown in Table 1. On the basis of standard method, the strain was assigned a code and identified as *Bacillus licheniformis* according to the method described in Bergey's Manual of Determinative Bacteriology (Buchanan and Gibbons, 1974). This strain (*B. licheniformis*) was selected for further studies (Figure 2).

Amylolytic activity (Iodine method)

Figure 3 shows the growth profile of *B. licheniformis* in

Horikoshi medium II during the 120 h cultivation. Maximum growth of the bacterium was obtained within 24 to 84 h of cultivation based on biomass dry weight (0.13 to 0.148 mg/mL). The activity of the enzyme reached a maximum within 36 h after inoculation (0.6867 U/mg/ml) and plateau off for the next two days. Beyond 96 h of growth, no increase in enzyme activity was recorded. The two profiles were paralleled and showed that the fermentation kinetics of α- amylase production by *B. licheniformis* might be classified as the growth associated type (Figure 3). Enzyme production was found to be concomitant with growth.

The result of the crude enzyme of *B. licheniformis might* also contain a mixture of amylolytic enzymes as well asα-amylase as the iodine assay may not be specific for α- amylase activity. DNS method which determined the reducing sugar was used as a specific method. The results of enzyme activity using DNS method (Figure 4) showed thesame results as Iodine method indicating that the enzyme might be α- amylase.

Optimisation of DNS assay conditions

The influences of pH, time of incubation, temperature and substrate concentration were measured to optimize DNS assay conditions. The results showed a broad peak optimum pH with ranges between 6.0 and 9.0 and some activity was gained (0.01 U/mg/ml) at pH 10.0. The maximum *B. licheniformis* crude enzyme activity (0.0237 U/mg/ml) was initially observed between pH 6.0 and 9.0 (Figure 5).

Table 1 Morphological and biochemical characteristics of the isolated amylolytic bacterial strain.

Characterization tests	Bacterial reaction
Gram reaction	+
Cell shape	Rod
Motility test	+
Spore	+
Growth in air	+
Catalase	+
Oxidase	-
Starch hydrolysis	+
Nitrate reduction	+
VP	+
Citrate utilization	+
Urease	+
Oxidative- fermentation test	F
Casein hydrolysis	+
Indole production	-
Anaerobic growth	+
Acid from sugar fermentation	
Glucose	+
Sucrose	+
Melibiose	+
Salicin	+
Raffinose	-
Cellobiose	+
Galactose	+
Xylose	+
Mannose	+
Growth at 50 C	+
Growth at 10% NaCl	-
Probable identity	*Bacillus licenformis*

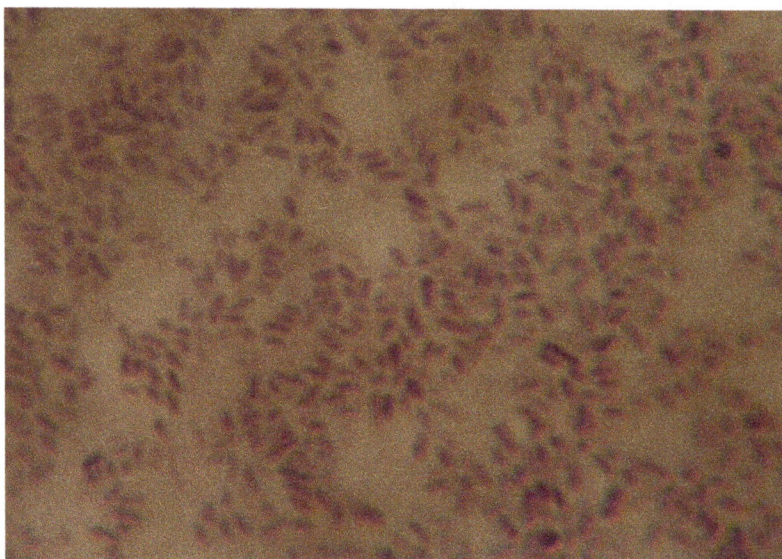

Figure 2. *B. licheniformis* (microscopic characteristics).

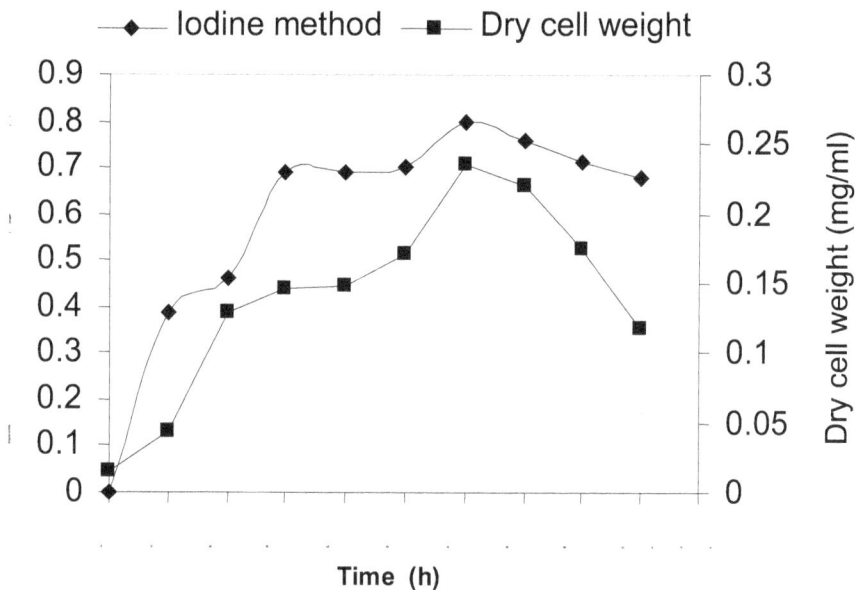

Figure 3. Biomass and crude enzyme activity profiles during the growth of *B. licheniformis* using hydrolysis method (Iodine method).

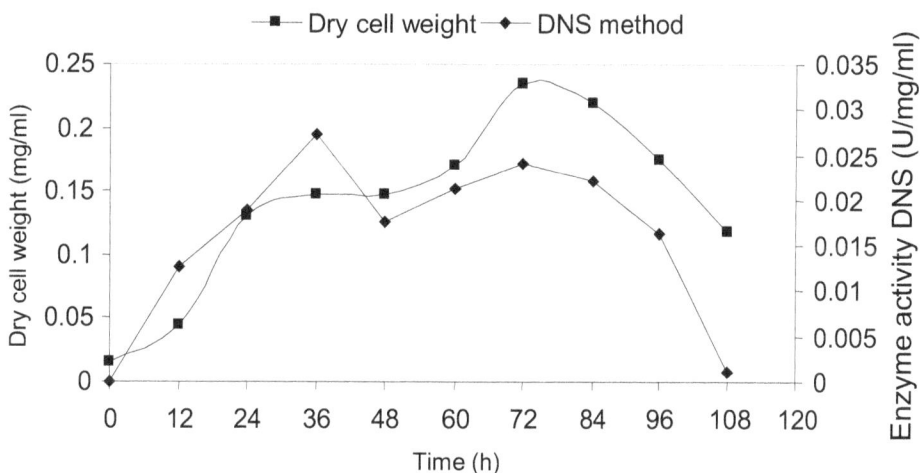

Figure 4. Time course of *B. licheniformis* crude enzyme activity and growth curve using saccharification method (DNS method).

Based on the substrate concentration, the extracted crude enzyme activity is shown to be linear between 0.5 to 1.0%. Hence, a practical assay substrate concentration is taken to be 1.0% of 0.7941 U/mg/ml of α- amylase activity (Figure 6).α- amylase from *B. licheniformis* was found to be active (0.018 to 0.017 U/mg/ml) in a broad temperature range of 50 to 90°C and showed a maximum activity (0.0252 U/mg/ml) at around 70°C (Figure 7). The enzyme is active even at high temperatures of up to 90°C.

In selected reaction conditions, namely at a temperature of 70°C, substrate concentration of 1% and pH at 9.0 were used as optimal conditions to assay for B.

licheniformis α- amylase activity using DNS method.

HPLC analysis

The result of the reaction mixture of glucose, maltose and *B. licheniformis* crude enzyme with soluble starch revealed that the retention time of glucose (Figure 8) and maltose were 8.841 and 2.588 respectively, where the area were 36909 and 11.5902 respectively, (Figure 9). *B. licheniformis* crude enzyme resulted in 8.561 retention time and an area of 279965 (Figure 10) which were nearly similar to the reading for standard glucose and

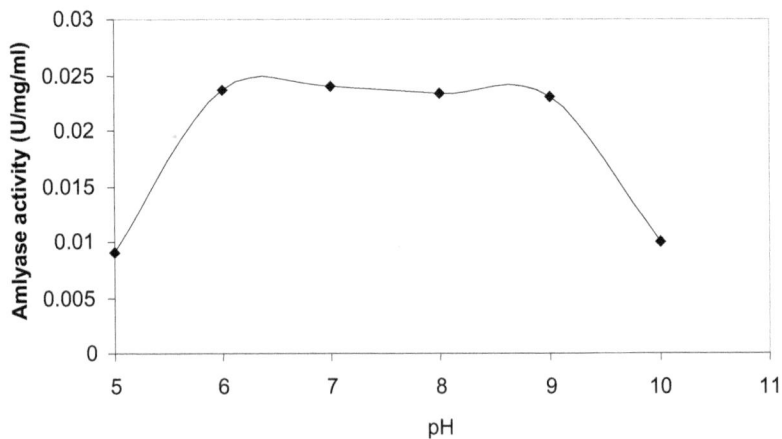

Figure 5. Kinetic evaluation of different pH on the production of α-amylase by *B. licheniformis*.

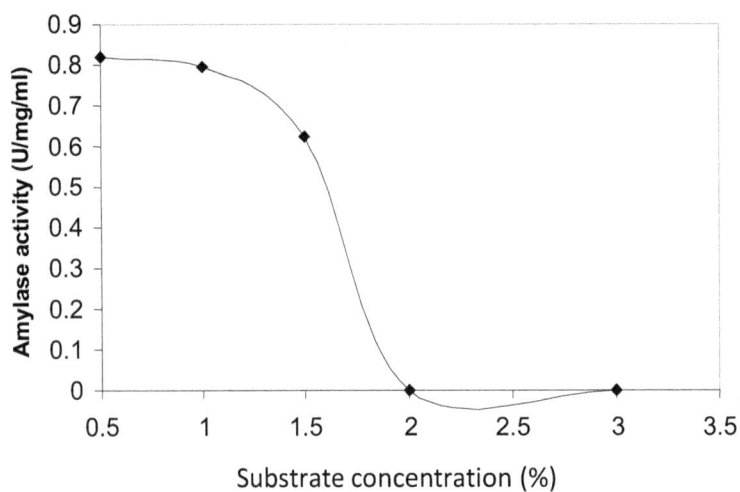

Figure 6. Kinetic evaluation of different substrate concentration on the production of α-amylase by *B. licheniformis*.

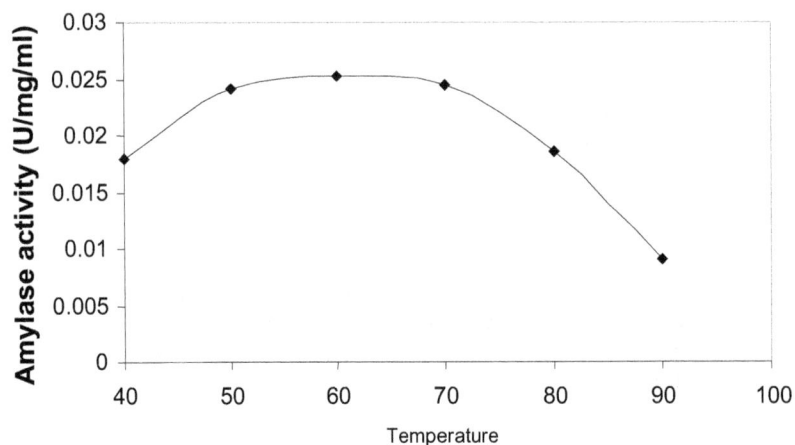

Figure 7. Kinetic evaluation of different temperature on the production of α-amylase by *B. licheniformis*.

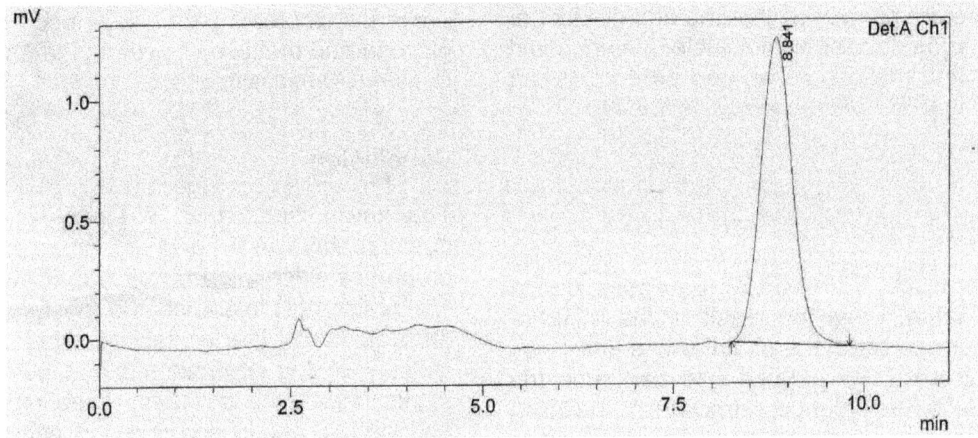

Figure 8. HPLC chromatogram of standard glucose.

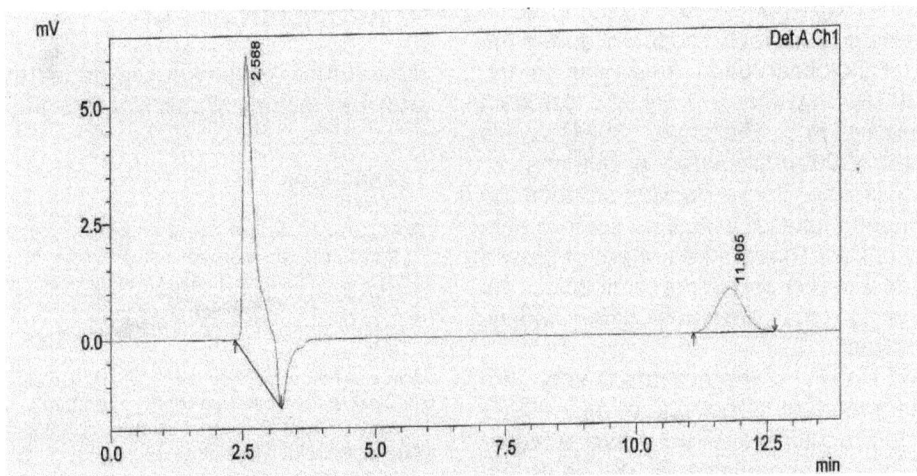

Figure 9. HPLC chromatogram of standard maltose.

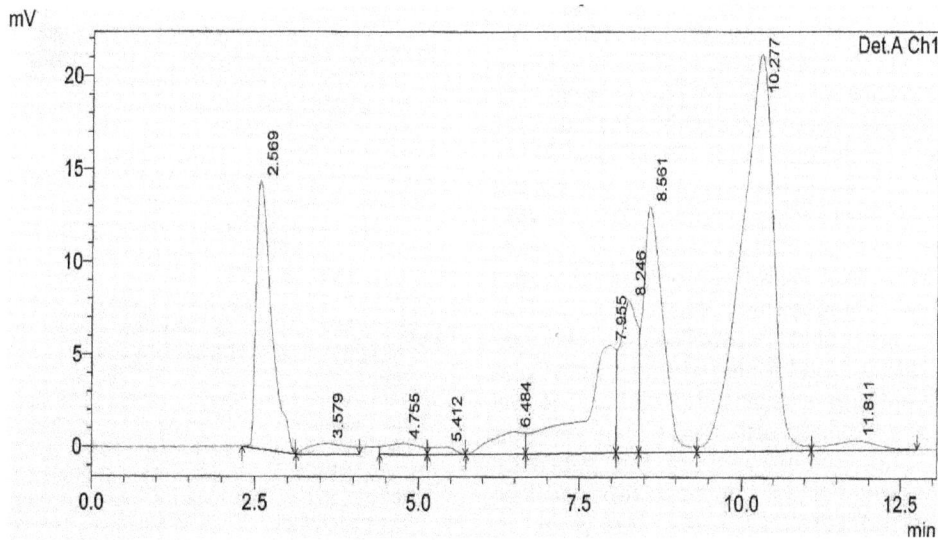

Figure 10. HPLC chromatogram of glucose produced from *B. licheniformis* crude α- amylase with soluble starch after 20 min incubation time at 70°C and pH 6.0.

maltose. It is therefore concluded that the end product of B. licheniformis crude enzyme with soluble starch could produce glucose and maltose as an end product. It can be clearly seen that B. licheniformis is mainly an α-amylase producer.

DISCUSSION

Many bacteria produce extra-cellular amylases during fermentation of starch. From the result of the morphological and biochemical tests, the strain was identified as B. licheniformis. Robyt (1984) stated that many bacteria including B. licheniformis produce extracellular amylases during the fermentation of starch. B. licheniformis is a ubiquitous bacterium of importance in the environment as a contributor to nutrient cycling due to its production of amylase and other enzymes. The biosynthesis of α-amylase by B. licheniformis appears to be related since the enzyme in this isolate is primarily produced during the exponential phase. This observation is similar to the pattern of amylase synthesis by Clostridum acetobutylicum (Annous and Blaschek, 1991) while Adeniran and Abiose (2009) produced a potential α-amylase from a fungi isolate. Some Bacillus produce the enzyme in the exponential phase, where as some others in the mid-stationary phase. Though the pattern of growth and enzyme profile of Bacillus spp. have similarities, the optimized conditions for the enzymes differ widely, depending upon the strain.

The production and stability of the enzyme is very sensitive to pH and temperature (Declerck et al., 2003). Slight changes in temperature and pH have adverse effect on the growth of microorganisms as well as on the produc-tivity of α-amylase (Anyangwa et al., 1993). The activity of B. licheniformis crude α- amylase is in a broad temperature range showing that the enzyme is active at high temperatures of up to 90°C which was quite high comparable with other thermostable amylase (Jin et al., 1990; Saxena et al., 2007). Ivanova et al. (1993) reported an optimum temperature of 90°C for B. licheniformis CUMC 305. This broad-temperature activity characteristic of the α-amylase produced lends the isolate B. licheniformis as a potential microbe to produce α-amylase required for high temperature applications. At elevated temperatures, they improve the solubility of starch, decrease the viscosity, limit microbial contaminants and reduce reaction time (Thippeswamy et al., 2006). This isolate showed that it is an alkaline Bacillus that produces alkaline enzyme at pH 9.0 and it also showed activity at pH 10. This is advantageous since the higher pH will lessen the tendency of gelatinized starch to retrograde (Yu et al., 1988). The concentration of the substrate resulted in 1% as optimum and it has been reported that the synthesis of carbohydrate-degrading enzymes in most species of Bacillus is subject to catabolic repression by readily metabolizable substrates such as glucose (Lin et al., 1998).

HPLC chromatogram confirmed the availability of glucose and maltose, indicating that the crude enzyme in B. licheniformis is mainly α- amylase.

Conclusion

The newly isolated B. licheniformis strain which produces α-amylase is novel and offers interesting hydrolytic properties since the enzyme was active between pH 7 to 10. Besides pH tolerance, the most striking feature of the enzyme is its thermostability. Also significant thermostability of the enzyme make it potential for industrial applications such as starch liquefaction for sweeteners and syrups, textile desizing and paper industries, which requires the process to be carried out in multiple steps at high temperature.

ACKNOWLEDGEMENT

The authors greatly acknowledge the Ministry of Science and Technology, Sudan for their financial support.

REFERENCES

Adeniran AH, Abiose SH (2009). Amylolytic potentially of fungi isolated from some Nigerian Agriculture waste. Afr. J. Biotechnol., 8: 667-672.

Adeyanju MM, Agboola FK, Omafuvbe BO, Oyefuga OH, Adebawo OO (2007). A thermostable Extracellular α- amylase from Bacillus Licheniformis isolated from cassava steep water. Biotechnology, 6(4): 473-480.

Annous BA, Blaschek HP (1991). isolation and characterization of Clostridum acetobutylicum mutants with enhanced amylolytic activity. Appl. Environ. Microbiol., 57: 2544-2548.

Anyangwa EM, Mapsev C, Musanage P, Elemva M (1993) The effect and removal of starch in the sugar refining industry. J. Int. Sugar, 95: 210-213.

Arnesen S, Eriksen SH, Olsen J, Jensen B (1998). Increased production of α- amylase from Thermomyces Lanuginosus by the addition of tween 80 Enz. Microbial. Technol., 23: 249-252.

Buchanan RE, Gibbons NE (1974). Bergey's manual of determinative bacteriology. Baltimore: The Williams and Wilkins Co.

Burhan A, Nisa U, Gokhan C, Omer C, Ashabil A, Osman G (2003). Enzymatic properties of a novel thermostable, thermophilic, alkaline and chelator resistant amylase from an alkaliphilic Bacillus sp. isolate ANT-6. Process Biochem., 38: 1397-1403.

Clark HE, Bordner GEF, Kabler PW, Huff CB (1958). Applied Microbiology. International Book Company, NY, pp: 27-53

Declerck N, Machius M, Joyet P, Wiegand G, Huber R, Gaillardin C (2003). Hyperthermostabilization of Bacillus Licheniformis α- amylase and modulation of its stability over a 50°C temperature range. Protein Eng., 16(4): 287-293.

DeMoraes LMP, Fillo SA, Ulhoa CJ (1999). Purification and some properties of an α- amylase glucoamylase fusion protein from Saccharomyces cerevisiae. World J. Microbiol. Biotechnol., 15: 561-564.

Eduardo Carlos de Souza T, Meire LLM (2000). Cultural conditions for the production of thermostable amylase by Bacillus sp. Braz. J. Microbiol., 31(4): doi: 10.1590/51517- 8382000000400011.

Fogarty WM, Dooylc EM, Kelly CT (1999). Comparison of the action pattern of two high maltose forming alpha-amylase on linear malto-oligosaccharides. Enzyme Microbial. Technol., 25(3-5): 330-335

Fuwa H (1954). A new method for micro determination of amylase activity by the use of amylose as substrate. J. Biochem., 41(5): 583-593.

Gupta R, Gigras P, Mohapatra HVK, Goswami and Chauhan. (2003).

Microbial amylases: a biotechnological perspective. Process Biochem., 38:1599- 1616.

Hagenimana V, Vezina LP, Simaed RE (1992). Distribution of amylasewithin sweet Potato (Ipomea batatas L) root tissue. J. Agric. Food Chem., 40: 1777-1783.

Hanan MI, Wan MWY, Aidil AH, Rosli Md Illias, Othman H, Othman O (2004). Optimization of Medium for the Production of β-Cyclodextrin glucanotransferase using Central Composite Design (CCD). Process Biochem., 40(2): 753-758.

Hag I, Ashraf H, Qadeer MA, Iqbal I (2003). Production of alpha amylase by Bacillus licheniformis using an economical medium. Bioresour. Technol., 87(1):57-61.

Herbert D (1961). The chemical composition of microorganisms as a function of their environment. Symp. Soc. Gen. Microbiol., 11: 391-416.

Horikoshi K (1999). Alkalophilic: Some applications of their product for biotehnology. Microbiol. Mol. Biol. Rev., 63(4): 735-750.

Ivanova VN, Dobreva EP, Emanuilova EL (1993). Purification and characterization of a thermostable alpha-amylase from B. Licheniformis. J. Biotechnol., 28: 277-289.

Jin F, Cheng X, Shi Y, Zhang C (1990). Isolation of new thermophilic aerobic bacteria which produce thermostable amylase. J. Gen. Microbial, 36: 415-424.

Karbalai-Heidari HR, Ziaee AA, Schaller J, Amoozegar MA (2007). Purification and characterization of an extracellular haloalkaline protease produced by moderately halophilic bacterium salinivibrio sp strain AF-2004 (Enzyme Microbial, Technol. 40: 266-272, doi: 10.1016/j.enzmictec. 2006.04.0006.

Kouame LP, Ahipo ED, Niamke SL, Kouame FA, Kamenan A (2004). Synergism of cockroach (Periplaneta americana) α-amylase and α-glucosidase hydrolysis of starch. Afr. J. Biotechnol., 3(10): 529-533

Lin LL, Chyaum CC, Hsu WH (1998). Production and properties of a raw- starch- degrading amylase from Thermofilic and alkalophilic Bacillus sp. TS-23. Biotechnol. Appl. Biochem., 28: 61-68.

McTigue MA, Kelly CT, Doyle EM, Fogarty WM (1995). The alkaline amylase of the alkalophilic Bacillus sp. IMD 370 Enzyme Microbial Technol., 17: 570-573.

Pandy A, Nigam P, Soccol CR, Soccol VT, Singh D, Mohan R (2000). Advance in microbial amylases; Review article. Biotechnol. Appl. Biochem., 31: 135-152.

Rao MB, Tanksale AM, Ghatge MS, Deshpande VV (1998). Molecular and Biotechnological aspect of Microbiol Protease. Microbiol.. Mol. Biol. Rev., 62: 597-635.

Reddy NS, Nimmagadda A, Sambasiva RKRS (2003). An overview of the microbial α-amylase family. Afr. J. Biotechnol., 2 645- 648.

Robyt JF (1984). Enzyme in the hydrolysis and synthesis of starch. In: Starch chemistry and technology. 2nd edi., Whistler, R. L., J. N. Br Miller and E. F. Paschall (Eds), 4: 88-90.

Sadhukhan R, Roy SK, Raha SS, Manna S, Chakrabarty SL (1992). Induction and regulation of amylase synthesis in a cellulolytic thermophilic fungus Myceliophthora thermophillia D14 (ATCC 48104). India J. Exp. Biol., 30: 482-486.

Saxena RK, Dutt L, Agarwal PN (2007). A highly thermostable and alkaline amylase from Bacillus sp. PN5. Bioresour. Technol., 98: 260-265.

Thippeswamy S, Girigowda K, Mulimami HV (2006). Isolation and identification of α- amylase producing Bacillus sp. From dhal industry waste. Indian J. Biochem. Biophys., 43(5): 295-298.

Van der Maarel MJEC, Van der Veen B, Uitdehaag JCM, Leemhuis H, Dijkhuizen L (2002). Properties and aplications of starch converting enzymes of α-amylase family. J. Biotecnol., 94:137-155.

Yu EKC, Hiroyuki A, Masanaru M (1988). Specific apha-cyclodextrin production by a novel thermostable cyclodextrin glycosyltransferase. Appl. Microbiol. Biotechnol., 28: 377-379.

15

Biochemical and molecular characterization of *Haemophilus influenza* isolated from Riyadh, Kingdom of Saudi Arabia

Afaf I. Shehata*, Amal Abdulaziz Al-Hazani, Hesham Al-Aglaan and Hanan O. Al Shammari

Department of Botany and Microbiology, College of Science, King Saud University, P. O. Box 2455, Riyadh 11451, Saudi Arabia.

The objective of this study focused on the prevalence of *Haemophilus influenza* to confirm the colonies of *H. influenza* on the basis of their growth requirements and serotype distribution. This study prepared 80 isolates of *H. influenze* isolated from five different sources (eye, ear, sputum (SP), lower genital tract (TA), and nasopharyngeal (NPA)) with different ages for infants and elderly persons. The phenotypic characteristics, which included the biotype, serotype, antibiogram and β-lactamase production, were applied by using APINH and minimum inhibitory concentration (MIC). Also, the study focused on the identification of selected serotype using PFGE analysis. The discussion of this study differentiates the age groups occurrence in the isolates, alongside with non-typeable strain versus the typeable ones and their percentages in the sample of isolates. This clustering of most strains in one PFGE pattern might be explained with the colonel population structure of the encapsulated *H. influenza*.

Key words: *Haemophilus influenza*, biotype, serotype, antibiogram, β-lactamase.

INTRODUCTION

Haemophilus influenzae is an important opportunistic pathogen that is rapidly evolving toward widespread distribution among the human population. It colonizes the nasopharynx of approximately 75% of healthy children and adults and becomes a leading cause for a variety of infectious disease such as acute otitis media (AOM), sinusitis, pneumonia, sepsis and meningitis. Although the overwhelming majority of the invasive *Haemophilus* species infections are caused by *H. influenzae* Serotype b, other Serotype as well as nontypeable influenzae (NTI) have been associated with invasive diseases. The pathogens of *H. influenzae* are classified into six encapsulated strains called typeable depending on the serologically distinct capsular polysaccharides (serotype a-f) and noncapsule or nontypeable strains. *H. influenzae* are obligate normal flora of the upper respiratory tract of humans. These pathogen/commensally whose carriage populations are likely to be restricted in size can in some hosts cause life threatening invasive diseases.

Different pathogenic potential which occur almost exclusively in the most Vulnerable population sector- children between 6 months and 6 years of age are usually caused by the encapsulated strain. Non-encapsulated strain is mainly a cause of acute otitis media in both children and adults with exacerbations of chronic bronchitis complicating chronic obstructive pulmonary disease in adults (Murphy and Apicella, 1987).

Biochemical characterization of *H. influenzae* have yield valuable epidemiological information's as specific biotypes have shown to associate different infections, to vary with the source of isolation as well as the antigenic properties and antimicrobial resistance patterns (Berhofer and Back, 1979).

Although most of *H. influenzae* strain recovered from systemic infection belongs to serotype b, the other serotypes have also been isolated from infection in all age groups. Moreover none capsulated strain proved to cause serious ones following breakdown of persons-host defense mechanism (Cerquetti et al. 2000). Antibiotic resistance has emerged to ampicillin, chloramphenicol and tetracycline which are antimicrobial agents traditionally used in treatment of *H. influenzae* infection *H. influenzae* producing β- lactamase have increased globally since it was first recognised in 1972 (Jorgensen, 1992).

*Corresponding author. E-mail: afafsh@ksu.edu.sa.

Chlormaphenicol was consequently recommended in the initial therapy of invasive disease caused by this organism (Ister et al., 1984). Unfortunately chloramphenicol resistance has thereafter been reported worldwide after the first chloramphenicol-resistance strains were isolated in 1976 (van Klingeren et al., 1977).

β- lactamase prevalence rate was found to vary considerably among the different capsulated strain type and the non-capsulated. It also differs with the geographical locality (Garica-Rodriguez et al., 1999) and with patient age.

Consequently future spread of antibiotic resistance in *H. influenzae* may be halted or delayed by a more restricted use of antibiotics guided by continuous epidemiological monitoring and predicated therapeutic outcome.

The widespread infections of *H. influenzae* in The Kingdom of Saudi Arabia among children's and adults have resulted in the initiation of studies on this pathogen. In the present study focus is made on the isolation of *H. influenzae* from patients hospitalized in Riyadh region (KSA) and characterizes these isolate by a biochemical and molecular technique.

This study aims to assess the relationship of biotypes for *H. influenzae* to serotyping, antimicrobial susceptibility age and antibiogram typing. Also this study will evaluate the bacterial applicability for identification of genotype of the bacterial isolates. The main objectives of this study are:

i) To confirm the colonies of *H. influenzae* on the basis of their growth requirements for hemin and Nicotinamide dinucleotid (NAD).
ii) Investigate the prevalence and serotype distribution of *H. influenzae* isolated from patients with lower respiratory tract infection (LRTIS). In order to provide reliable data on-the prevalence of *H. influenzae* serotypes.
iii) Determine the prevalence, age-group distribution, serotyping, antibigram typing and susceptibility patterns of invasive *H. influenzae* in KSA children and adults.
iv) Identify the selected serotypes of the isolated bacteria on the basis of genetic diversity through molecular typing.

MATERIALS AND METHODS

Clinical specimens

The 80 *H. influenzae* isolates were collected from five sources that are eye, ear, sputum, lower genital tract, and nasopharyngeal, by the patients at the Clinical Microbiology Laboratory of the Riyadh Military Hospital 2007-2008. *H. influenzae* strains were grown on sterile Brain Heart infusion broth and the subculture was made by the use of chocolate blood agar that was ready for biotyping.

Identification of *H. influenza* isolate

Bacterial culture of *H. influenzae* is performed on nutrient agar, Chocolate agar, plate with added X & V factors at 37°C in an enriched 5% CO_2 incubator (*OXOID Wade Road, Basingastoke,*

Hampshire, RG24 8PW, England). Gram-stained and microscopic observation of a specimen of *H. influenzae* will show Gram-negative, coccobacilli, with no specific arrangement. The APINH kit is a miniaturized rapid system for the identification of *Haemophilus* species which enable the performance of 12 identification tests (enzymatic reactions or sugar fermentations, detection of a penicillinase. Antimicrobial susceptibility testing confirmad by detection β-lactamase. The Kirby-Bauer disk diffusion method, or culture susceptibility test, is applied. *H. influenzae* agglutinating sera are intended for use in the qualitative slide agglutination test to identify serologically the type antigen of pathogenic strain of *H. influenzae* (types *a* to *f*) for epidemiological and diagnostic purposes. Biochemical testes (catalase and oxidase) and pulsed-field gel electrophoresis (PFGE) were applied. The preparation of genomic DNA suitable for PFGE begins for cultures by suspending them in Tris-EDTA (TE) buffer (usb Corporation Cleveland. OH USA) 100 mM and adjusted to a turbidity of 0.5 using a Dade Behring MicroScan turbidity meter.DNA in plugs for PFGE was then digested using the Apal- (Promega Corporation.2800 Woods Hollow Road. Medison, WI53711-5399 U.S.A) restriction enzyme.

RESULTS

Descriptive epidemiology

A total of 80 invasive isolate of *H. influenzae* were identified , 36 of 80 isolates were nontypeable strain, 25 was a type f strain, 6 a type e strain, 7 a type a, 2 a type b strain, 2 a type c strain, and 2 a type d strain. Patients with invasive *H. influenzae* disease ranged in age from 1 day to older than 50 years old. 53 (65%) occurred in children less than 2 years of age; 12 (22%) occurred in age ranging from 3-19 years; the remaining 15 occurred in adults ranging from 20 years to older than 50 years old, (Figure 1). There are 7 serotypes in the 80 isolates. 7(8.75%) were serotype "a", 2 (2.5%) serotypes "b", 25 (31.25%) serotype "f", 2 (2.5%) serotype "d", 2 (2.5%) serotype "c", 36 (45%) NT, and 6 (7.5%) serotype "e", (Table 1). Moreover, all isolate were biotypeable as types I through V, and all encapsulated strain were serotype specific. No encapsulated biotype or non-capsulated biotypes VI, VII, and VIII organism were isolated. Of the serotypeable strain (96.25%) belonged to biotype II,III, and IV. 4 of 7 type A were typed as biotype II, 1 as biotype I and 2 biotype III. 2 of 6 type e were typed as III, 2 as IV ,1 as biotype II ,and 1 as biotype V. 18 of 26 type f were typed as biotype II and 7 as biotype III. 18 of 36 NT were typed as II, 15 as biotype III and 3 as type IV. 1of 2 biotype D was biotype II and 1 as biotype V. Serotype b organism belonged to biotype III and the 2 serotype c isolate were biotype II, Table 1.

Biotype *and H. influenza*

H. influenza may be subdivided into eight biotype (I through VII) each on the basis of three biochemical reactions that are also important in the differentiation between some of the *Haemophilus* SP; indole production, urease, and ornithine decarboxylase activities. Biotype of *H. influenzae* shows a relationship to the source of

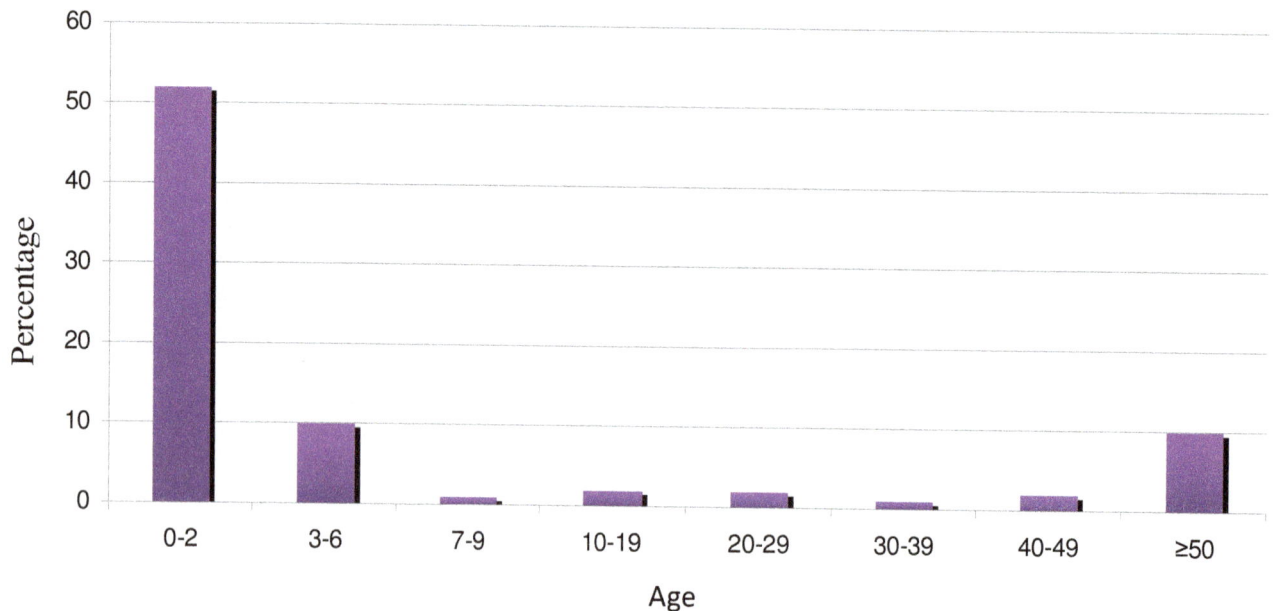

Figure 1: Age distribution of 80 with *H. influenzae* isolate.

Table 1. Relationship between biotypes and serotypes of *H .influenzae* isolate from different side of body from children and elderly.

Serotype	No. of isolate	No. of strain with following biotype				
		I	II	III	IV	V
A	7	1	4	2	—	—
B	2	—	—	2	—	—
C	2	—	2	—	—	—
D	2	—	1	—	—	1
E	6	—	1	2	2	1
F	25	—	18	7	—	—
NT	36	—	18	15	3	—
Total	80	1	46	26	5	2

isolation. Moreover, the biotypes of 80 *H. influenza* strains were determined. The results are presented in Table 2.

The most prevalent biotype was biotype II (57.5%) taken from the eye with a high isolate of 20 (43%) and the lowest one was taken from the ear with a low isolate of 4 (8%), followed by biotype III (32.5%) taken from the eye with a high isolate of 12 (46%) and the lowest one taken from the ear with a low isolate of 1 (4%). This was then followed by biotypes IV (6.25%), V (2.5%) and I (1.25%), respectively.

Serotyping and *H. influenza*

Although serologic typing of *H. influenzae* may be used to establish for isolate *H. influenze* as being any one of the six serotypes "a" to "f", it is usually only used to identify type "b" strain. All *H. influenzae* from cases of invasive infection should be serotyped to rule in/out type "b" infections. Testing can be performed using a slide agglutination test. Table 3 shows the distribution of serotypes *H. influenze* strains and different clinical specimens. From Table 3, most of the two common serotypes in all clinical specimens were NT (43% occurrence), and f which occur at 43 and 31%, respectively. The least common serotypes were a (9% occurrence), and e (7.5% occurrence) that were taken from 4 different clinical specimens that are SP, NPA, TA and EYE. Moreover, serotypes c and d occur only 2.5% in all specimens. Most of serotypes were isolated from EYE (43.7%), NPA (23.7%), SP (13.75%), and TA (11.25%) while only 11.25% serotypes were isolated from the ear. invasive infection should be serotyped to rule in/out type "b" infections.

Table 2. Distribution of biotype by source of isolate in patients.

Source	Biotype					
	I	II	III	IV	V	Total
SP	–	5	4	1	1	11
NPA	–	8	9	1	1	19
TA	–	9	–	–	–	9
EYE	1	20	12	2	–	35
EAR	–	4	1	1	–	6
Total	1	46	26	5	2	80

Table 3. Distribution of serotype *H. influenze* strains in different clinical specimens.

Serotype	Side of isolate					
	SP	NPA	TA	EYE	EAR	Total
A	2	1	3	1	0	7
B	1	0	0	1	0	2
C	0	0	1	1	0	2
D	0	1	1	0	0	2
E	2	2	1	1	0	6
F	3	3	2	15	2	25
NT	3	12	1	16	4	36
Total	11	19	9	35	6	80

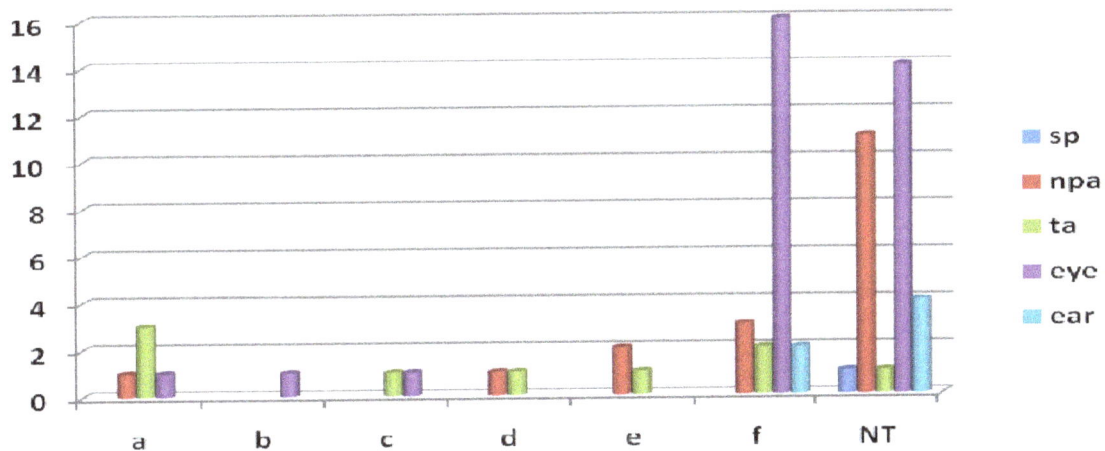

Figure 2. Distribution of serotype *H. influenze* strain in different clinical specimens.

Testing can be performed using a slide agglutination test. Table 3 shows the distribution of serotypes *H. influenze* strains and different clinical specimens. From Table 3, the most two common serotypes in all clinical specimens were NT (43% occurrence), and f which occur at 43 and 31%, respectively. The least common serotypes were a (9% occurrence), and e (7.5% occurrence) that were taken from 4 different clinical specimens that are SP, NPA, TA and EYE. Moreover, serotypes c and d occur only at 2.5% in all specimens. Most of serotypes were isolated from EYE (43.7%), NPA (23.7%), SP (13.75%), and TA (11.25%) while only 11.25% sero-types were isolated from the ear (Figure 2). Figure 3 shows that children are infected by *H. influenze* more than adult. The most serotypes infected children are NT (38.7%), "f" (28.6%) while the least infected serotypes are "b" (only 1.25%) comparing to serotypes "a" (6.25%) and "c" (2.5%). On adult, the most infected serotype was NT (25%) while "b" (1.25%) was the least infected one. Moreover, there are no serotypes "c" and "d" among adult

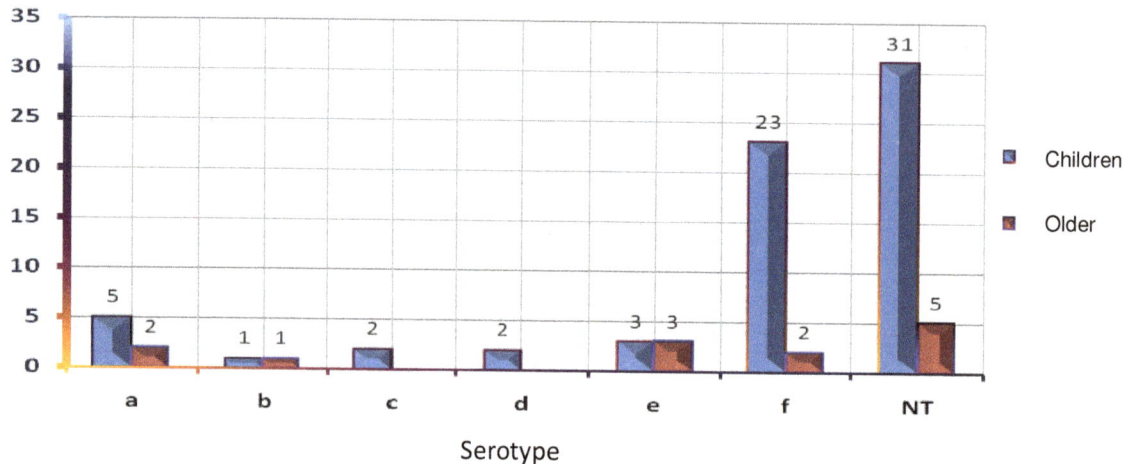

Figure 3. The comparison of serotypes of *H. influenza* between children and elderly.

Table 4. The comparison of serotypes of *H. influenza* between children and elderly.

Group	Serotype							
	a	b	c	d	e	f	NT	Total
Children	5	1	2	2	3	23	31	79
Elderly	2	1	0	0	3	2	5	13
All isolate	7	2	2	2	6	25	36	80

and only 2.25% among children.

The most serotypes infected children are NT (38.7%), f (28.6%) while the least infected serotypes are b (only 1.25%) comparing to serotypes a (6.25%) and c (2.5%).
On adult, the most infected serotype was NT (25%). While b (1.25%) was the least infected one. Moreover, there are no serotypes c and d among adult and only 2.25% among children. Looking at the different sources of isolating *H. influenze* among children. Table 4 shows that the *H. influenze* is mostly isolated from the EYE (49.2%).

Figure 4 shows that children are infected by *H. influenze* more than adult. The most serotypes infected children are NT (38.7%), "f" (28.6%) while the least infected serotypes are "b" (only 1.25%) comparing to serotypes "a" (6.25%) and "c" (2.5%). On adult, the most infected serotype was NT (25%) while "b" (1.25%) was the least infected one. Moreover, there are no serotypes "c" and "d" among adult and only 2.25% among children (Table 5).

Antibiotic susceptibility

Table 6 shows the relation between antibiotic and the isolated specimens. 20 of the 80 *invasive H. influenzae* produced a β-lactamase ampicillin resistant, of the 10 β-lactamase ampicillin resistant were typeables and 10 of

the β-lactamase ampicillin resistant were nontypeable. The most effective antibiotics on *H. influenze* are AUG and AZI. All the isolated serotypes are sensitive to AUG expect F. For AZI, two serotypes are resistant to it which are F and NT. On the other hand for the antibiotic TE; there are 8 (10%) isolated bacteria that are resistant to it. Antibiotic C and CRM have a strong resistant from the bacteria with 13.75% of the total number of isolated bacteria from the 80 patients (Figure 5).

The following result were obtained from the isolate, the high sensitive was to Aug 95% with MIC50 0.5 µg/ml and MIC90 2 µg/ml, and AZI 93.7% with MIC50 and MIC90 were 2 µg/ml. However, out of 80 isolate of *H. influenzae* strains 30% resistant to Amp with MIC50 0.5 µg/ml AND mic90 >8 µg/ml, Te resistant came next where 18.5% with MIC50 1 µg/ml AND MIC90 4 µg/ml. Table 7. The following result were obtained from the isolate in children, the high sensitive was to Aug and Azi 97% with MIC50 1 µg/ml and MIC90 2 µg/ml (Figure 6).

However, out of 67 isolate of *H. influenzae* strain in children 22% were resistant to Amp with MIC50 0.5µg/ml and MIC90 >8µg/ml. The resistant came next where 13% with MIC50 1µg/ml and MIC90 8 µg/ml, Table 8. It found that 33% of the TA isolate and 26.3% of the NPA were Ampicillin resistant. For the EYE was 23%, Ear 17%, and SP 10% resistant to Ampicillin. Isolate tested against CRM showed resistance in 44% in TA, 1 8% SP, 10.5% NPA, and EYE was 5.7%. For C, it has been found 22%

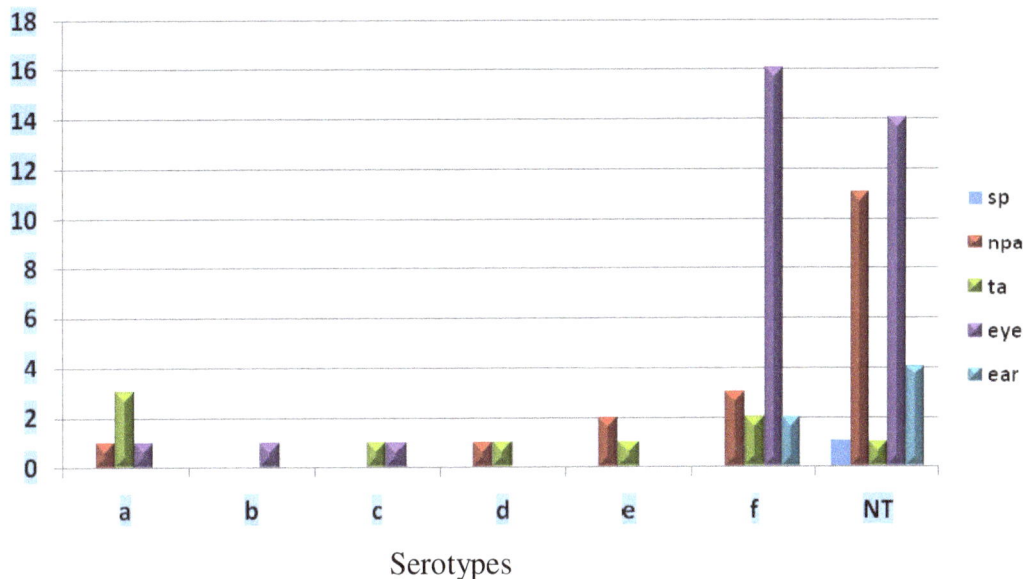

Figure 4. Comparison of serotypes of *H. influenza* in children

Table 5. The comparison between serotypes and sources of isolate of *H. influenzae* in children.

Side of isolate	Serotype						
	a	b	c	d	e	F	NT
SP	0	0	0	0	0	0	1
NPA	1	0	0	1	2	3	11
TA	3	0	1	1	1	2	1
Eye	1	1	1	0	0	16	14
Ear	0	0	0	0	0	2	4

Table 6. MIC 50%, MIC 90% range interpretation of 80 *H.influenzae* invasive isolate.

Antibiotic	MIC mg/l		% of isolate	
	MIC50%	MIC90%	SEN	RES
AMP	0.5	>8	75	25
CRM	1	8	82.50	17.5
TE	1	4	81.25	18.5
AUG	0.5	2	95	5
C	1	8	88.70	11.10
AZI	2	2	93.70	6.25

isolate from TA, 17% of isolate from EAR, NPA 15%, and EYE 14%. Isolate from SP were no resistance in C. The lowest resistant 3% isolate from EYE by using AUG and AZI. Isolate from EAR and NPA were showed no resistant by using AUG and AZI, (Figure 7). Moreover, it was found that isolate from SP is of higher resistant at 18% by using CRM, AZI, the lowest resistant was 0% by using AUG, C. Isolate from NPA were found to be higher in resistant at 26.3% in AMP, the lowest resistant 0% in AUG and AZI. Source Isolate from TA, the lowest resistant 0% by using AZI and higher resistant 44% in CRM. Source isolate from the EYE is of lowest resistant at 3% in AZI and higher at 23% in AMP. Finally, source of isolate from Ear, high one were 17% in AMP. C, and TE. The lowest resistant was 0% by using CRM, AUG, and AZI, Figure 8.

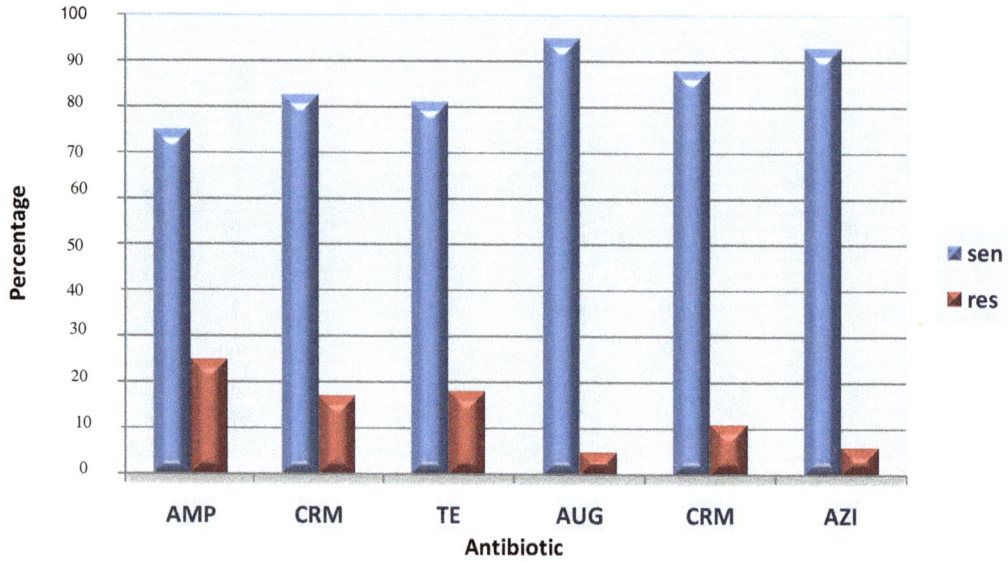

Figure 5. MIC 50%, MIC 90% range interpretation of 80 *H. influenzae* invasive isolate

Table 7. MIC 50%, MIC 90% range interpretation in children.

Antibiotic	MIC mg/l		% of isolate	
	MIC50%	MIC90%	SEN	RES
AMP	0.5	>8	77	22
CRM	1	8	88	12
TE	1	8	86	13
AUG	1	2	97	3
C	1	8	89	10
AZI	1	2	97	3

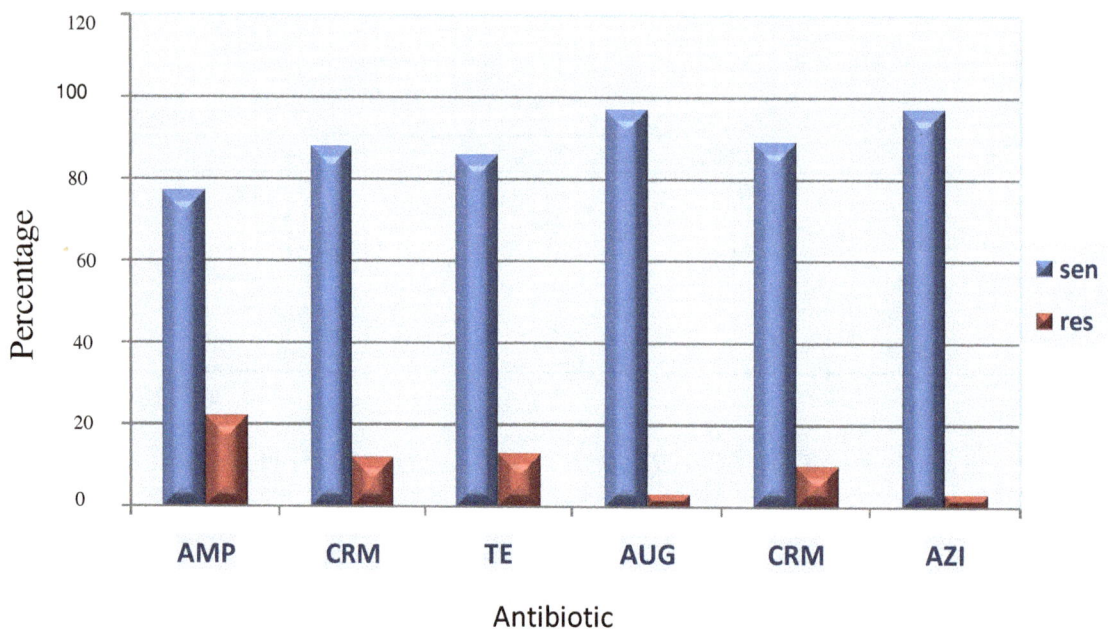

Figure 6. MIC50%, MIC90% range interpretation in children.

Table 8. Antibiotic sensitive & resistant of *H. influenzae* isolate by MIC test.

Source of isolate	No.	AMP (%)		CRM (%)		TE (%)		C (%)		AUG (%)		AZI (%)	
		SEN	RES	SEN	RES	SEN	RES	SEN	RES	SEN	RES	SEN	RES
SP	11	90	10	82	18	90	10	100	-	100	-	82	18
N PA	19	37.60	26.40	89.40	10.50	79	21	84	15	100	-	100	-
TA	9	67	33	56	44	89	11	78	22	89	11	100	-
EAR	6	83	17	100	-	83	17	83	17	100	-	100	-

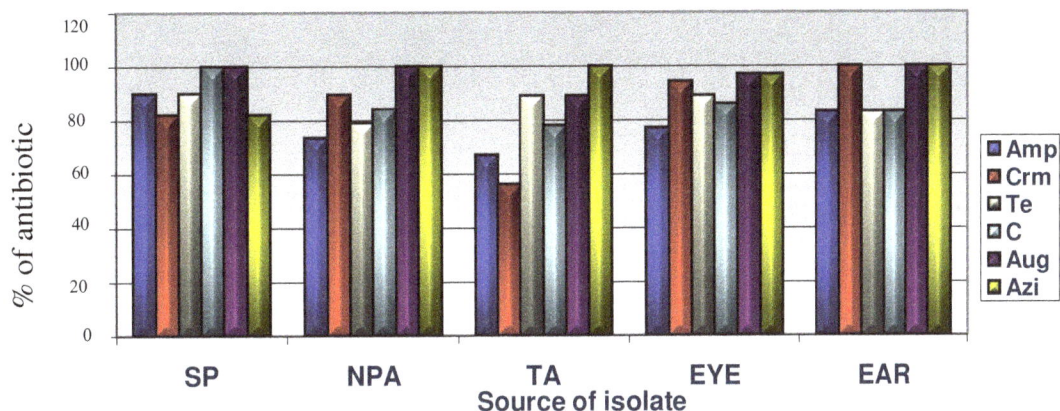

Figure 7. Antibiotic sensitive of *H. influenzae* isolate by MIC test.

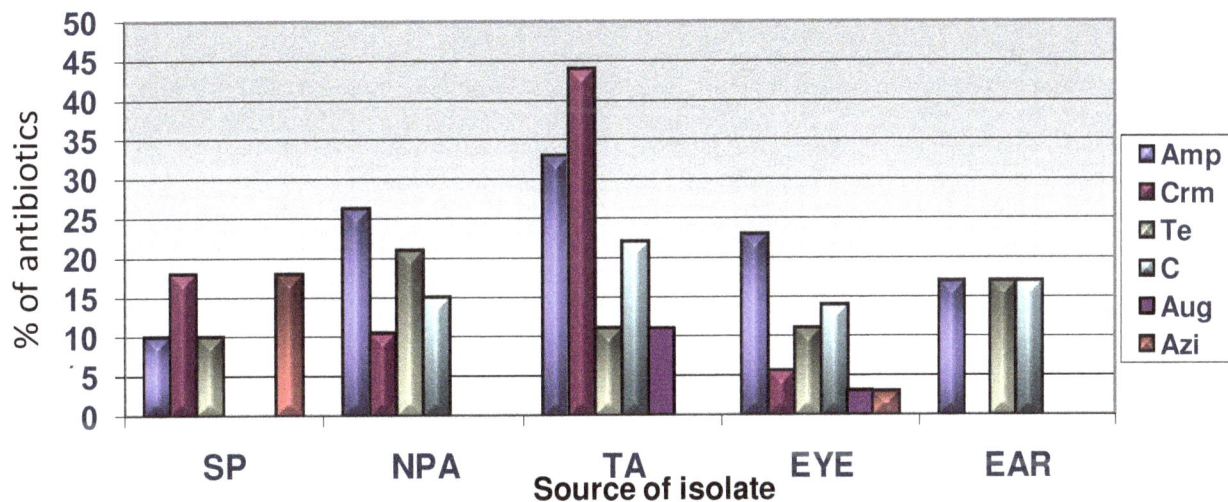

Figure 8. Antibiotic resistant of *H. influenzae* isolate by MIC test.

The susceptibility test was observed on *H. influenza* by AMP and AUG was more than C and TE using MIC and Kirby Bauer methods (Table 9 and Figure 9).

Pulse-field gel electrophoresis (PFGE)

PFGE was performed and compared with the fingerprints of genomic DNA for the typeable *H. influenzae* (serotype a to f). Many distinct restriction patterns were identified among the 6 strains available for PFGE, reflecting the expected genetic diversity among *H. influenzae* strains (Figure 10).

DISCUSSION

This study has been conducted to get a pilot profiling

Table 9. Comparative antibiotic susceptibility result using Kirby Bauer disc diffusion method and MIC.

Susceptibility method	% Resistance			
	AMP	AUG	C	TE
MIC	24	2.5	15	15
Kirby Bauer	27	5	15	17

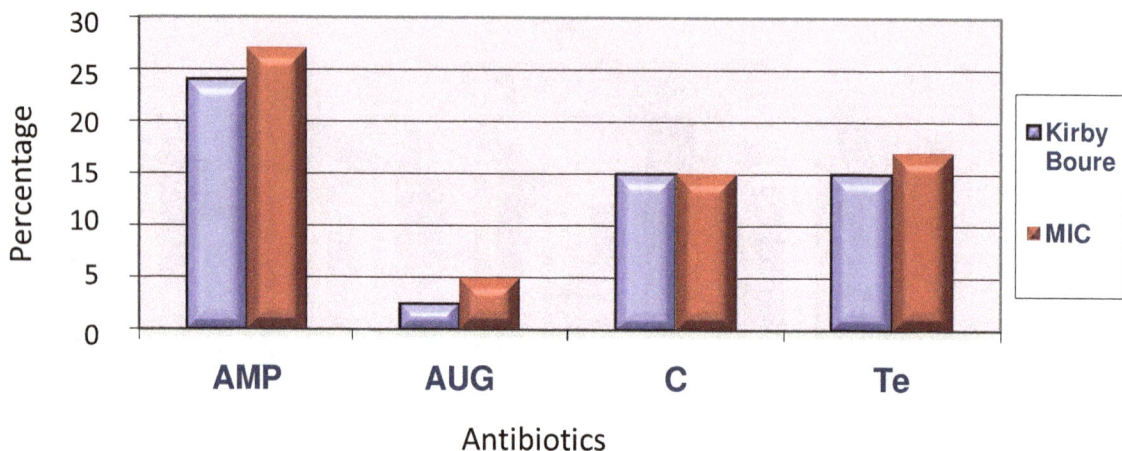

Figure 9. Comparative antibiotic susceptibility result using Kirby Bauer disc diffusion method.

biotype, serotype and antibiotic susceptibility. *H. influenzae* strain prevalent in Riyadh area through studying clinical specimens received in routine bacteriology laboratory of Riyadh Military hospital for the period from 2007-2008. The 80 *H. influenzae* isolates are collected from 5 sources that are eye, ear, sputum, lower genital tract, and nasopharyngeal. *H. influenzae* isolates ere almost equally prevalent in both sex males and females (Vidya and Devarajan, 2001.).

Distinct age predilection was observed where Patients with invasive *H. influenzae* disease ranged in age from 1 day to adult older than 50 years. 53 (66.25%) occurred in children less than 2 years of age; 12 (15%) occurred in age ranging from 3-19 years; the remaining 15 occurred in adults ranging from 20 years to older than 50 years old (Figure 1). The least prevalence was found in middle age groups varying from 2-6% of *H. influenzae.* in many other studies 86, 88, 90, and 95% invasive *H. influenzae* infection have been found to occur under 5 years of age and 9% in adults (Anderson, 1983; Wang et al., 2008). This emphasizes the fact that *H. influenza* type b (Hib) disease should be recognized as potentially contagious in young children and elderly. Based on the result of three tests-indole production, urease activity, and ornithine decarboxylase activity *H. influenzae* strain have been identified and characterized into 8 biotype I-VIII. Biochemical characterization of *Haemophilus* sp. has lead to important epidemiologic information and certain biotypes have been found paralleled with different origins of isolate

(Barry et al., 1991). Almost of the total isolate belonged to biotype II 57.5% followed but two folds less frequently 33.75% equally by each biotype I and III. Biotype IV and V constituted only 6.25 and 2.5% of stain respectively. Rarely were biotypes VII and VI met with. The most prevalent biotype was biotype II (57.5%) that were taken as the highest isolate from eye 20 (43%) and lowest one was taken from the ear 4(8%) , followed by biotype III (32.5%) that were taken as the highest isolate from eye 12 (46%) and lowest one taken was from ear (4%) (Table 2). Biotype distribution varied considerably in different infection. Biotype II was principally isolated from sputum (sp), eye, next from ear, nasopharyngeal (NPA) and genital tract (TA). Of interest was the fact that in all specimens this biotype was equally representing almost 50% of all and any infection. For instance biotype I and serotype b are commonly associated with meningitis in children, and biotypes II and III are commonly associated with upper respiratory tract infections (Gratten, 1983; Olsen et al., 2005). Serotyping of *H. influenzae* isolate differentiated them into 55% as typeable (THi) and 45% as nontypeable *H. influenzae* (NTHi) using slide agglutination test (Table1).Co agglutination test of nontypeable strains of *H. influenza* showed no false-positive reactions. The use of co agglutination reagents for serological identification of clinical isolates of *H. influenza* proved more advantageous than conventional slide agglutination in several respects. The reactions were rapid, sensitive, and easy to visualize and interpret (Farmer, and Tilton. 1980;

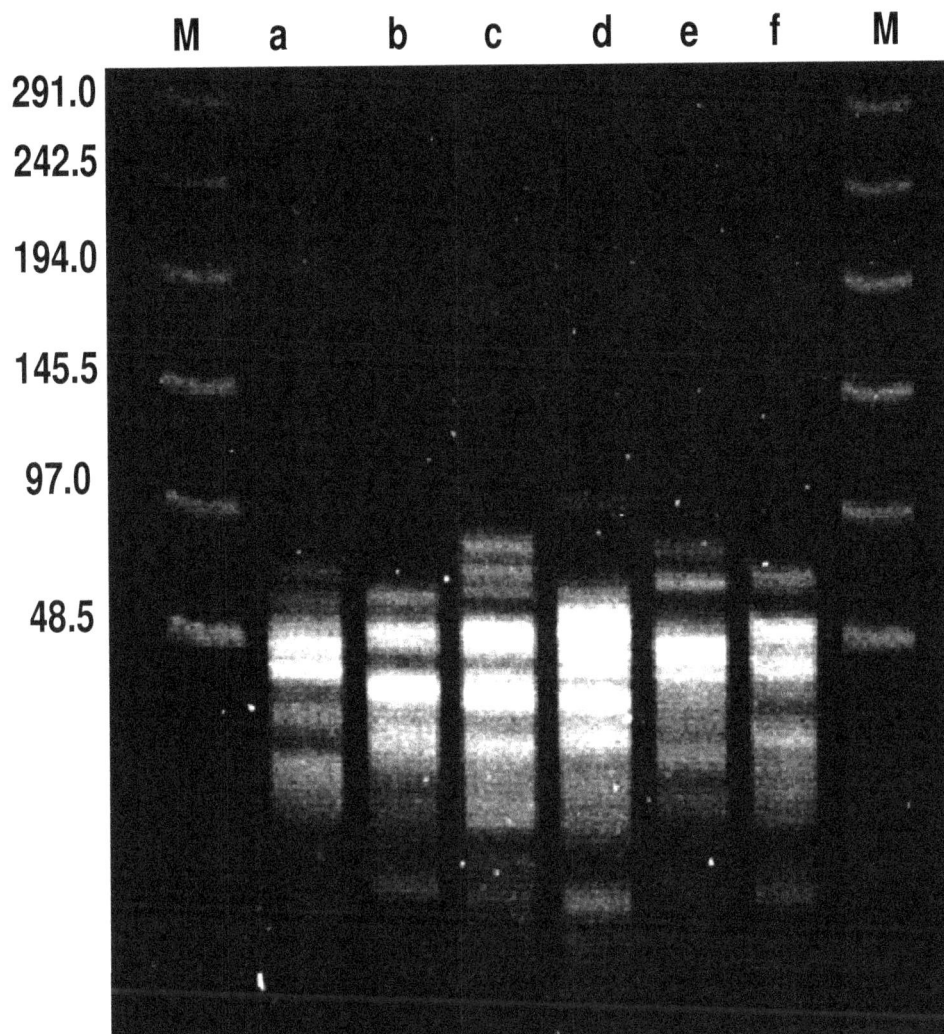

Figure 10. Pulsed-field gel electrophoresis (PFGE) patterns of Apal-digested chromosomal DNAs of *Haemophilus influenzae* isolates. Lan M, Lan serotype a, Lan serotype b, Lan serotype c, Lan serotype d, Lan serotype e, Lan serotype f, Lan M.

Silverman et al., 1977). *H. influenza* asymptomatically colonizes the nasopharynx of healthy individuals, and causes systemic disease and mucous membrane infections. Eight biotypes and six serotypes are used as epidemiological markers for studying the pattern of colonization of *H. influenzae* and to identify the strains of bacterium commonly known to be pathogenic (Alrawi et al., 2002; Amita et al., 2006). In acute infection, most cases of primary invasive diseases, secondary chronic pulmonary infection with serious complication as well as meningitis with predisposing factor as sinusitis or otitis media were found to be due to nontypeable *H. influenzae* (Bilton et al., 1995; Meats et al., 2003.).

This study focused on the distinctive relationship between serotype prevalence and age. Capsulated strain was found to be solely restricted to the first two decodes of life but in a remarkable humble percentage. On the other hand nontypeable *H. influenzae* strain apart from

being extensively encompassing all age group were exclusively prevalent beyond decade of life (Table 4).

Distribution of serotypes *H. influenze* strains and different clinical specimens, the most two common serotypes in all clinical specimens were nontypeable (NT) (43% occurrence), and serotype f that occur 43 and 31%, respectively. The least common serotypes were serotypes a (9% occurrence) and e (7.5% occurrence) that were taken from 4 different clinical specimens (SP, NPA, TA and eye), while the major serotype a (43%) was found in TA and the major serotype e (33%) was found in SP and NPA. Moreover, serotypes c and d occur only at 2.5% in all specimens. Most of serotypes were isolated from eye (43.7%), NPA (23.7%), SP (13.75%), and TA (11.25%) while only 11.25% serotypes were isolated from the ear (Table 3) shows that the *H. influenze* is mostly isolated from the eye (49.2%). Moreover, the most serotypes in children are Serotype f (23.8%) and NT (20.8%). Looking at the isolated sources in

H. influenze is less common in the following sources: TA, ear and SP for children. Furthermore, only nontypeable (NT) was found on all the sources for children (Table 5). The study pointed out that, *H. influenza* serotype "f" was the principal serotype encountered in almost 32.5% of capsulated strain and as the exclusive serotype in eye and genital specimens. High relationship has been found between *H. influenzae* strain serotypes and biotypes. Biotype II was the principal one in *H. influenzae* serotype "f" & "a". biotype III were equally belonging to serotype "a"& "f" (Table 1). In fact a substantial correlation has been found between biotypes and serotypes of *H. influenzae* strain (8). in agreement with the fact that the serotypes constitute separate evolutionary lineaging. Intensively was biotype I belonging to serotype "a", "b", & "f"; biotype II to serotype "c"; and biotype IV to serotype "d" or e "(22). About 85% of *H. influenzae* strains were β-lactamase negative and 25% β-lactamase positive. The majority of the β-lactamase positive strains were ampicillin highly resistant, chloramphenicol and tetracycline 35 and 45% resistant respectively and cefotaxime 90% sensitive. In contrast 90% of β-lactamase negative *H. influenzae* strains were all among the sensitive strain to the tested antibiotic (Table 6). Other prevalence rate of β-lactamase production approached (25%) or showed distinct difference as in isolate from the Middle East being 65.3% in Kuwait, 79.5% in Oman, 82.9% in Saudi Arabia and united Arab Emirates abd 95.7% in Europe in Turkey (26). In the United States, the proportion of BLNAR isolate remains very low while during the last decade the proportion of β-lactamase negative isolate has rapidly increased in Japan (Hasegawa et al., 2004).

In this study β-lactamase positive has *H .influenzae* strain were 10% nontypeable, for all 80 *H. influenzae* isolates were studied for their *in vitro* susceptibility against 6 antibiotics known to be first choice treatment in *H. influenzae* infection. These were ampicillin, chloramphenicol, azithromycin, tetracycline, cefuroxime, and amoxicillin using 1 method run in MIC (Table 6). Influenzae isolate were found susceptible to amoxicillin clavulanate in wide scale conducted study in 4 European countries in France, Germany, Italy and Spain (Blosser et al., 2002). In the present work almost all *H. influenzae* (89%) were sensitive to chlorophenicol (Tables 7, and 8). In fact wild type strain of *H. influenzae* is known to be susceptible to chloramphenicol and to the macrolide groups of antimicrobials. However, because of the spread of conjugative plasmids, a vast proportion of clinical isolates have become among others, chloramphenicol resistant (Jacobs et al., 1999).

For *H. influenza* isolate, tetracycline and cefuroxime showed a lesser effect than the rest other tested antibiotics where 81.25 and 82.5% of strain were sensitive to respectively. Resistance to amoxicillin was negligible (5%) by susceptibility test (Table 7).

In France resistance of *H. influenzae* to tetracycline and chloramphenicol was fond to be less than 10 and 2%

respectively in 1997 (Abdurahman et al., 2000). In Portugal 20% of *H. influenzae* was proved to be resistant to ampicillin (Shibl, 2001). Data from the Alexander Project 1999 for Saudi Arabia indicated that 27.9 and 29% af *Haemophilus* isolate were ampicillin and trimethoprime sulphamethoxazole resistant respectively (Felmingham and Washington, 1999).

Regarding the specimen site antibiotic relationship studied in the present work, it was found that Ciproloxacin and fotaxime were universally potent regardless of their specimen origin (Table 9); although& Chlorophenicol meningitis showed 100% potency in SP as expected. However, Tetracycline and Ampicillin almost had the same efficiency on *H. influenzae* isolate from SP (90%). However, prevalence of ampicillin resistant *H. influenzae* β-lactamase positive strain increased conreased considerably (Hasegawa et al., 2003) in both typeable and non typeable strain posing a serious problem in Europe and USA (10) and Middle East (Shible and Gaillot, 1994). However Ampicillin resistant prevalence, showed large different between individual countries, where it was highest in Spain 30.6% and lowest in Germany 1.6%. Moreover, the prevalence of chloramphenicol resistance remains below 1% in USA and European countries, yet resistant has become a major problem in Spain 24.9% and Belgium 10.9% and are often additionally resistant to ampicillin, tetracycline (Machka et al., 1988). This study shows, in β-lactamase positive strain, a high level of resistant to ampicillin was displayed with MIC90 of >8 µg/ml and resistance to chloramphenicol and tetracycline with MIC90 of 4 or >4 µg/ml (Table 7). *H. influenzae* genotype was identified by PFGE.

Briefly, the bacterial effective resistance to antibiotic or its susceptibility were identified and discussed in the sample of isolates. Out of the 80 isolates, there were 53 (65%) occurred in children less than 2 years of age. 12 (22%) occurred in age ranging from 3-19 years. The remaining15; occurred in adults ranging from 20 years to older than 50 years old. During the study period, a total of 80 invasive isolates of *H. influenza* were identified, 36 of 80 isolates were nontypeable strain, 34 isolates were typeable. There were 6 serotypes in the 80 isolates. There percentages found to be, 7 (8.75%) were serotype "a", 2 (2.5%) serotypes "b", 25 (31.25%) serotype "f", 2 (2.5%) serotype "d", 2 (2.5%) serotype "c", 36 (45%) nontypeable, and 6 (7.5%) serotype "e".

Nontypeable (38, 7%) was the most serotypes that infect children , ranging from one day to ten years old, serotype "f" (28.6%) while the least infected serotypes was "b" (only 1.25%) comparing to serotypes "a" (6.25%) and "c" (2.5%). For adult, the most infected serotype *was* nontypeable (25%) while "b" (1.25%) was the least infected one. Looking at the different sources of isolated *H. influenza* among children were mostly isolated from the eyes (49.2%). Moreover, the most serotypes in children were "f" (23.8%) and nontypeable (20.8%).The order of the most prevalent biotype for all isolates were biotype II (57.5%), then biotype III (32.5%), biotype IV (6.25%), biotype V (2.5%) followed by biotype I (1.25%). (Studying the effect of antibiotic showed there were 20 of the 80 invasive *H. influenza* produced a β-lactamase ampicillin resistant , of the 10 β-lactamase ampicillin resistant were typeables and 10 of the β-lactamase ampicillin resistant were nontypeable. The rate of β-lactamase producers was extensively among nontypeable of *H. influenza*. Resistance wise, the most effective antibiotics on *H. influenze* found to be amoxicillin (AUG) and azithromycin (AZI). All of the isolates serotypes found to be sensitive to amoxicillin (AUG) expect F. For: azithromycin (AZI), two serotypes found resistant to it which are type "f" and nontypeable (NT). On the other hand for the antibiotic Tetracycline (Te), there were 8 (10%) isolated bacteria found to be resistant to it. Antibiotic chloramphenicol (C) and cefuroxime (CRM) have a strong resistant from the bacteria with 13.75% of the total number of isolated bacteria from the 80 patients. All isolate exhibited the serotype a, b, c d, e and f capsular serotypes. PFGE with Sma1 restriction enzyme digestion generated not well- resolved profiles, as several very close fragments of 97.0-48.5 kb were obtained (Figure 10). Following the PFGE profiles were easier to compare the isolate with other kb such as Lan 1,3,4,5. The isolate in Lan serotype a, Lan serotype b, Lan serotype c, Lan serotype d, Lan serotype e, Lan serotype f were ranging between 97.0 and 48.5 kb. According to criteria reported by Tenover (1995) and Wang et al. (2008), all isolated were related to *H. influenzae* serotype a to f. According to epidemiologic data, patient with strains that showed indistinguishable information that did not appear to in PFGE patter which cannot be observed in any common risk factor. The result of PFGE in data needed to be assured by marking with more than one restriction enzyme and Multiple PCR analysis.

REFERENCES

Abdurahman EM, Ismael NA, Dixon RA (2000). Antimicrobial resistance and prevalence of β-lactamase in *Haemophilus influenzae* isolate a surveillance study of patients with respiratory infection in Saudi Arabia. Diagn. Microbiol. Infect. Dis., 36: 203-208.

Alrawi AM, Chern KC, Cevallos V, Lietman T, Whitcher JP, Margolis TP, Cunningham ET (2002). Biotypes and serotypes of *Haemophilus influenzae* ocular isolates. Br. J. Ophthalmol., 86: 276-277

Amita J, Pradeep K, Shally A (2006). High ampicillin resistance in different biotypes and serotypes of *Haemophilus influenzae* colonizing the nasopharynx of healthy school-going Indian children. J. Med. Microbiol., 55: 133-137.

Anderson P (1983). Antibody response to *Haemophilus influenzae* type b and diphtheria toxin induced by conjugates of oligosaccharides of the type b capsule with the nontoxic protein. Infect. Immun., 39: 233-238.

Barry AL, Jorgensen JH, Hardy DJ (1991). Reproducibility of disc susceptibility tests with *Haemophilus influenza*. Antimicrob. Chem., 26: 295-301.

Berhofer TR, Back AE (1979). Biotypes of *Haemopilus influenzae* encountered in clinical laboratories. J. Clin. Microbial., 10: 168-174.

Bilton D, Pye A, Johson M (1995). The isolate and characterization of nontypeable *H. influenzae* from the sputum of adult cystic fibrosis patients. Eur. Respir. J., 8: 948-53.

Blosser MRS, Karlowsky JA, Critchely LA (2002). Antimicrobial surveillance of *Haemophilus influenzae* in the United states during 2000-2001 leads to detection of clonal dissemination of β- lactamase negative and Amoicillin resistant strain. J. Clin. Microbiol, 40: 1063-1066.

Cerquetti M, Atti M, Renna G (2000). characterisation of non-type b *Haemophilus influenzae* strain isolated from patients with invasive disease. The HI study group. J. Clin. Microbiol., 38(12): 46-52.

Farmer SG, Tilton RC (1980). Immunoserological and immunochemical detection of bacterial antigens and antibodies: In E. H. lennette, A. Balows, W. H. Hausler, and J. P. Truant (ed.), Manual of clinical microbiology, 3rd ed. American Society for Microbiology, Washington, D.C., pp. 528 -529

Felmingham D, Washington J (1999). Trends in the antimicrobial susceptibility of bacterial respiratory tract pathogens-finding of the Alexander Project 1992-1996. J. Chemother., 11: 5-21.

Garica-Rodriguez JA, Baqiero F, De Lomas G (1999). Antimicrobial susceptibility of 1422 *Haemophilus influenzae* isolates from respiratory tract infection in spain. Result of a 1 year (1996-1997) multicenter surveillance study. Infect. Abstr., 27: 265-267.

Gratten M (1983). *Haemophilus influenzae*, biotype VII. J. Clin. Microbiol., 18: 1015-1016.

Hasegawa K, Chiba N, Kabayashi R, Murayama S (2004). Rapidly increasing prevalence of β-lactamase non producing ampicilin resistant *Haemophilus influenzae* patients with meningitis. Antimicrobial Agents Chemother., 48: 1509-1514

Hasegawa K, Yamamoto K, Chiba N, Kobayashi R, Nagai K, Jacobs MR (2003). Diversity of ampicillin-resistance genes in *Haemophilus influenzae* in Japan and the United States. Microb. Drug Resist. Spring, 9(1): 39-46

Ister GR, Conner JS, Glode MP (1984). Increasing ampicillin resistance rates in *Haemophilus influenzae* meningitis. Am. J. Dis. Child, 138: 366-369.

Jacobs MR, Bajaksouzain S, Zilles A (1999). Susceptibilities of streptopneumonia, and *Haemophilus influenzae* to 10 oral antimicrobial agents based on pharmacodynamic. Antimicrob. Agents Chemother., 43(139): 1091-1908.

Jorgensen JH (1992) Update on mechanisms and prevalence of antimicirobial resistance in *Haemophilus influenzae*. Clin. infect. Dis., 14: 11-23.

Machka K, Barveny I, Daberant H (1988) Distribution and resistance pattern of *Haemophilus influenzae* a European cooperative study. Eur. J. Clin. Microbiol., 7(140): 14-24.

Meats E, Feil EJ, Stringer S, Cody AJ, Goldstein R, Kroll JS, Popovic T, Spratt BG (2003). Characterization of encapsulated and noncapsulated *Haemophilus influenzae* and determination of phylogenetic relationships by multilocus sequence typing. J. Clin. Microbiol., 41: 1623-1636

Murphy TF, Apicella MA (1987). Nontypable *Haemophilus influenzae* review of clinical aspects, surface antigen, and the human immune response to infection. Rev. Infect. Dis., 9: 1-15.

Olsen SJ, Dejsirilert S, Sangsuk L, Chunsutiwat S, Dowell SE (2005). Frequent *Haemophilus influenzae* type b colonization in rural Thailand. Pediatr. Infect. Dis. J., 24: 739-742.

Shibl A (2001). Comparison of the *in vitro* activity of commonly prescrbed antibiotics againt *Haemopilus influenzae* isolate from the Middle East. Clin. Microbiol. Infect., 7(1): 1-394.

Shible AM, Gaillot O (1994). Susceptibility of clinical significal *Haemophilus influenzae* strain to oral antimicrobial agent in Saudi Arabia. Chemotherapy, 40(138): 300-403.

Silverman M, Stratton D, Diallo A, Egler L (1977). Diagnosis of acute bacterial pneumonia in Nigerian children. Arch. Dis. Child. 52: 925-931.

van Klingeren B, Van Embden JDA, Dressens KM (1977). Plasmid mediated chlramphenicol resistamnce in *Haemophilus influenzae*. Antimicrob. Agents Chemother, 11: 383-387.

Vidya R, Devarajan MD (2001). *Haemophilus influenzae* infection. E Med. J., 2: 80.

Wang CC, Kuo HY, Chiang DH, Tsai CC, Lin ML, Chan YJ. (2008). Invasive *Haemophilus influenzae* disease in adults in Taiwan. J. Microbiol. Immunol. Infect., 41(3): 209-214.

16

Identification of single chain Fv antibody fragment against *Helicobacter pylori*

Cesar Pedroza-Roldan[1, 2]*, Oscar Zavala-Tapia[1], Leny J. Alvarez-Araujo[1], Claudia Charles-Niño[2], Angel G. Diaz-Sanchez[3] and Raymundo Rivas-Caceres[1]

[1]Instituto de Ciencias Biomédicas, Universidad Autónoma de Ciudad Juárez. Anillo Envolvente del Pronaf y Estocolmo s/n. C.P.32300 A.P. 1595-D. Cd. Juárez, Chihuahua, México.
[2]Instituto de Investigaciones Biomédicas, Universidad Nacional Autónoma de México, AP 70228, Ciudad Universitaria, México, D.F. 04510, México.
[3]Departamento de Bioquímica, Facultad de Química, Universidad Nacional Autónoma de México, AP 70228, Ciudad Universitaria, México, D.F. 04510, México.

***Helicobacter pylori* is one of the most common causes of infectious diseases around the world. Although combined antibiotic treatment for the infection exist, the development of therapies such as immunotherapy or an effective vaccine can be helpful for containment and prevention of the infection. Phage display technologies offer a simple way for selecting specific antibody fragments against a specific target. In this work, we report the construction of a phage display antibody library against *H. pylori* strain N2. The library is constituted of 2.4 × 10⁴ transformants. A total of 94 clones were randomly selected and screened against the N2 strain, where 22% of them recognized whole-cell extracts of *H. pylori*. Moreover, other 94 clones were screened against the J99 strain, where 12% of the clones recognized this strain. The scFv-E1N2 was selected and it showed high reactivity in ELISA experiments. Furthermore, it reacted with an antigen of 27 kDa in both strains of *H. pylori* tested in our study. In conclusion, this scFv library may be used for the selection of antibodies for the development of an immunodiagnostic test. Also, it may provide insights in the identification of antigens for the development of a new vaccine or immunotherapy.**

Key words: Phage display libraries, scFv antibodies, *Helicobacter pylori*.

INTRODUCTION

Helicobacter pylori is one of the most common causes of infectious diseases around the world (Aguilar et al., 2001). Besides causing gastric and peptic ulcers, the presence of *H. pylori* in the human stomach also represents a risk of developing gastric cancer and lymphoma MALT (mucosa-associated lymphoid tissue) (Owens and Smith, 2011). Studies on patients with low degree of lymphoma MALT have confirmed a high occurrence of *H. pylori* (Wotherspoon, 2000). The bacteria is associated with the proliferation of gastric epithelial cells and this epithelial cells and this proliferation can be reverted

through its complete eradication, suggesting that tumor cells remain in response to the infection (Maruta et al., 2005). Although combined antibiotic therapy for treating the infection exist, the development of therapies such as immunetherapy or an effective vaccine can be helpful for containment and prevention of the infection. Cao J. and coworkers demonstrated that phage-displayed scFv exhibited a bactericidal effect and inhibited the growth of *H. pylori*; additionally, pretreatment of the bacteria with these recombinant phages before oral inoculation prevented colonization of the mouse stomachs (Cao et al., 2000). On the other hand, as it has been seen with other pathogens, the development of an effective vaccine against *H. pylori* would be of enormous benefits, especially in developing countries where the infection reaches up to 90% among population (Frenck and Clemens,

---

*Corresponding author. E-mail: cpedroza46@gmail.com.

2003). However, some vaccine candidates against *H. pylori* tested in mouse models did not demonstrate protection in human clinical trials (Czinn and Blanchard, 2011). Meanwhile, many important efforts have been made for the identification of proteins and epitopes related to protection against these bacteria. Because of this, several methods have been used for helping in the development of vaccines such as bioinformatic tools, proteomics (Chakravarti et al., 2000), and phage display technology. Using the last technology, Li et al. (2010). identified two different mimotopes which induced humoral immune response against the protein urease B derived from *H. pylori* after immunization in mice. The antiserum induced by these mimotopes clearly inhibited the enzymatic activity of the urease B (Li et al., 2010). In addition, this protein was identified by antibody phage display as a potential target for the immune system and this protein has been used as a vaccine component in human clinical trials (Czinn and Blanchard, 2011).

A more in-depth understanding about the process of infection and its relation with the immune systems is necessary. The identification of potential epitopes or proteins with the capacity to stimulate an effective immune response is critical to eliminate the infection. Furthermore, the identification of new potential antibodies might be particularly relevant as a complement to current therapies against *H. pylori*. In the current report, we describe the construction of a scFv antibody library against *H. pylori* strains J99 and N2. From this library, we isolated and characterized a monoclonal scFv fragment E1N2 which demonstrated high reactivity against *H. pylori*. This library can be helpful for the identification of antibodies for its use as immunotherapy or immunodiagnostics; moreover, it would be useful for the identification of potential epitopes/mimotopes which might be included as a component in future immunogens or vaccines.

MATERIALS AND METHODS

Helicobacter pylori culture conditions

The *Helicobacter pylori* strains J99 and N2 were a kind gift of Dr. Guadalupe Ayala (Centro de enfermedades infecciosas, Instituto Nacional de Salud Publica de México). Bacteria were grown in tryptic soy agar supplemented with 5% sheep blood and 20 µg/ml of vancomycin under a microaerobic atmosphere (10% CO_2) at 37°C for 48 h. The bacteria were checked for the presence of urease, catalase, and oxidase activities in addition with stain characteristic as previously described (Oliva et al., 2000).

Ethics statement

The Local Ethics Committee, (Instituto de Ciencias Biomédicas, UACJ) approved the study.

Chicken immunization

Two chickens White Leghorn were immunized with lysed *H. pylori* N2 strain three times at 14-day intervals. The immunogen (20 and 200 µg, respectively) was mixed with Freund´s incomplete adjuvant

(IFA) for priming and with complete Freund´s adjuvant (CFA) for boost injections in the pectoral area. Pre- and post-immunization sera were tested in ELISA for the presence of specific anti-*H. pylori* antibodies. Briefly, a 96-well plate (Nunc, Roskilde, Denmark) was coated with 2 µg/ml of lysed *H. pylori* strain N2 in phosphate buffer (pH 7.2) overnight at 4°C. The plate was washed with PBS containing 0.2% Tween 20 and blocked with PBS containing 2% of serum albumin. After washing, chicken sera diluted 1:100 in PBS-BSA 1% was added, and the plate was incubated for 1 h at 37°C. After washing, goat anti-chicken IgY conjugated with HRP (Abcam, USA). After incubation of 1 h at 37°C, the plate was washed and 100 µl of ABTS single solution (Zymed) was added. The OD reading at 405 nm was registered using a microplate reader.

Construction of anti-*Helicobacter pylori* scFv antibody phage display library

The mRNA was extracted from the splenocytes of immunized chickens using RNA STAT-60 (Tel-Test Inc, USA). The kit RT-PCR Superscript II (Invitrogen, USA) was used for cDNA synthesis. All procedures were done according to the manufacturer´s instructions. Oligonucleotides used for the amplification of the heavy (V_H), light (V_L) chain and ScFv construction were described previously (Andris-Widhopf et al., 2000). Briefly, the reaction mixture for V_L amplification was prepared with 10 µl of PCR buffer 10x, 8 µl of dNTP´s (2.5 mM), 3 µl of the oligonucleotide CSVK and CKJo-B (20 pmol/µl), 2 µl of cDNA and 5 µl of Taq DNA polymerase (1U/µl). The primers CSCVHo-F and CSCG-B were used for V_H amplification in the same conditions. Assembly reaction to join V_L and V_H genes were done using the primers CSC-F and CSC-B. PCR products were gel purified using Wizard SV Gel and PCR Clean-Up System (Promega, USA). The PCR products were digested using Sfi I (New England Biolabs, USA) for 12 h and ligated into the similarly digested vector pComb 3X (Andris-Widhopf et al., 2000). The ligation was precipitated with isopropanol and then it was introduced into 50 µl *Escherichia coli* X1-blue competent cells by heating and chilling. Transformed cells were cultured in 10 ml of super broth media with ampicillin (50 µg/ml) and tetracycline (10 µg/ml). Next, phagemid-containing bacterial colonies were infected with the helper phage VCSM13 (6 × 10^{11} PFU/ml) (Invitrogen, USA) and the volume of the culture media adjusted at 100 ml with ampicillin and kanamycin (50 µg/ml). The culture was grown overnight in agitation.

Selection of scFv antibodies against *H. pylori* by biopanning

Selection and amplification procedures for the scFv library were described previously (Pedroza-Roldan et al., 2009). Briefly, phages were recovered by double precipitation with PEG/NaCl. The precipitated phages were resuspended in 400 µl of PBS-BSA 1%. The panning was done as reported previously (Solorzano-Vargas et al., 2008). For this, a 96-well plate was coated overnight with 2 µg/ml of lysed *H. pylori* strains J99 and N2 separately. Plates were washed and blocked with BSA 3% and incubated for 1 h at 37°C, and then the phages were added at a concentration of 3 × 10^{11}/ml. After washing, bound phages were eluted using 50 µl of trypsin (10 mg/ml). The eluted phages (individually for J99 and N2) were used to repeat another round of selection. Four rounds of selection were performed.

Selection and production of monoclonal scFv antibodies

Phagemid DNA was extracted from the third (strain N2) and fourth (strain J99) selection rounds and introduced into *E. coli* TOP10F competent cells. Individual colonies (94 for N2 and 94 for J99, respectively) were randomly selected and cultured in a 96-well plate and incubated overnight at 37°C. Next day, a 5 µl aliquot of each sample from the overnight culture were added into another 96-well plate and incubated for 6 h at 37°C. After this time, 5 µl (0.5 M) of Isopropyl β-D-1-thiogalactopyranoside (IPTG) were added to each well and cultured for 22 h at 37°C. Supernatants were stored at -20°C. Protein extraction from bacteria used in this study was performed as follows: pellets were resuspended in PPD buffer (100 mM NaCl, 25 mM Tris HCl pH 8.0, 0.02% NaN$_3$ and 1% triton X-100). The resuspended pellet was sonicated in ice four times for 15 s each time.

The immunoassay was conducted as follows; two micrograms of lysed Helicobacter pylori strains N2 and J99 were coated separately in a 96-well plate overnight at 4°C. Plates were blocked with PBS/BSA 3% for 1 h at 37°C, after washing, 50 µl of the super-natant was added. After washing, goat anti-HA conjugated with HRP (Roche Diagnostics, USA) was added and incubated for 1 h at 37°C. After incubation for 1 h at 37°C, plates were washed and 100 ul of ABTS single solution (Zymed) were added. The OD reading at 405 nm was registered using a microplate reader.

Characterization of the monoclonal scFv antibody E1N2

To characterize the capacity of the scFv EIN2 to recognize H. pylori, the clone E1N2 was cultured in 30 ml of SB and induced by adding 300 µl of IPTG 0.5 M in the culture medium for 16 h at 37°C. The scFv antibody was purified from the supernatant and cell-associated fraction by a Ni-NTA agarose system (Qiagen, USA) according to the manufacturer's instructions. Bradford's assay was used to quantify proteins extracts.

A 96-well plate was coated with 2 µg/ml of lysed Helicobacter pylori strain N2 or Klebsiella sp, E. coli, Enterobacter cloacae, Salmonella typhimurium, Proteus vulgaris, and Staphylococcus sp. in PBS overnight at 4°C. The plate was washed with PBS containing 0.2% Tween 20 and blocked with PBS containing 2% of BSA. After washing, E1N2 scFv antibody (10 µg/ml) was added, and the plate was incubated for 1 h at 37°C. As a control, a non-related antibody fragment that binds a hapten was used. After washing, goat anti-HA conjugated with HRP (Roche Diagnostics, USA) was added and incubated for 1 h at 37°C. Next, the plate was washed and 100 µl of ABTS single solution (Zymed) was added. The OD reading at 405 nm was registered using a microplate reader. The supernatant and sonicated bacteria were separated in SDS-PAGE under reducing conditions and transferred onto PVDF membrane (Osmonics 0.45µM). The membrane was blocked with PBS-BSA 3% for 1 h at 4°C. Goat anti-HA conjugated with HRP diluted 1:5000 in PBS-Tween-BSA 0.2% was added and incubated for 2 h at 4°C. On the other hand, to identify the molecular weight of the antigen recognized by the scFv antibody E1N2; 10 µg of lysed H. pylori strains J99 and N2 were separated in SDS-PAGE and transferred onto PVDF membrane. The membrane was blocked with PBS-BSA 3% for 1 h at 4°C and then 2 µg/ml of the purified antibody was added and incubated for 2 h at 4°C. After washing, goat anti-HA conjugated with HRP diluted 1:5000 in PBS-Tween-BSA 0.2% was added and incubated for 1 h at 4°C. The membranes were developed using the ECL western blotting detection system (Amersham Biosciences, UK).

DNA sequencing

Four clones were selected for DNA sequencing. The sequencing process was made in an ABI PRISM 3100 sequencer (Applied Biosystems, USA) using the primers g-back and OMPSEQ described previously (Andris-Widhopf et al., 2000).

Statistical analysis

All the experiments were reproduced at least 2 times. Data were expressed as mean ± standard deviations of the means (SD). Tukey's multiple comparison test was performed using GraphPad Prism v.5.0 for Windows. P-value less than 0.05 was considered significant.

RESULTS

Library construction

To achieve the construction of the library, two white leghorn chickens were immunized separately with 20 and 200 µg of whole-cell lysate of H. pylori strain N2. Chicken two which received 200 µg had a better response in comparison with chicken 1 (20 µg) as shown in Figure 1. As shown in Figure 1, chicken were immunized three times with 20 and 200 µg of lysed H. pylori in the pectoral area in the presence of CFA. Sera were tested in immunoassays experiments for the presence of anti-H. pylori antibodies. Chicken 2, which presented high levels of specific anti-H. pylori antibodies, was sacrificed and the splenocytes extracted for cDNA synthesis. The V genes were amplified from mRNA. A molecular weight of 400 pb and 350 pb were found in an agarose gel for V_H and V_L, respectively (Figure 2). The V_H and V_L were spliced with each other (Figure 2) and the obtained scFv gene repertoires were cloned into pComb 3X phagemid to generate a phage-displayed scFv antibody library constituted of 2.4×10^4 transformants. PCR analysis of 24 randomly selected clones showed that all carried full length inserts (data not shown).

Selection of scFv antibodies by biopanning and monoclonal scFv screening

In order to identify specific monoclonal scFv antibodies against H. pylori, phagemid DNA from the eluted phages of the third round for J99 and fourth round for N2 were used to transform E. coli TOP10F cells. The transformed cells express the scFv antibody without fusion to the coat protein III of the phage. A total of 188 clones (94 for N2 and 94 for J99) were screened to evaluate their reactivity. For the N2 scFv library, 22% of the selected clones recognized the antigen in ELISA assays (Figure 3a), compared to 12% for the J99 scFv library (Figure 3b). As a consequence of the screening, 10 clones were selected, 5 from the N2 library (10GN2, E1N2, 9EN2, 11FN2, and 11AN2) and 5 from the J99 library (10HJ99, 11CJ99, 11DJ99, 10EJ99, and 6HJ99). Next, we decided to determine whether the selected scFv antibodies might be the same or were different. To this end, we employed restriction fragment length polymorphism procedure. We isolated plasmid DNA and amplified the scFv by PCR; the PCR products were digested with Bst N1 and separated in an agarose gel. The scFv antibody genes 6HJ99,

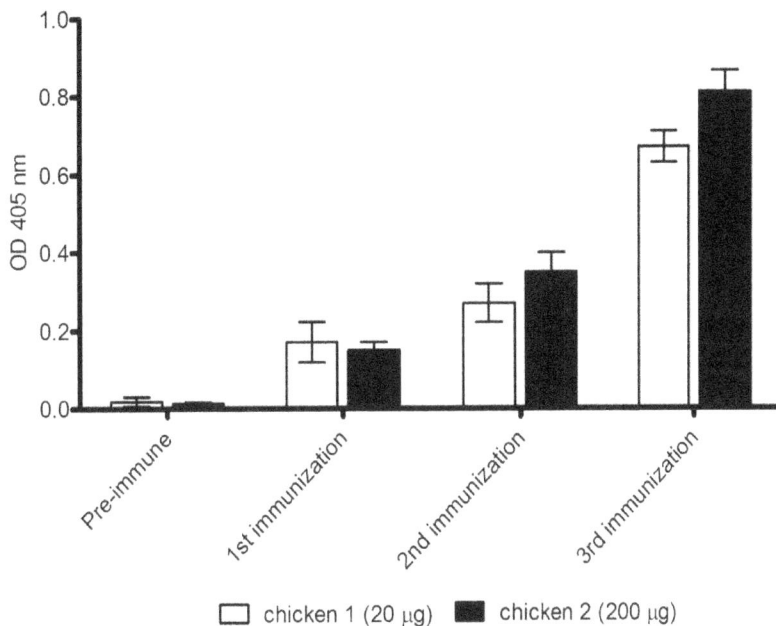

chicken 1 (20 µg)   chicken 2 (200 µg)

Figure 1. Immune response induced in chickens by immunization with whole-cell lysate of *Helicobacter pylori* N2 strain. Chicken were immunized three times with 20 and 200 µg of lysed *H. pylori* in the pectoral area in the presence of CFA. Sera were tested in immunoassays experiments for the presence of anti-*H. pylori* antibodies. Data shown are the mean ± SD of two independent experiments performed by triplicate.

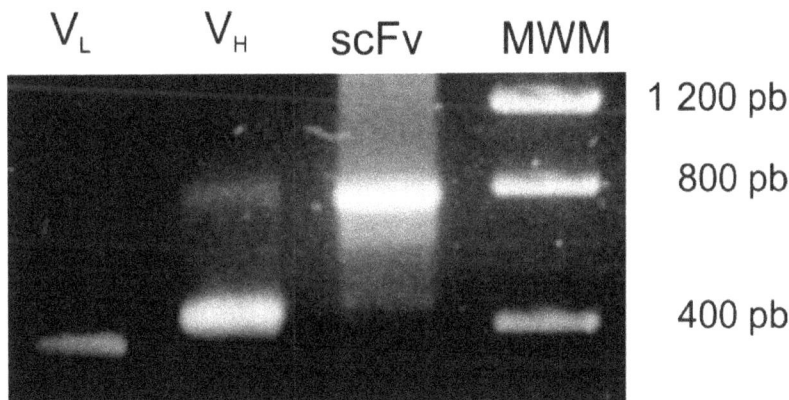

Figure 2. Amplification of the V genes repertories by PCR. V_L and V_H genes were amplified from the splenocytes of the chicken 2. V_H and V_L were spliced in order to obtain the scFv gene repertories. MWM, Low mass ladder.

11DJ99, 10EJ99 and E1N2 presented different restriction patterns; the other scFv genes presented the E1N2 restriction pattern (data not shown).

Characterization of the scFv antibody E1N2

We tested the capacity of the scFv antibody E1N2 to react against *H. pylori*. To this end, we expressed and purified the scFv antibody. The scFv E1N2 demonstrated

a significant sensitivity on the detection of the bacteria with respect with the control (P < 0.05) (Figure 4a). Moreover, the recognition of *H. pylori* by the scFv antibody E1N2 in ELISA was higher in comparison with non-related bacteria strains (P < 0.05) (Figure 4b). Next, we purified the antibody fragment from culture supernatant and from cell-associated fraction; these extractions were separated by SDS-PAGE and transferred onto membranes for Western blot analysis. A great proportion of the antibody was found accumulated in the bacterial extract

a

b

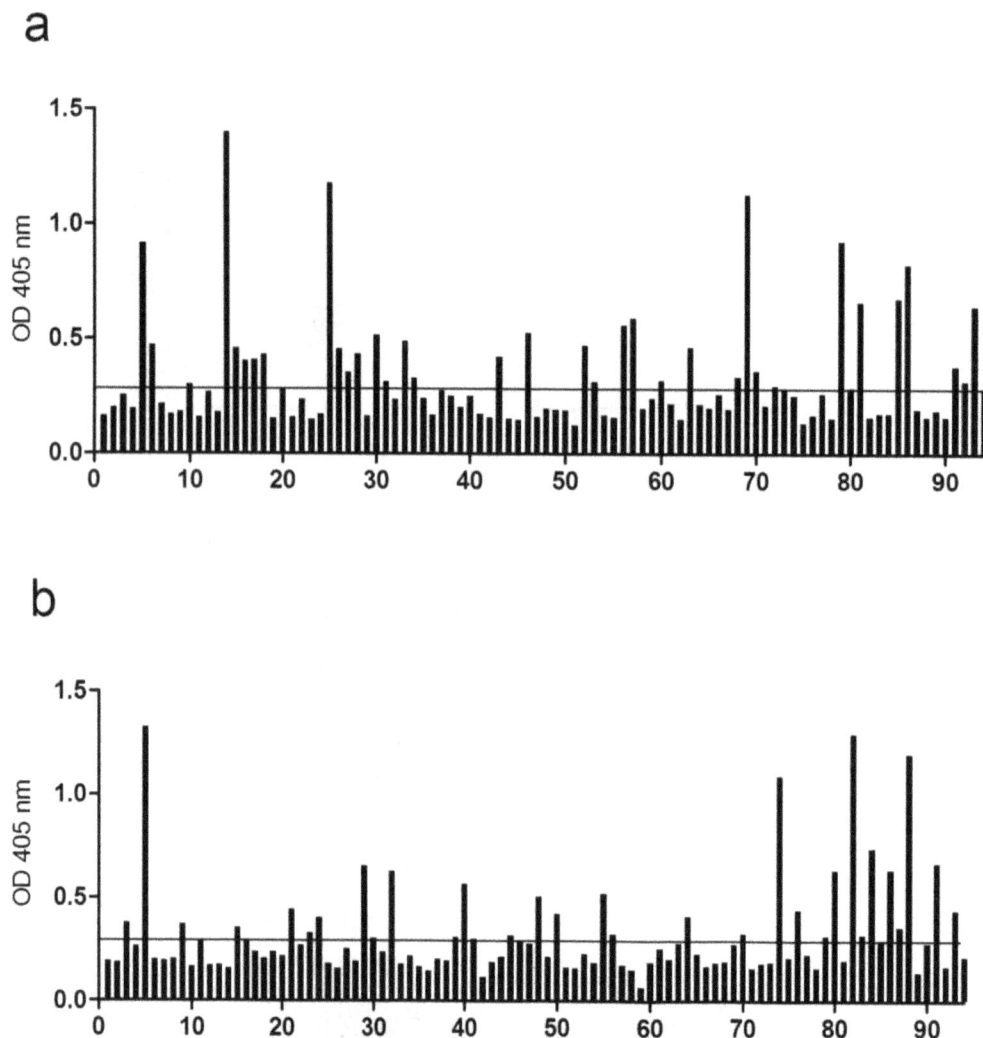

Figure 3. Isolated scFv clones from enriched phages after 3 and 4 rounds of selection against *H. pylori* J99 and N2 strain, respectively. Monoclonal scFvs were expressed without phage fusion and screened for binding activity in ELISA. a) 22% of the clones bound to the *H. pylori* strain N2 lysate and 12% bound to the *H. pylori* strain J99 lysate.

in the bacterial extract as a dimer with a MW of 46 kDa, whereas most of the monomer conformation was found with a MW of 26 kDa in the supernatant (Figure 5a). On the other hand, to determine the molecular weight of the antigen recognized by the scFv antibody E1N2, whole protein from *H. pylori* (J99 and N2) were separated by SDS-PAGE and transferred onto PVDF membranes. The scFv antibody E1N2 recognized an antigen with a MW of 27 kDa. Interestingly, the identified protein was shared by both strains as shown in Figure 5b. No reactivity was found with proteins from *E. coli*. The four clones selected by restriction patterns (6HJ99, 11DJ99, 10EJ99, and E1N2) were sequenced. They showed a high diversity in the complementary determining regions (Figure 6). The sequences presented a 93% of identity with the germline sequence of the chicken.

DISCUSSION

H. pylori is one of the major challenges in the field of public health; therefore, the appropriate diagnosis and treatment can diminish the risk of development of gastric ulcers, lymphoma malt and eventually cancer. With this in mind, we constructed a scFv library against *H. pylori* from which we isolated and characterized the scFv-E1N2 antibody. This antibody demonstrated a significant reactivity against *H. pylori* strains N2 and J99 (Figure 4a). Furthermore, the antibody fragment demonstrated a specific recognition of the bacteria (Figure 4b). We produced the scFv E1N2 antibody in *E. coli*, and found it as a dimer in sonicated bacteria (Figure 5a). On the other hand, protein purification from the supernatant showed that the majority of the antibody could be found as a

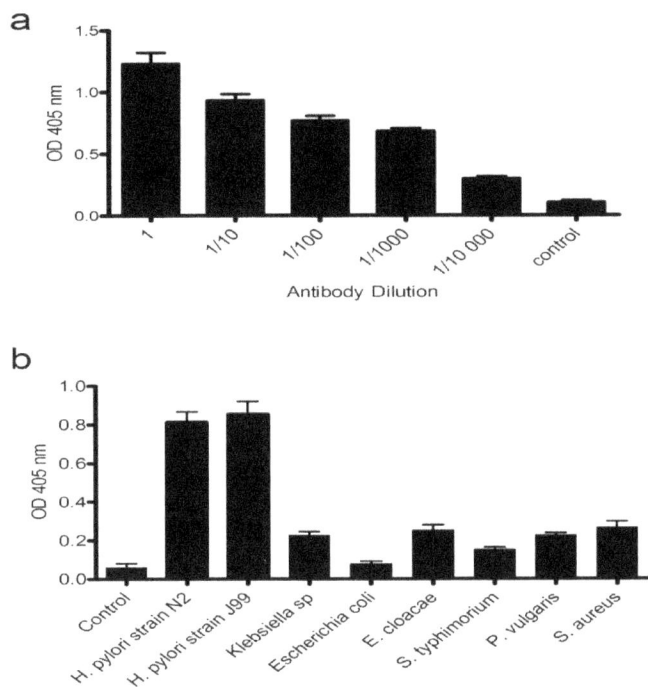

Figure 4. Reactivity of the E1N2 scFv antibody. (a) E1N2 antibody fragment was serially diluted from a stock with a concentration of 10 µg/ml, the dilutions were added in a previously coated 96-well plate with *H. pylori* N2 strain; after washing, goat anti-HA HRP was added. (b) For testing specificity, a 96-well plate was coated with sonicated *H. pylori* strain N2, J99 and other non-related bacteria. ScFv-E1N2 was added to each well; after washing goat anti-HA HRP was added. ABTS single solution was added and the OD reading at 405 nm was registered. As a control, a non-related antibody fragment that binds a hapten was used. Data shown are the mean ± SD of three independent experiments by triplicate.

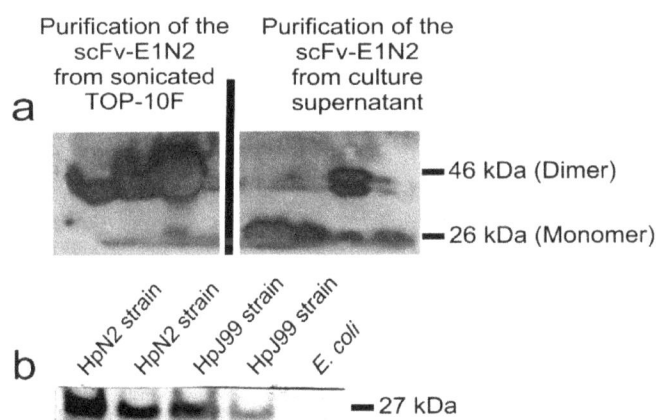

Figure 5. SDS-Page and Inmunoblotting analysis of purified scFv-E1N2 antibody and sonicated extracts of *H. pylori*. a) scFv-E1N2 antibody was recovered and purified from sonicated TOP-10F and culture supernatant by Ni-NTA agarose system. The antibody was separated and transferred; after washing, the membrane was incubated with anti-HA HRP. b) Sonicated *H. pylori* N2 and J99 strains were separated and transferred. Membrane was incubated with Ni-NTA purified E1N2 antibody; after washing, anti-HA HRP antibody was added. All the membranes were revealed with the ECL western blotting detection. *Escherichia coli* was used as a control.

monomer (Figure 5a). This conformation taken by the antibody fragment is because of the length of the linker that joins V_L and V_H chains; in this study the linker used was composed of 7 aminoacids (GQSSRSS) (Andris-Widhopf et al., 2000). Previous reports have shown that shortening of the linker (1 to 7 aminoacids) decrement stability of monomers, increasing the capacity of the scFv to form dimers, trimers, tetramers, and higher molecular mass multimers (Dolezal et al., 2000). Moreover, reduction of the linker could be a strategy to increment the avidity and stability of the antibody fragment; however, these types of modification do not increment affinity for the target (Cai-Qun et al., 2010; Dolezal et al., 2000).

H. pylori employ urease protein for living in a highly acidic gastric milieu (Sachs et al, 2006). The native urease of *H. pylori* is a hexameric molecule with a molecular mass of approximately 540 kDa. This enzyme is composed of two subunits, one of them called UreB with a molecular mass of 60 kDa and a small one called UreA with a molecular mass of 26.5 kDa (Olivera-Severo, 2006;Berdoz and Corthesy, 2004). Species identified to date, produce large amounts of urease, which may account for 15% of the bacterial protein composition and is highly immunogenic (Czinn and Blanchard, 2011). The scFv-E1N2 antibody had reactivity against an antigen of approximately 27 kDa in both strains of *H. pylori* (Figure 5b). Because of this, we may speculate that the antigen recognized by the scFv-E1N2 antibody is related with the subunit A of ureasa protein; however, more experiments are needed to characterize the antigen.

Because this pathogen is able to acquire resistance to pharmacological treatment (Benhar, 2001;Weeks et al., 2011), there is a special interest in the development of immunotherapies, as alternatives for the treatment of infectious microorganisms as viruses and bacteria. In previous reports, phage displayed scFv antibodies recognized a 60 kDa protein which was related with the urease protein B (UreB; 60 kDa) (Houimel et al., 2001; Reiche et al., 2002). On the other hand, Cao J and co-workers found a scFv antibody that recognized an antigen not identified with a molecular weight of 30 kDa. Interestingly, an *in vivo* experiment with these recombinant phages exerts a bactericidal effect and inhibited bacterial growth in six different strains. Moreover, previous treatment of the bacteria with these phages reduced the colonization of the bacteria in mouse stomachs (Cao et al., 2000). Even though the main mechanism by which there was an inhibition of the bacteria was not identified in that work, other reports have shown that is possible to reduce the capacity of replication by antibody neutralization of the proteins urease B (60 kDa) and heat-shock protein 60 (Li et al., 2010;Yamaguchi et al., 1997). It would be interesting to determine if the scFv-E1N2 antibody can be able to diminish the capacity of colonization of the bacteria in mouse stomachs. Because *H. pylori* colonizes under the mucus layer of the gastric mucosa, it is well protected against conventional antibodies (Cao et al., 2000).

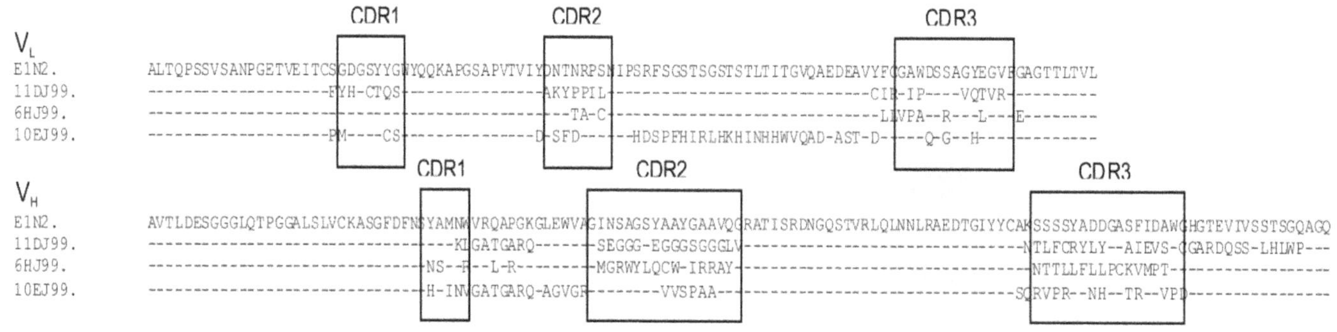

Figure 6. Sequence comparison of the four binders scFv antibody fragments. Clones E1N2, 11DJ99, 6HJ99, and 10EJ99 that showed different pattern of restriction with the enzyme *Bst*N I, were sequenced. The figure shows the V genes (V$_L$ and V$_H$) and complementary determining regions (CDRs). Dashes denote sequence identity.

Nystrom J and coworkers demonstrated that albeit high levels of *H. pylori* specific serum antibody titers, there is not a significant reduction of the colonization until the development of a Th1 profile simultaneously with an IgA specific antibody response (Nystrom and Svennerholm, 2007). This and other reports suggest the importance of IgA antibody for blocking bacterial adhesion to the mucus layer (Berdoz and Corthesy, 2004). For this reason, it might be possible to convert the scFv-E1N2 into a polymeric secretory IgA as it has been reported previously (Berdoz and Corthesy, 2004). This modification might give the antibody more stability into the gastric environment.

Currently, the detection of *H. pylori* antigens in feces is a valid method to identify whether a person is infected or not. Some diagnostic tests as ImmunoCard STAT, HpSA and Diagnostec *H. pylori* antigen EIA Kit has a sensibility and specificity comparable with other invasive and non-invasive methods (Calvet et al., 2010). The scFv-E1N2 or another scFv antibody selected from the library might be used for the development of a low cost immunodiagnostic with the primary approach to detect in challenging conditions as human feces and biopsies. Anti-*H. pylori* scFvs might be also being used for the identification of potential proteins, epitopes or mimotopes to being incurporated in future vaccines as it was reported previously (Gevorkian et al., 2004; Manoutcharian et al., 2004). Because this scFv library was constructed from immunized chickens with whole lysate of *H. pylori*, it might be possible to find scFv antibodies against many *H. pylori*-related proteins. In conclusion, these scFv antibodies might be used for the progressive development of complementary immunotherapies and immunodiagnostics, in addition to its use for the identification of additional *H. pylori*-related antigens.

REFERENCES

Aguilar GR, Ayala G, Fierros-Zarate G (2001). Helicobacter pylori: recent advances in the study of its pathogenicity and prevention. Salud Publica Mex, 43: 237-247.

Andris-Widhopf J, Rader C, Steinberger P, Fuller R, Barbas CF (2000). Methods for the generation of chicken monoclonal antibody fragments by phage display. J. Immunol. Methods, 242: 159-181.

Benhar I (2001). Biotechnological applications of phage and cell display. Biotechnol. Adv., 19: 1-33.

Berdoz J, Corthesy B (2004). Human polymeric IgA is superior to IgG and single-chain Fv of the same monoclonal specificity to inhibit urease activity associated with *Helicobacter pylori*. Mol. Immunol., 41: 1013-1022.

Cai-Qun Bie, Dong-Hua Y, Xu-Jing Liang, Shao-Hui T (2010). Contruction of non-covalent single-chain Fv dimers for hepatocellular carcinoma and their biological functions. World J. Hepatol., 2(5):185-191.

Calvet X, Lario S, Ramirez-Lazaro MJ, Montserrat A, Quesada M, Reeves L, Masters H, Suarez-Lamas D, Gallach M, Miquel M, Martinez-Bauer E, Sanfeliu I, Segura F (2010). Accuracy of monoclonal stool tests for determining cure of *Helicobacter pylori* infection after treatment. Helicobacter, 15: 201-205.

Cao J, Sun Y, Berglindh T, Mellgard B, Li Z, Mardh B, Mardh S (2000). Helicobacter pylori-antigen-binding fragments expressed on the filamentous M13 phage prevent bacterial growth. Biochim. Biophys. Acta, 1474: 107-113.

Chakravarti DN, Fiske MJ, Fletcher LD, Zagursky RJ (2000). Application of genomics and proteomics for identification of bacterial gene products as potential vaccine candidates. Vaccine, 19: 601-612.

Czinn SJ, Blanchard T (2011). Vaccinating against *Helicobacter pylori* infection. Nat. Rev. Gastroenterol. Hepatol., 8: 133-140.

Dolezal O, Pearce LA, Lawrence LJ, McCoy AJ, Hudson PJ, Kortt AA (2000). ScFv multimers of the anti-neuraminidase antibody NC10: shortening of the linker in songle-chain Fv fragments assembled in V$_L$ to V$_H$ orientation drives the formation of dimers, trimers, tetramers and higher molecular mass multimers. Protein Engineering, 13(8):565-574.

Frenck RW, Jr Clemens J (2003). Helicobacter in the developing world. Microbes. Infect., 5: 705-713.

Gevorkian G, Petrushina I, Manoutcharian K, Ghochikyan A, Acero G, Vasilevko V, Cribbs DH, Agadjanyan MG (2004). Mimotopes of conformational epitopes in fibrillar beta-amyloid. J. Neuroimmunol., 156: 10-20.

Houimel M, Corthesy-Theulaz I, Fisch I, Wong C, Corthesy B, Mach J, Finnern R (2001). Selection of human single chain Fv antibody fragments binding and inhibiting *Helicobacter pylori* urease. Tumour. Biol., 22: 36-44.

Li Y, Ning Y, Wang Y, Peng D, Jiang Y, Zhang L, Long M, Luo J, Li M (2010). Mimotopes selected with a neutralizing antibody against urease B from *Helicobacter pylori* induce enzyme inhibitory antibodies in mice upon vaccination. BMC. Biotechnol., 10: 84.

Manoutcharian K, Acero G, Munguia ME, Becerril B, Massieu L, Govezensky T, Ortiz E, Marks JD, Cao C, Ugen K, Gevorkian G (2004). Human single chain Fv antibodies and a complementarity determining region-derived peptide binding to amyloid-beta 1-42.

Neurobiol. Dis., 17: 114-121.

Maruta F, Sugiyama A, Ishizone S, Miyagawa S, Ota H, Katsuyama T (2005). Eradication of *Helicobacter pylori* decreases mucosal alterations linked to gastric carcinogenesis in Mongolian gerbils. J. Gastroenterol., 40: 104-105.

Nystrom J, Svennerholm AM (2007). Oral immunization with HpaA affords therapeutic protective immunity against H. pylori that is reflected by specific mucosal immune responses. Vaccine, 25: 2591-2598.

Oliva G, Romero I, Ayala G, Barrios-Jacobo I, Celis H (2000). Characterization of the inorganic pyrophosphatase from the pathogenic bacterium *Helicobacter pylori*. Arch. Microbiol., 174: 104-110.

Olivera-Severo D, Wassermann GE, Carlini CR (2006). Ureases display biological effects independent of enzymatic activity. Is there a connection to diseases caused by urease-producing bacteria?. Brazilian J. Med. Biol. Res., 39:851-861.

Owens SR, Smith LB (2011). Molecular Aspects of H. pylori-Related MALT Lymphoma. Patholog. Res. Int., 2011: 1931-49.

Pedroza-Roldan C, Charles-Nino C, Saavedra R, Govezensky T, Vaca L, Avaniss-Aghajani E, Gevorkian G, Manoutcharian K (2009). Variable epitope library-based vaccines: Shooting moving targets. Mol. Immunol., 47:270-282.

Reiche N, Jung A, Brabletz T, Vater T, Kirchner T, Faller G (2002). Generation and characterization of human monoclonal scFv antibodies against *Helicobacter pylori* antigens. Infect. Immun., 70: 4158-4164.

Sachs G, Kraut JA, Wen Y, Feng J, Scott DR (2006). Urea transport in bacteria: Acid acclimation by gastric *Helicobacter* spp. J. Membrane Biol., 212: 71-82.

Solorzano-Vargas RS, Vasilevko V, Acero G, Ugen KE, Martinez R, Govezensky T, Vazquez-Ramirez R, Kubli-Garfias C, Cribbs DH, Manoutcharian K, Gevorkian G (2008). Epitope mapping and neuroprotective properties of a human single chain FV antibody that binds an internal epitope of amyloid-beta 1-42. Mol. Immunol., 45: 881-886.

Weeks JN, Boyd KL, Rajam G, Ades EW, McCullers JA (2011). Combined immunotherapy with intravenous immune globulin and P4 peptide rescues mice from post-influenza pneumococcal pneumonia. Antimicrob. Agents Chemother, 55:2276-2286

Wotherspoon AC (2000). A critical review of the effect of *Helicobacter pylori* eradication on gastric MALT lymphoma. Curr. Gastroenterol. Rep., 2: 494-498.

Yamaguchi H, Osaki T, Taguchi H, Hanawa T, Yamamoto T, Fukuda M, Kawakami H, Hirano H, Kamiya S (1997). Growth inhibition of *Helicobacter pylori* by monoclonal antibody to heat-shock protein 60. Microbiol. Immunol., 41: 909-916.

17

Identification and pathogenicity of phytopathogenic bacteria associated with soft rot disease of girasole tuber in Egypt

Ismail M. E.[1], Abdel-Monaim M. F.[2]* and Mostafa Y. M.[3]

[1]Department of Plant Pathology, Faculty of Agriculture, El-Minia University, Minia, Egypt.
[2]Plant Pathology Research Institute, Agricultural Research Center, Giza, Egypt.
[3]Department of Horticulture, Faculty of Agriculture, El-Minia University, Minia, Egypt.

Six bacterial isolates were isolated from naturally infected tubers of girasole plants (*Helianthus tuberosus* L. cv. Balady) showing soft rot, which were collected from the experimental farm of the Faculty of Agriculture, Department of Horticulture, El-Minia University, during 2010 and 2011 growing seasons. Pathogenicity tests showed the virulence of the isolated bacteria in girasole tubers which harbored the pathogen; the bacteria were characterized as rod-shaped, Gram negative, α-methyl-d-glucoside medium, reducing substances from sucrose, phosphatase activity and deep cavities on pectate medium. Otherwise, diagnostic tests suggested that the pathogen was *Erwinia carotovora* ssp. *carotovora*. The isolated bacteria caused soft rot of wounded tubers when inoculated into tissues. The bacterial isolates were compared for their degree of pathogenicity as well as for differences in the specific symptoms induced in the different hosts. The tested isolates could infect several host range such as fruits of apricot, apple, olive, lemon, squash, eggplant, potato tubers, cloves of bulbs, garlic and onion; root of radish, carrot, sweet potato and rape, were infected by the tested isolates. On the other hand, no symptoms were exhibited on pods of bean and cowpea, faba bean, fruits of pepper and tomato. The extracts of experimentally diseased girasole tubers were active in pectinase at pH 6 and also caboxymethyl cellulose on pH 6 compared to enzyme activities in healthy tissues. The bacterial isolates also increased the total and reducing sugars in infected than healthy tissues.

Key words: Girasole, *Erwinia carotovora* ssp. *carotovora*, pectinase and caboxymethyl cellulose, total and reducing sugars.

INTRODUCTION

Jerusalem artichoke is grown primarily for tubers which can be eaten fresh or raw, cooked in appetizing ways similar to Irish potatoes, or pickled. Tubers are used to fatten cattle, sheep and hogs. Stems and leaves are rich in fats, protein and pectin, and make good forage and silage. The alcohol fermented from the tubers is said to be of better quality than that from sugar beets. It is good weed eradicator, as it makes so dense a shade that few other plants can compete.

It is good in ridding fields of quack grass (Margaritis and Bajpai, 1982).

Post harvest diseases caused by bacteria affect quality and availability of fruit and vegetable (Wells et al., 1993). Bacterial pathogens involved in this respect include the species of soft-rotting *Erwinia*, *Pseudomonas*, *Xanthomonas*, *Cytophaga* and *Bacillus* (Liao and Wells, 1987; Lund, 1983).

---

*Corresponding author. E-mail: fowzy_2008@yahoo.com.

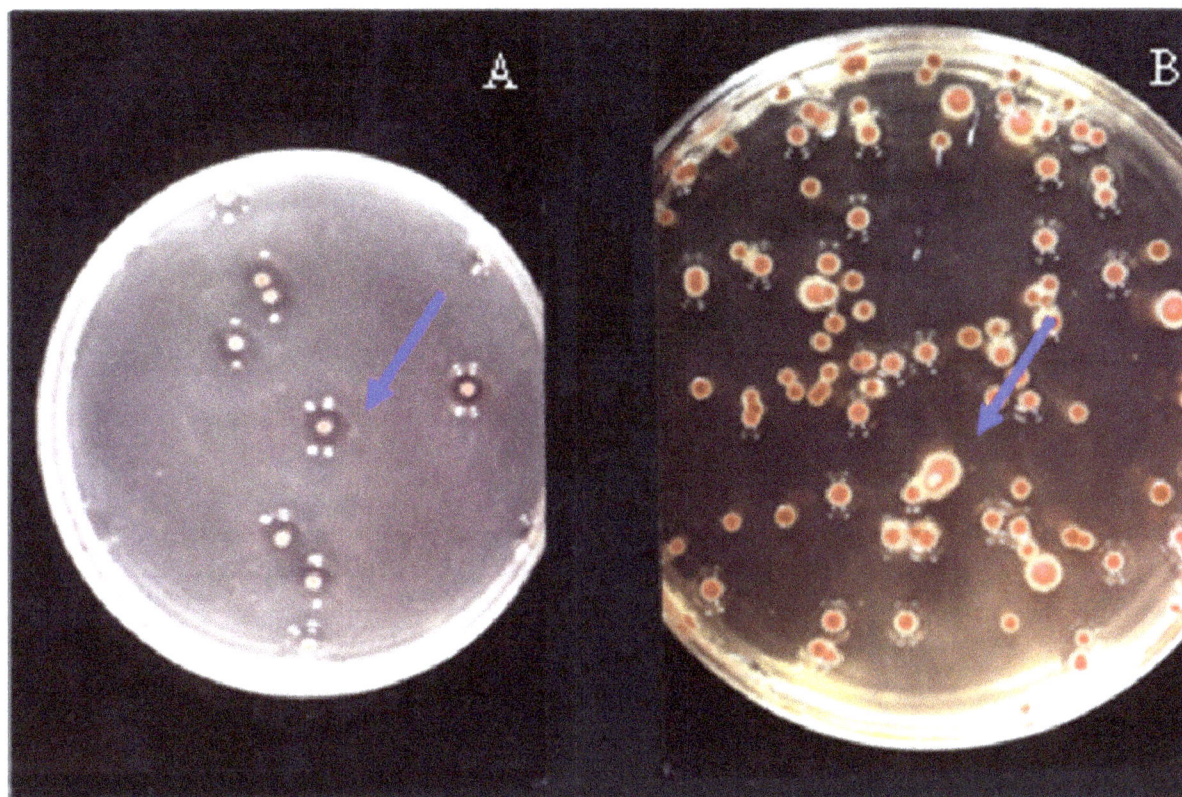

Figure 1. a) Cavity formation by soft-rot bacteria after incubation at 27°C for 24 h on the CVPM medium; b) recovery of soft-rot bacteria from artificially inoculated tubers after 24 h.

Plant diseases caused by plant pathogens are a complicated process because a number of factors play a part. However, direct involvements of pectic and cellulitic enzymes produced by the pathogen in pathogensis were reported (Gaber et al., 1990; Walker et al., 1994). Bacteria soft rot caused by *Erwinia carotovora* ssp. *carotovora* (Van Hall) Dye is one of the most important and widespread bacterial disease of a wide variety of plants either in the field or during storage (Hajhamed et al., 2007).

The objective of this investigation is to isolate and identify the pathogenic agent involved or associated with soft rot disease of girasol tubers at El-Minia governorate. Furthermore, the cell wall degradation enzymes in pathogenesis were discussed.

MATERIALS AND METHODS

Isolation

Infected girasole tubers showing typically developed soft rotting (Figure 1) were subjected for isolation. Samples of girasole tubers rot (cv. Balady) were collected from experimental farm of Faculty of Agriculture, Department of Horticulture, El-Minia University, during 2010 growing season; the isolation of the microorganism(s) associated with these symptoms was conducted. Diseased tubers were firstly washed with tap water then surface sterilized with 3% sodium hypochlorite solution (NaOCl) for 3 min then washed thoroughly 3 times with sterilized distilled water; the rotted tissues of tuber was put into sterilized mortar and homogenized then left to stand for 20 min then streaked into plates containing crystal violet pectate modified (CVPM) medium (Ahmed et al., 2000). The plates were incubated at 27±1°C for 48 to 72 h. Only bacterial colonies in deep cavities (Figure 2) were subcultured onto King's B medium and nutrient agar (NA) medium and stored on slants till they were used.

Identification of the causal pathogen

Six bacterial isolates - EC1, EC2, EC3, EC4, EC5 and EC6 - were identified by studying their morphological, physiological and biochemical characters as recommended by Breed et al. (1974), Sneath et al. (1986), Lelliott and Stead (1987), and Klement et al. (1990).

Pathogenicity tests on girasole tubers

The pathogenicity of the bacterial isolates was determined by inoculating intact unblemished healthy tubers. Each isolate was used to inoculate 5 tubers. This was done by making a 1 cm wound

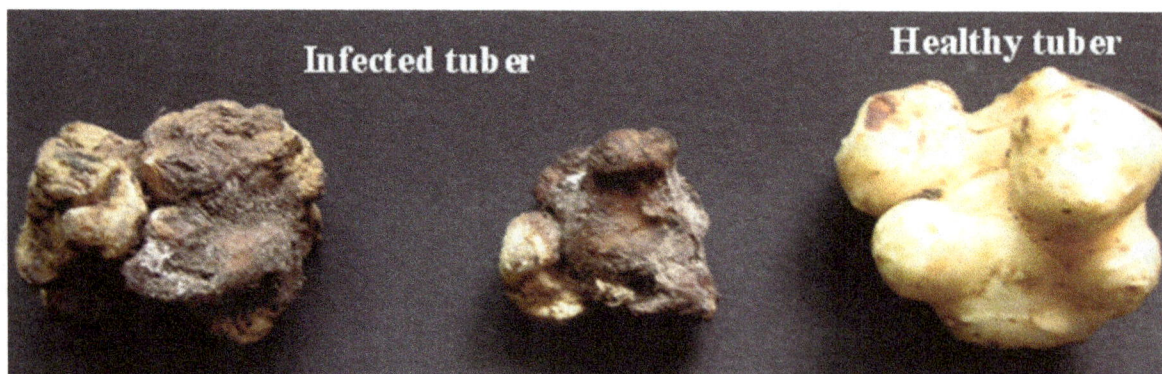

Figure 2. Naturally infected soft rot on girasole tubers cv. Balady infected by *E. carotovora* ssp. *carotovora*.

Table 1. List of plant species tested for their reaction to *Erwinia carotovora* sub sp. *carotovora* pathogen.

Hosts (common name)	Scientific name	Family name	Variety	Part organ
Apricot	*Prunus aremeniaca*	*Rosaceae*	Canino	Fruits
Bean	*Phaseolus vulgaris*	Leguminosae	Contender	Pods
Carrot	*Daucus carota*	Umbelliferae	Chantinay	Storage roots
Cowpea	*Vigna unguiculata*	Leguminosae	Black eye	Pods
Cucumber	*Cucumis sativus*	Cucurbitaceae	Balady	Fruits
Eggplant	*Solanum melogena*	Solanaceae	Black Beauty	Fruits
Lemon	*Citrus limon* (L.) Burm	Rutaceae	Balady	Fruits
Tobacco	*Nicotiana tabacum*	Solanaceae	Samsun	Leaves
Pepper	*Capsicum frutesences*	Solanaceae	Romy	Fruits
Radish	*Raphanus sativus*	Carucifera	Balady	Storage roots
Onion	*Allium cepa*	Amaryllidaceae	Giza 20	Storage onions
Squash	*Cucurbita pepo*	Cucurbitaceae	Eskandarani	Fruits
Sweet potato	*Ipomea batatas*	**Convolvulaceae**	Balady	Storage roots
Potato	*Solanm tuberosum*	Solanaceae	Diamant	Tubers
Tomato	*Lycopersicon esculentum*	Solanaceae	Super strain B	Fruits
Turnip	*Brassica rape*	Carucifera	White globe	Storage roots

in the middle of the tuber, and then inoculating by smearing the inside of the wound with an entomological needle filled with 48 h old cultures of the bacterial isolates grown on NA medium individually. Inoculated tubers (cv. Balady) were kept in sterilized boxes containing piece of sterilized distilled water-saturated cotton to insure high humidity and then incubated at 25±1°C. Seven days later, rot quantity and rot severity were assayed. The amount of rotten tissue produced in each tuber was determined and the percentage of rotten tissue was calculated and taken as a criterion of the pathogenicity to each isolate. Every tuber was weighed before and after removing the rotten portion, and calculation was done via the following formula put forward by Kelman and Dickey (1980):

Rot severity = (W1-W2)/W1 ×100

Where, W1= weight of whole tuber and W2= weight of tuber after removal of the rotten tissue.

Host range

The highly pathogenic isolate EC1 of the causal pathogen was inoculated into 16 plant species as listed in Table 1. Five plants were used in each treatment. Control treatments were similarly tested with sterile water only and kept at the same conditions. Disease severity was recoded after 7 days.

Assessment of some hydrolytic enzymes (cellulase and pectinase) in diseased and healthy girasole tubers

Assessment of pectinase and cellulase enzymes were assayed in tissue extracted from diseased and healthy tubers taken from the subjected plants during pathogenicity test.Half gram of either healthy and/or infected rot tissues were extracted and macerated separately with sterilized mortar containing 5 ml of 0.05 M phosphate buffer (pH 6). The homogenated tissue extracts were

Table 2. Morphological, biochemical and physiological characters of bacterial isolates.

Test	Bacterial isolates						Bradbury (1986)
	Ec1	Ec2	Ec3	Ec4	Ec5	Ec6	
Shape	Rod	Rod	Rod	Rod	Rod	Rod	Rod
Motility	+	+	+	+	+	+	+
Gram reaction	+	+	+	+	+	+	-
Pigment on $CaCO_3$ agar	-	-	-	-	-	-	-
Sporulation	+	+	+	+	+	+	-
Potato slices	-	-	-	-	-	-	+
Aerobiosis	+	+	+	+	+	+	F
Gelatin liquefication	+	+	+	+	+	+	+
Catalase production	+	+	+	+	+	+	-
Levan production	-	-	-	-	-	-	-
Indole formation	+	+	+	+	+	+	-
Tolerance 5, and 7% NaCl	+	+	+	+	+	+	+
Maximum temperature (°C)	35	35	35	35	35	35	30
Utilization of sugars from Arabinose	-	-	-	-	-	-	?
Galactose	+	+	+	+	+	+	+
Glucose	+	+	+	+	+	+	?
Lactose	+	+	+	+	+	+	+
Fructose	+	+	+	+	+	+	D
Insitol	+	+	+	+	+	+	+
Maltose	+	+	+	+	+	+	-
Mannitol	+	+	+	+	+	+	?
Mannose	+	+	+	+	+	+	?
Sorbitol	+	+	+	+	+	+	?
Trehalose	+	+	+	+	+	+	+
Celliobiose	+	+	+	+	+	+	+
Xylose	+	+	+	+	+	+	?
Raffinose	+	+	+	+	+	+	+
Sucrose	-	-	-	-	-	-	-

+ = all isolates are positive, (+) = weakly reaction, - = negative reaction, D = isolates differed, F = facultative anaerobic and ? = isolates not tested.

filtered through two layers of cheese-cloth, cooled to temperature near zero then centrifuged at 5000 rpm for 20 min. The clarified enzyme preparation of healthy and infected tissues was directly subjected to the viscometrical assessment according to Mahadevan and Sridhar (1982).

Total carbohydrate and reducing sugars in healthy and artificially inoculated girasole tubers

Total carbohydrates

The phenol-sulphuric acid method was used for determining the total sugars in clarified tissue extract as described by Hodge and Horfreir (1962).

Determination of reducing sugars

Determination of reducing sugars was performed according to the methods of Somogyi (1952).

Statistical analysis

All experiments were performed twice. Analyses of variance were carried out using MSTAT-C program version 2.10 (1991). Least significant difference (LSD) was employed to test for significant difference between treatments at P≤0.05 (Gomez and Gomez, 1984).

RESULTS AND DISCUSSION

Isolation and identification of the causal organisms

Six isolates of creamy-white bacteria were isolated from girasole plants showing typical tuber rot symptom (Figure 1). Regardless of some slight differences in certain characteristics, all bacterial isolates appeared to be representative of *E. carotovora* ssp. *carotovora* (Table 2) according to the description of Bergey's Manual of

Table 3. Pathogenicity of *E. carotovora* sub sp. *carotovora* on girasole tubers.

Isolates	Rot weight (gm)	Rot severity(% rotted tissues)
Ec1	1.40[a]	92.0
Ec2	0.99	41.3
Ec3	1.23	74.6
Ec4	0.89	23.6
Ec5	0.68	21.3
Ec6	1.07	57.3
L.S.D. at 5%	0.12	5.88

[a] Mean of five replicates; calculated as percentage of rotted tissues.

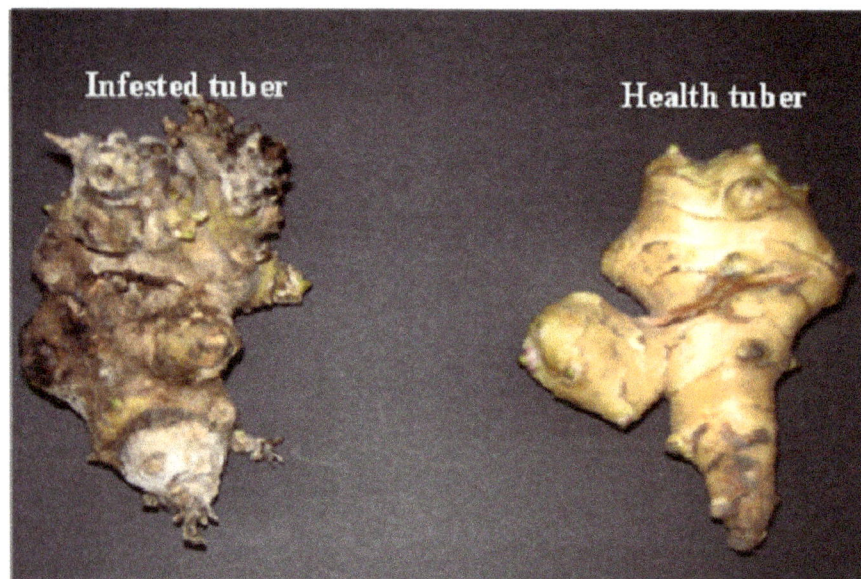

Figure 3. Artificially infested soft rot on girasole tubers cv. Balady infected by *E. carotovora* sub sp. *carotovora*.

Determinative Bacteriology (1974) and Bergey's Manual of Systematic Bacteriology (Sneath et al., 1986). However, the tests were carried out as described by Lelliott and Stead (1987), and Klement et al. (1990). The results of the present work revealed that all the tested bacterial isolates are rod-shaped, motile, gram negative and non-spore forming, and they grow at 35°C. They are facultative anaerobic, negatively react with phosphatase production, indol formation and H_2S production, while they positively react with gelatine liquefaction, rot of potato and carrot slices; however, they grow in the presence of 5% NaCl, and reduce nitrate reduction. All the tested isolates produce deep cavities semi-selective (CVPM) medium. Otherwise, the bacterial isolates utilized glucose, galactose, fructose, cellubiose, lactose, mannitol, raffinose, trehalose, mannose and xylose but they did not utilize arabinose, maltose, sorbitol, sucrose and methy glucoside. Comparison of the characters of the isolated bacteria with those reported by Dye et al. (1980) and Dickey (1981) could be identified as *E. carotovora* ssp. *carotovora*.

Pathogenicity tests

Data presented in Table 3 indicate that all bacterial isolates under investigation were able to infect girasole tubers and induce soft rot, although they varied in severity of rot they initiated. Inoculation with any of these isolates showed disease symptoms appearing as soft rot at wounded sites, and eventually collapsed within two weeks. However, the control plants remained unaffected. Soft rot symptoms (Figure 3) sites of inoculation were obvious 5 to 10 days after inoculation, whereas from 10

Table 4. Symptoms expression with 6 bacterial isolates of *E. carotovora* ssp. *carotovora* on different hosts.

Hosts bacterial isolates	Site of inoculation	Erwinia carotovora ssp. carotovora isolates					
		Ec1	Ec2	Ec3	Ec4	Ec5	Ec6
Apricot	Fruits	+[a]	+	+	+	+	+
Bean	Pods	-	-	-	-	-	-
Carrot	Storage root	+	+	+	+	+	+
Cowpea	Pods	-	-	-	-	-	-
Cucumber	Fruit	+	+	+	+	+	+
Eggplant	Fruits	-	-	-	-	-	-
Lemon	Fruits	+	+	+	+	+	+
Tobacco	Leaves [b]	+	+	+	+	+	+
Pepper	Fruits	-	-	-	-	-	-
Radish	Storage roots	+	+	+	+	+	+
Onion	Leaves	+	+	+	+	+	+
Squash	Fruits	+	+	+	+	+	+
Sweet potato	Storage roots	-	-	-	-	-	-
Potato	Tubers	+	+	+	+	+	+
Tomato	Fruits	-	-	-	-	-	-
Turnip	Storage root	+	+	+	+	+	+

[a] Data are means of 5 replicates per treatment; [b] hypersensitive reaction (HR).

Table 5. Effect of extract of diseased girasole tubers on percentage of viscosity of 1% citrus pectin solution during incubation for 3 h at room temperature.

Time(min)	% Loss in viscosity of 1% citrus pectin					
	Erwinia carotovora sub sp. carotovora isolates					
	Ec1	Ec 2	Ec 3	Ec 4	Ec 5	Ec 6
0.0	6.8 [a]	5.4	5.4	2.4	2.4	2.0
30	19.7	8.2	13.2	14.5	6.5	7.1
60	33.2	11.8	20.8	17.8	15.8	17.7
120	38.0	21.4	30.4	21.3	17.3	23.6
180	38.0	21.4	30.4	21.3	17.3	23.9

[a] Values are mean of 3 replicates.

to 15 days, the tubers were collapsed. Amount of rotting, also rated from 22.2 and 42.2% after 21 days from incubation. Also, the obtained results indicate that isolate Ec1 and Ec3 could be considered as highly pathogenic, whereas other isolates were weak virulent. Several authors reported that *Erwinia chrysanthemi* and *E. carotovora* ssp. *carotovora* were isolated from different plants and caused soft rot diseases (Liu et al., 2002; Scortichni and Ascenzo, 2003; Hajhamed et al., 2007).

Host range

Results in Table 4 show that all isolates produced soft

rot on most different plant tested. On the other hand, plants such as pods of cowpea, bean and fruits of eggplant, pepper, sweet potato and tomato are not affected by inoculated bacteria.

Production of pectolytic and cellulolytic activity by *E. chrysanthemi in vivo*

All tested isolates were active in secreting pectolytic and cellulolytic enzymes in tuber tissues of girasole plants after 10 days of inoculation (Tables 5 and 6), whereas the isolate Ec1 (more virulent) was higher after 180 min than their activities with the weakly virulent isolate (Ec5). These results confirmed those reported by Saleh (1995),

Table 6. Effect of extract of diseased girasole tubers on percentage loss of viscosity of 1% carboxymethyl cellulose (CMC) solution during incubation for 3 h at room temperature.

Time(min)	% Loss in viscosity of 1% carboxymethyl cellulose (CMC) solution					
	Erwinia carotovora sub sp. *carotovora* isolates					
	Ec1	Ec 2	Ec 3	Ec 4	Ec 5	Ec 6
0.0	22.7[a]	15.5	16.7	11.1	0.0	10.9
30	35.4	22.1	23.7	23.0	12.2	17.8
60	43.0	28.7	39.5	27.5	20.3	22.4
120	57.4	36.1	42.6	32.6	28.9	33.4
180	57.4	36.1	43.90	32.9	28.9	33.6

[a] Values are mean of 3 replicates.

Table 7. Total carbohydrate and reducing sugars in healthy and diseased tissue extracts of girasole plants inoculated with bacterial isolates.

Treatment	Total carbohydrate (mg/g fresh weight)	Reducing sugars(mg/g fresh weight)
Control (healthy tissue)	36.22 ± 1.4[a]	20.42 ± 3.3
Isolate Ec1	95.44 ± 3.4	17.82 ± 3.0
Isolate Ec2	44.13 ± 3.2	10.19 ± 2.0
Isolate Ec3	82.85 ± 2.3	13.23 ± 2.4
Isolate Ec4	41.17 ± 3.0	11.22 ± 5.0
Isolate Ec5	35.27 ± 1.0	9.12 ± 4.1
Isolate Ec6	39.12 ± 3.0	7.33 ± 3.5

[a] Data are means of 3 replicates ± SD.

Ouf et al. (1997), and Galal et al. (2002). They reported that the pectolytic activity of the enzymes were higher in infected tissues than in the healthy ones.

Activity of these enzymes was higher in infected tissue than in healthy ones. Data indicated that the highest activity was shown after 2 h incubation at room temperature. These results are generally in line with those reported by previous investigators (Ouf and El-Sadek, 1997; Ouf et al., 1997). Similar results were reported for *Bacillus subtilis* and *E. chrysanthemi* causing soft rot of carrot roots (Kararah et al., 1985; Saleh and Gabr, 1989; Saleh, 1995; Saleh and Stead, 2003). Severin et al. (1985) reported that *E. carotovora* pv. *carotovora*, *E. carotovora* ssp. *atroseptica*, *E. chrysanthemi* ssp. *chrysanthemi* and *Xanthomonas campestris* pv. *pelargonii* (the causal pathogens of soft rot of potato, dahlia and pelargonium, respectively) were able to produce pectinase(s) and cellulase(s) enzymes.

Generally, the data in Table 7 indicate that total carbohydrates were much higher in inoculated tissue extracts than in healthy ones particularly with isolate Ec1 (more virulent). Similar trends were obtained with reducing sugars.

Data presented by Saleh (1995) indicate a similar effect of the pathogen (*B. subtilis* and *B. pumilus*) on total carbohydrates in infected tissues.

REFERENCES

Ahmed ME, Mavridis A, Rudolph K (2000). Detection of latent contamination of potato tubers by *Erwinia spp.* using a semi-selective agar medium. Phytomedizin, Mitteilung. Deutsch. Phytomedizin. Gesellsch, 30(1): 15-16.

Breed RS, Murray EGD, Smith NR (1974). Bergey's Manual of Determinative Bacteriology 8th ed. The Williams and Williams Company, Baltimore, p. 1268.

Dickey DS (1981). Erwinia chrysanthemi: Reaction of eight plant species to strains from several hosts and to strains of other Erwinia species. Phytopathology, 71: 23-29.

Dye DW, Bradbury JF, Goto M, Hayward AC, Lelliott RA, Schroth MN (1980). International standards for naming pathovars of phytopathogenic bacteria and a list of pathovar names and pathotype strains. Rev. Plant Pathol., 59: 163-168.

Gaber MR, Saleh OI, Abo El-Fotouh E (1990). Pectolytic, cellulolytic and proteolytic activities of two isolates of *Botryodiplodia theobramae*. Annals Agric. Sci., Fac. Agric., Ain Shams Univ., Cairo, Egypt. 35(1): 445-457.

Hodge JE, Horfreir BT (1962). Determination of reducing sugars and carbohydrates (In: Methods in Carbohydrates Chemistry,. Academic Press, New York, 1: 380-394.

Galal AA, Abdel-Gawaed, TI, El-Bana AA (2002). Post – harvest decay of garlic cloves caused by Bacillus polymyxa and Fusarium moniliforme. Egypt. J. Microbiol., 36: 71-88.

Gomez KA, Gomez AA (1984). Statistical Procedures for Agricultural Research. Wiley, Interscience Publication New–York, p. 678.

Hajhamed AA, Abd El-Sayed W. M, Abou El-Yazied A, Abd-El-Ghaffar NY (2007). Suppression of bacterial soft rot disease of potato. Egyptian J. Phytopath., 35(2): 69-80.

Kararah MA, Barakat FM, Mikhail MS, Fouly HM (1985). Pathophysiology in garlic cloves inoculated with Bacillus subtilis, Bacillus pumilus, and Erwinia carotovora. Egyptian J. Phytopathol. 17(2): 131-140.

Kelman A, Dickey RS (1980). Soft rot or "carotovora" group. In: Schaad. N.W. (ed) Laboratory "Guide gor Identification of Plant Pathogenic Bacteria APS St" Paul Mim. USA, pp. 31-35.

Klement Z, Rudolph K, Sands DC (1990). Methods in Phytobacteriology Akademiai, Kiado, Budapest, p. 568.

Lelliott RA, Stead DE (1987). Method in Plant Pathology, Vol. II. Method for the Diagnosis of Bacterial Diseases of Plants,. Blackwell Scientfic Publication, Oxford, London, Edinburg, pp. 1-216.

Liao CB, Wells JM (1987). Diversity of pectolytic, fluorescence pseudomonas causing soft rot of fresh vegetables at produce market. Phytopathology, 77: 673- 677.

Liu HL, Hsu ST, Tzeng KC (2002). Bacterial soft rot of chrysanthemum cuttings: characteristics of the pathogen and factors affecting its occurrence-Plant Pathol. Bull., 11(3):157-164.

Lund BM (1983). Bacterial spoilage.. In: Postharvest Pathology of Fruits and Vegetables. (C. Demis, ED. Academic press, London), p. 219-257.

Mahadevan A, Sridhar R (1982). Methods in Physiological Plant Pathology, Sivakami publications, India, p. 119.

Margaritis A, Bajpai P (1982). Continuous ethanol production from Jerusalem artichoke tubers. II. Use of immobilized cells of Kluyveromyces marxianus. Biotechnol. Bioengineering, 24: 1483-1493.

Ouf MF, El-Sadek SAM (1997). Bacterial stem rot of dieffenbachia in Egypt. Egyptian J. Microbiol., 32(2): 269-281.

Ouf MF, El-Sadek SAM, Abdel-Latif MR, Abd El-Aziz N (1997). Pectolytic and cellulolytic activity of enzymes by Erwinia dieffenbachia, causing stem soft rot on dieffenbachia plants. Egyptian J. Microbiol., 32(1): 99-115.

Saleh OI (1995). Identification of phytopathogenic bacteria associated with post harvest disease of garlic cloves in relation to cell wall degrading enzymes. Egyptian J. Microbiol., 30: 177-202.

Saleh OI, Stead D (2003). Bacterial soft rot disease of pea in Egypt. Integrated control in protected crops, Mediterranean climate IOBC WPRS Bulletin,. Editura Ceres, Bucharst, Romania, 26(10): 115-120, 1-218.

Saleh OI, Gabr MR (1989): Studies on core rot of carrot in relation to cell wall- degrading. Minia J. Agric. Res. Dev., 11: 1713-1722.

Scortichini M, Dascenzo D (2003). New bacterial diseases striking ornamentals. Colture Protelle 32(5): 81-84.

Severin V, Kupferberg S, Zurini I (1985). Plant Pathogenic Bacteria,. Editura Ceres, Bucharst, Romania. pp. 1-218.

Sneath PHA, Mair NS, Sharpe ME, Holt JG (1986). Bergey's Manual of Systematic Bacteriology. Williams Wilkins Co., Baltimore, p. 1123.

Somogyi M (1952). Notes on sugar determination. J. Biological Chemistry 195: 18-23.

Walker DS, Reeves PJ, Salmond GP (1994). The major secreted cellulose, CelV, of Erwinia carotovora sub sp. carotovora is an important soft rot virulence factor. MPMI, 7(3): 425-431.

Wells JE, Butterfield JE, Revear LG (1993). Identification of bacteria associated with postharvest diseases of fruits and vegetables by cellular fatty acid composition: An expert system for personal computers. Phytopathology, 83: 445-455.

18

Retrospective and clinical studies of Marek's disease in Zaria, Nigeria

Musa I. W.[1]*, **Bisalla M.**[2], **Mohammed B.**[3], **Sa'idu L.**[4] **and Abdu P. A.**[1]

[1]Department of Veterinary Surgery and Medicine, Ahmadu Bello University, Zaria, Kaduna State, Nigeria.
[2]Department of Veterinary Pathology and Microbiology, Ahmadu Bello University, Zaria, Kaduna State, Nigeria.
[3]Department of Veterinary Public Health and Preventive Medicine, Ahmadu Bello University, Zaria, Kaduna State, Nigeria.
[4]Veterinary Teaching Hospital, Ahmadu Bello University, Zaria, Kaduna State, Nigeria.

A ten year retrospective study (2000-2009) of Marek's disease (MD) was conducted in the Veterinary Teaching Hospital, Ahmadu Bello University, Zaria. 3,039 different poultry diseases were recorded, MD represents 4.9%. 63% of MD was recorded in birds 10 to 20 weeks old, 9.3% in birds above 30 weeks and 2.0% in birds below ten weeks. MD occurred mainly during the pre-rainy to rainy seasons with a progressive yearly increase. Strong association existed between season and MD ($p < 0.05$), risk estimate was also relevant (OR-2.4). A clinical analysis of birds affected with MD over the years revealed uneven growth and progressive weight loss as major complaints; ruffled feathers and whitish-yellow diarrhea were the major observable signs. Major gross lesions were severe emaciation, thickened proventriculus and flabby heart with loss of coronary fats. Histopathology revealed focal to diffuse neoplastic lesions in the affected organs. MD mainly affected chickens aged 10 to 20 weeks. The disease is endemic and on the increase in Zaria and its environs. We recommend that an effective MD vaccination technique and schedule be established for this region, purchase of point of lay chickens for production should be discouraged and standard biosecurity measures must be enhanced in hatcheries and farms to prevent primary exposure.

Key words: Marek's disease, occurrence, diagnosis, Zaria.

INTRODUCTION

Consumption of poultry and poultry products in most developing countries has grown by 5.8%, faster than that of human population growth; as a result, the poultry industry has over the years recorded dramatic changes (David, 2000). There is no doubt that high rate of intensification may suffer management defects by deteriorating standards, interplay between environment, host and organism genes, slow responses to prevention and treatment that may ultimately subject a large number of birds to further suffering with a consequent high rate of disease spread (David, 2000; Frank, 2001). Almost all

industrialized nations had at one time experienced losses due to Marek's disease (MD) in their poultry industry and a crude estimate of the cost of Marek's disease is said to be in the range of several billion US dollars (Calnek and Witter, 1997; David, 2000; Frank, 2001; Ionica and Comand, 2009; Katherine et al., 2011). Marek's disease is common in intensively managed commercial layer type chickens (Cauchy and Coudert, 1986; Dong et al., 2006; Graham, 1976; Stephen et al., 2008), though few reports have been documented in broilers and geese in some countries (Payne, 2004; Ionica and Coman, 2009). It is a transmissible lymphoproliferative disease of primarily chicken that is caused by an alpha herpes virus and characterized by malignant tumour formations in internal organs, skin, eye and peripheral nerves (Calnek and Witter, 1997; David, 2000; Dong et al., 2008; Graham,

*Corresponding author. E-mail: ibwazkalt@yahoo.co.uk.

1976; Stephen et al., 2008). The Marek's disease virus (MDV) is cell associated in body organs and tumors, it replicates and exists as enveloped free form in the feather follicles making feathers particularly dander, dust and litter materials loaded with MD virus, thus facilitating virus transmission by air borne route (Calnek and Witter, 1997; Adene and Akpavie, 2004; Frank, 2001). Susceptible chickens infected with the pathogenic MDV suffer cytolysis of the lymphoid organs and a concomitant immunosuppression (Gordon, 1976; Frank, 2001). Such birds mainly die as a result of tumour development in the visceral organs and peripheral nerves (Frank, 2001). Though morbidity and mortality vary with virus strain (Graham, 1976; Cauchy and Coudert, 1986; Dong et al., 2008; Ionica and Coman, 2009), it is a major economic risk for poultry flocks, not only because of its worldwide distribution, but because it majorly affects young adults about to be utilized for meat or egg production, thereby reducing the profitability of an affected flock (Fatumbi and Adene, 1984; Frank, 2001).

Single vaccination against MD either in ovo or at day of hatch in genetically susceptible stock has been effectively used to control the occurrence of clinical disease in the recent pass (Gordon, 1979; Powel, 1981; Frank, 2001). But more recently, double vaccination against MD has also been reported to be more effective (Payne, 2004). However, the evolution of very virulent MD virus strains, inevitable exposure of flocks to other immunosuppressive diseases or conditions, exposure of birds to Marek's disease virus before vaccinal immunity develops and inappropriate vaccine handling and vaccination procedures have resulted to "vaccine break and failure" leading to increasing number of MD outbreaks despite consistent vaccination in many countries (Gordon, 1979; Frank, 2001; Payne, 2004). Another problem associated with MD control through vaccination is evaluating vaccine induced protection following vaccine administration (Frank, 2001).

The poultry industry in Nigeria is fast developing but with few hatcheries concentrated in some parts of the country (Fatumbi and Adene, 1984; Geidam et al., 2006; Olabode et al., 2009). Many outbreaks of poultry diseases have been directly or indirectly associated with high poultry farm concentrations (Cauchy and Coudert, 1986; Calnek and Witter, 1997; Frank, 2001). Such areas pose great risk of disease transmission even in the face of vaccination (Calnek and Witter, 1997). Few studies conducted on MD in Nigeria appear not to adequately address increasing MD outbreaks nationwide (Fatumbi and Adene, 1984; Adene and Akpavie, 2004; Akpavie, 2005; Olabode et al., 2009). A very low prevalence was recorded in Ibadan (Fatumbi and Adene, 1984), a 2.14% prevalence rate was recorded for avian neoplastic diseases (MD inclusive) in Ilorin, Nigeria (Geidam et al., 2006), with the highest year of occurrence in 1998. From the available literature, this is the first paper that describes adequately the status of MD in Zaria, Nigeria

as of 2009.

MATERIALS AND METHODS

Study area

This study was conducted in Zaria which is located between longitude 11°07'N and latitude 7°44'E within the Sudan Savannah Zone. The average rainfall ranges from 1000 to 1250 mm, diurnal temperature range of 17 to 33°C and vegetation cover of predominantly trees and grasses.

Sample collection

Carcases or live birds submitted to our poultry clinic that originated from mainly Kaduna State and occasionally from other neighbouring States (Kano, Katsina, Zamfara, Sokoto, Kebbi, Niger and Federal Capital Territory (FCT) Abuja) were used for the study. Each carcass was subjected to routine post-mortem examination paying attention to changes in size, colour and the transactional appearances that may be suggestive of neoplastic growth (Fatumbi and Adene, 1984). Two or more peripheral nerves were grossly examined and processed histopathologically to differentiate MD from avian leukosis (AL) (Fatumbi and Adene, 1984). Samples of liver, spleen, kidneys, proventriculus, skin, eye, muscle, schiatic nerve, gonads and bursas of fabricius either grossly affected by tumours or not were collected for histopathology examination. These samples were taken to Histopathology Laboratory of the Faculty of Veterinary Medicine, Ahmadu Bello University, Zaria for processing. Flock histories, clinical and pathological lesions observed in reported outbreaks stood as our bases to tentatively diagnose MD, while histopathology findings complemented the diagnoses.

Histopathology

Tissue samples collected were fixed in 10% neutral buffered formalin for 3 days, embedded in paraffin and sectioned with microtome (5 μm). These were stained with haematoxylin eosin (HE) and examined under light microscope at 25 and 40x magnifications for evidence of histological changes (Fatumbi and Adene, 1984; Baker et al., 2001; Ionica and Coman, 2009).

Data collection

A ten-year clinic record (2000-2009) of the Veterinary Teaching Hospital, Ahmadu Bello University, Zaria was utilised for this study. Data on season, year, month, breed and age were extracted. A case was considered as a farm that reported outbreak of a disease and diagnosed based on the history, clinical signs, post mortem findings and laboratory results. The age of birds in this study were categorised as follows: (i) 1 to 10 weeks (ii) 10 to 20 weeks (iii) 20 to 30 weeks (iv) >30 weeks while seasons in Zaria and environs were categorised as dry (January-March), pre-rainy (April-June), rainy (July-August), pre-dry (October-November) (Abdu et al., 1992). The monthly distribution of MD for the period under study (2000-2009) was reduced using the 10-year data to one year using the 12 months ratios to moving average method (Sa'idu et al., 2004).

Data analysis

Data generated were analysed statistically using Win-Episcope and

Table 1. Yearly distribution of MD and other poultry diseases in Zaria.

Year	MD	NMD	Year spec. rate	OR	95% CI on OR
2000	4	213	1.9	1.8	0.53- 6.46
2001	6	350	1.7	1.4	0.41- 4.47
2002	5	394	1.2	1	
2003	4	150	2.7	2.1	0.56- 7.93
2004	5	214	2.3	1.8	0.53- 6.43
2005	8	182	4.4	3.5	1.12- 10.73
2006	12	184	6.5	5.1	1.78- 14. 80
2007	19	296	6.4	5.1	1.87- 13.70
2008	39	509	7.7	6.0	2.36- 15.46
2009	48	397	12.1	9.5	3.75- 24.18
2001	6	350	1.7	1.4	0.41- 4.47

*Comparison point; MD- Marek's disease; NMD- Non-Marek's disease; spec. specific.

Table 2. Monthly distribution of MD and other poultry disease in Zaria.

Month	MD	NMD	Month spec. Rate	OR	95% CI on OR
Jan	11	211	5.2	2.4	0.94-5.97
Feb	11	152	7.2	3.2	1.28-8.25
Mar	12	249	4.8	2.2	0.87-5.37
Apr	17	261	6.5	2.9	1.25-6.93
May	15	294	5.1	2.3	0.96-5.48
Jun	25	310	8.1	3.6	1.61-8.14
Jul	9	326	2.8	1.2	0.47-3.25
Aug	10	359	2.2	1	
Sep	16	339	4.7	2.1	0.90-5.01
Oct	8	213	3.7	1.7	0.63-4.60
Nov	7	174	2.9	1.3	0.42-4.00
Dec	9	151	6.0	2.7	1.03-7.16
Total	150	3039			

*Point of comparison; MD- Marek's disease; NMD- Non-Marek's disease; spec. specific.

SPSS statistical package (version 16.0) for windows.

RESULTS

Table 1 shows a total of 150 MD cases out of 3,039 poultry diseases recorded over a period of ten years (2000-2009) amounting to a 4.9% MD prevalence in Zaria, Nigeria. The incidence of MD was lowest (1.7) in the year 2001 and progressively increased to (12.1%) in the year 2009.

The pre-rainy (2.1%) and rainy (1.4%) seasons in this study appear to have more cases of MD (Table 3). Two high monthly incidences occurred in February (7.2%) and June (8.1%) (Table 2). The favourable age for MD occurrence was seen to be 11 to 20 weeks (68%) (Table 4). Pearsons chi-square test established significant association between season and the disease occurrence ($p<0.05$), while estimated relative risk showed the pre-rainy season to be most significant (odds ratio-2.4) using the pre-dry season as reference point (Table 3, $X^2 = 18.8$; df = 3). All the cases involved commercial improved breeds of chickens and no case of MD was recorded in the indigenous breeds of chickens or even wild birds. Table 5 shows the common and rare clinical and pathological lesions associated with Marek's disease in the study area. History revealed that farmers usually began to suspect a problem in their flocks when they noticed significant non-uniform growth and light weight in the flocks despite adequate feeding. From history, vaccination against MD was almost always questionable because farmers believed that MD vaccination was the

Table 3. Seasonal distribution of MD in Zaria.

95% CI on OR	OR	Season specific rate	NMD	MD	Season
0.67-2.12	1.2	0.9	603	25	Dry(Jan-Mar)
1.46-3.90	2.4	2.1	746	62	Pre-rainy(Apr-Jun)
0.78-2.21	1.3	1.4	879	40	Rainy(Jul-Sep)
	1	0.8	661	23	Pre-dry(Oct-Dec)

Table 4. Age and breed specific rates of MD.

Age (weeks)	MD	Age/breed specific rate
1-10	3	2.3
11-20	102	68.0
21-30	31	20.7
>30	14	9.3
Breed		
Improved	150	100
Local	0	

responsibility of the hatchery, while few farmers revaccinated irrespective of whether MD vaccine was given at the hatchery or not. The consistent clinical signs observed where whitish-yellow diarrhoea and ruffled feathers; nervous signs which is characteristic of MD were rarely seen. The gross lesions observed in most cases were severe emaciation, flabby heart, distended gall bladder with bile and thickened proventriculus (Figures 1 to 3). Tumour growths in affected organs were also rarely seen. Microscopically, focal to diffuse infiltration or proliferation of immature and mature lymphocytes in the affected organs were consistent lesions (Plates 1 and 2).

DISCUSSION

Since the 1960s, increasingly virulent strains of MDV have caused continued poultry industry production losses worldwide (Graham, 1976; Cauchy and Coudert, 1986; Payne, 2004; Dong et al., 2008; Stephen et al., 2008; Katherine et al., 2011). In developing countries in addition, improper handling and administration of MD vaccines contribute to continuous outbreaks. This is because Marek's disease vaccine especially the cell associated type is the most sensitive vaccine requiring storage at -196°C until use, and administration of reconstituted vaccine must be within the first two hours (Frank, 2001). However, the prevalence of MD is generally considered to be low and rarely exceeds 5% (Fatumbi and Adene, 1984; Frank, 2001; Dong et al., 2008). A study conducted in 2009 established a 2.14% prevalence rate of avian neoplasm in Ilorin, Nigeria (Olabode et al., 2009). Within the same year in this study,

a prevalence rate of 4.9% for MD is established in Zaria. We have observed in addition, a progressive increase of MD incidences over the years with the lowest incidence (2.5%) in 2000 and highest (12.1%) in 2009. Accordingly, it has further been documented that for most countries of the world, MD outbreaks have been on the increase despite vaccination (Frank, 2001; Adene and Akpavie, 2004).

One other major issue of continuous MD outbreaks in the developing countries is that of vaccine handling. A prevalence study of MD in indigenous and exotic chickens had 2 peaks of high prevalence in 1977 and 1983 in Nigeria, while 3 times rise of MD outbreaks occurred between 1976 and 1982 in the US (Adene and Akpavie, 2004). In our study, we noticed 2 peaks of high monthly incidences in February (7.2%) and June (8.1%) in Zaria. In support of this, the incidence of MD continued to increase at intervals of few years worldwide despite increasing efforts to control it due mainly to MD virus mutations leading to its increasing virulence (Payne, 2004; Katherine et al., 2011). This allows MD virus mutants to break through the protection afforded by the vaccine in use (Payne, 2004).

From this study, the highest likelihood of MD occurrence in Zaria is in the pre-rainy season which may be due to the fact that in the study area the dry season that precedes the pre-rainy is characterised by cold temperatures and high wind velocity which usually favour virus survival and the most important route (air borne) of MDV transmission (Fatumbi and Adene, 1984; Frank, 2001) and since MD has an incubation period of at least 3 weeks to several months, the exposure of most birds to MDV may have occurred during the pre-dry to dry seasons followed by the manifestation of lesions in the pre-rainy and rainy seasons. In addition, the pre-rainy season in the study area is characterised by high ambient temperatures which is a serious stressful factor to birds and may have devastating effects on the birds' immune system.

The age group mainly affected by MD in this study is 11 to 20 weeks (68%). This is in agreement with the findings of many authors where 12 to 24 weeks were the age range mostly affected by clinical MD manifestations and MD rarely occurred above 24 weeks (Fatumbi and Adene, 1984; Cauchy and Coudert, 1986; Adene and Akpavie, 2004; Akpavie, 2005). This observation may not be unconnected with the long incubation period of MD

Table 5. Clinical and pathological features of MD in randomly selected poultry farms located in Zaria and environs.

Case features	Case location (farms)									
	A	B	C	E	F	G	H	I	J	K
Case history										
Uneven growth	+	+	+	+	+	+	+	+	+	+
Light weight	+	+	+	+	+	+	+	+	+	+
MD vaccination status	+	?	?	?	?	?	+	?	?	?
Clinical signs										
Whitish-yellow diarrhoea	+	+	+	+	+	+	+	+	+	+
Nervous signs	-	-	+	+	-	-	-	-	+	-
Ruffled feathers	+	+	+	+	+	+	+	+	+	+
Inconsistent signs	+	+	+	+	+	+	+	+	+	+
Gross lesions										
Emaciated carcass	+	+	+	+	+	+	+	+	+	+
Distended gall bladder	+	+	+	+	+	+	+	+	+	+
Enlarged/thickened proventriculus	+	+	+	+	+	+	+	+	+	+
Flabby heart with loss of coronary fats	+	+	+	+	+	+	+	+	+	+
Tumours in affected organs	+	-	-	+	+	+	-	+	-	-
Inconsistent lesions	+	+	+	+	+	+	+	+	+	+
Histopathology										
Focal to diffuse organ infiltration/proliferation of lymphocytes, lymphoblasts, plasma cells	+	+	+	+	+	+	+	+	+	+

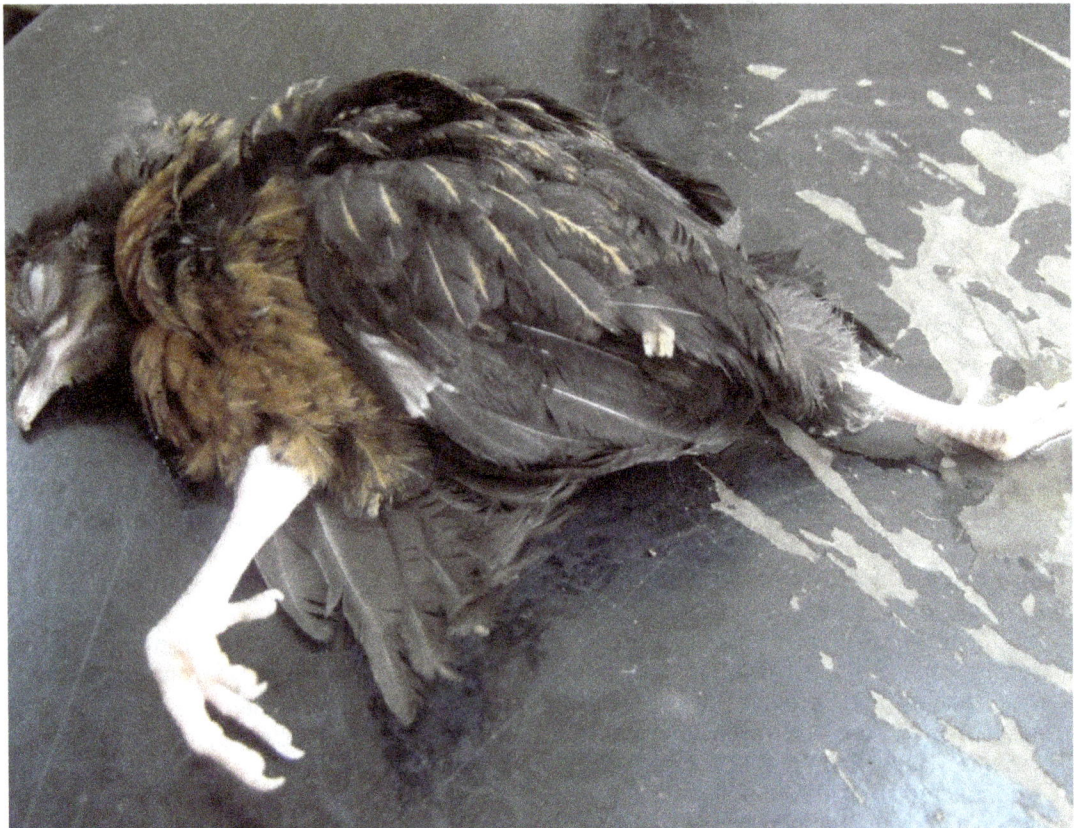

Figure 1. Marek's disease infected 6 week-old commercial chicken showing leg paralysis and ruffled feathers.

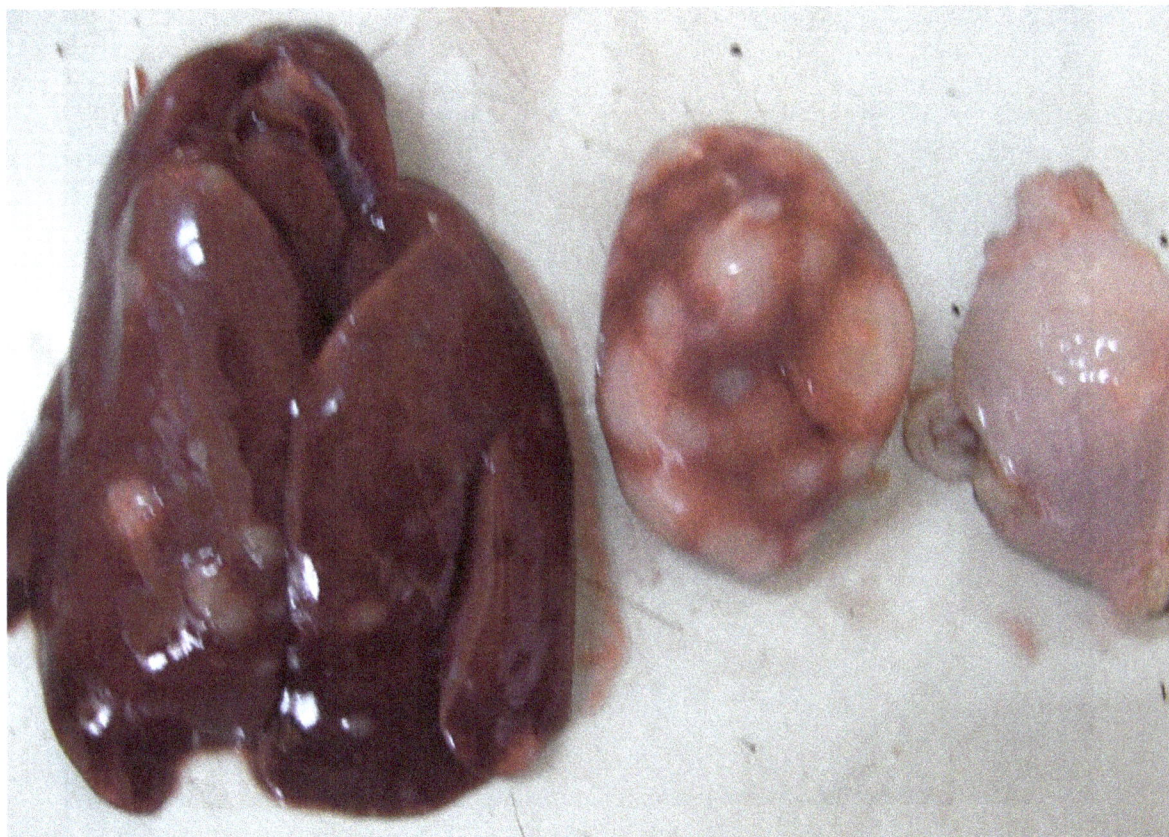

Figure 2. From left to right: whitish nodular neoplastic growths in the liver and spleen and enlarged proventriculus in MD infected 17 week-old commercial layer chicken.

Figure 3. MD affected 3 week-old pullet showing bilateral leg paralysis.

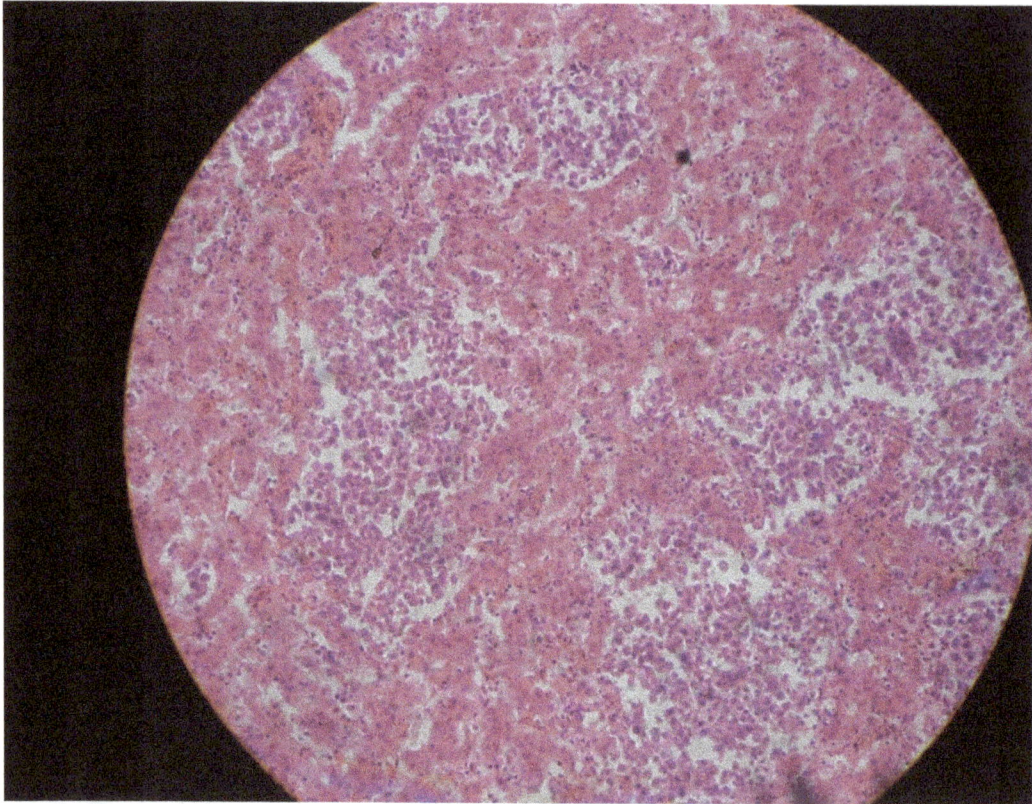

Plate 1. Diffuse infiltration of immature lymphocytes in the liver of MD infected chicken.

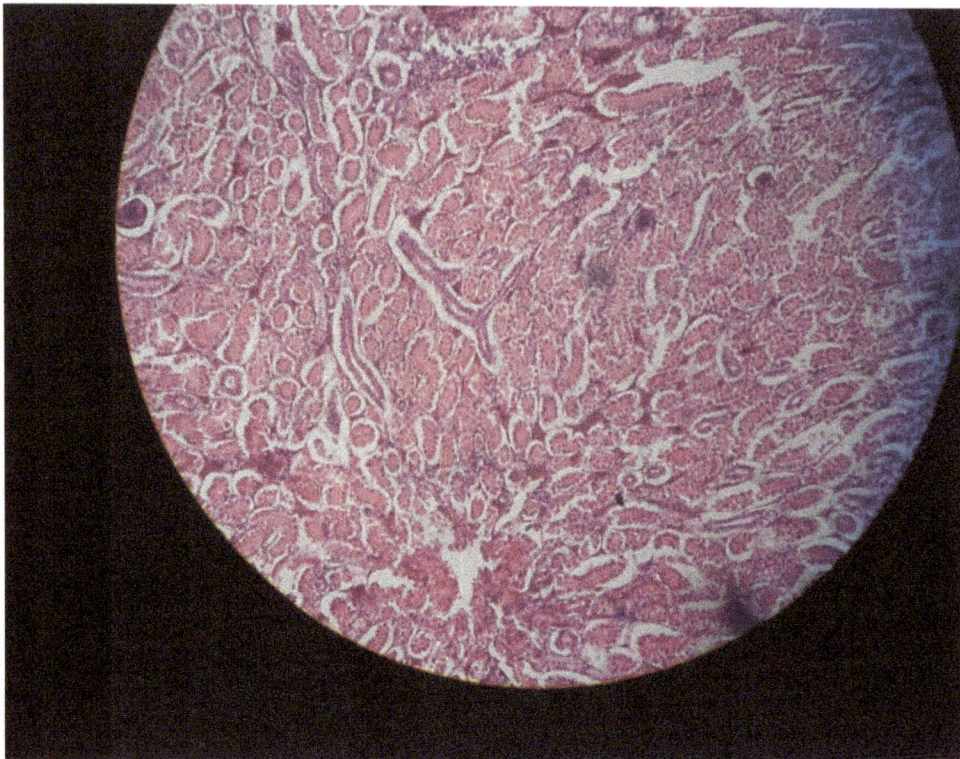

Plate 2. Diffuse infiltration of lymphocytes in the kidney of MD infected chicken.

(weeks to several months) being a chronic cancerous disease, so birds exposed at an early age may begin to show clinical disease at a much later age. Also, poultry are kept on low nutritive but high energy grower feed within this age range (9 weeks to point of lay) in most farms worldwide. Therefore, bird's immune response is not at its best to protect against this avian cancer and most other immunosuppressive conditions. In addition, the MD virus is reported to lie latent until a later age when birds are beginning to lay (Fatumbi and Adene, 1984; Frank, 2001).

All the cases of MD recorded in this study were in exotic egg producing type breeds (100%) and non in broilers or indigenous breeds or wild bird species. Meanwhile Nigerian indigenous chickens have only been shown to be seropositive despite their significance in the nation's poultry population (Adene and Akpavie, 2004). Should this be the situation, one will suspect that Nigerian indigenous breeds may possess some levels of genetic resistant to MD, and or MD may be only significant under intensive management systems since Nigerian indigenous breeds are mainly kept on extensive system of management. Accordingly, it has earlier been reported that certain exotic breeds of chickens have been shown to resist MDV (Cauchy and Coudert, 1986; Frank, 2001; Adene and Akpavie, 2004). From the available literature, there has not been any documented report of MD in broilers in Nigeria. Few reports and seasonal incidences of MD were observed in broiler flocks in other few countries (Calnek and Witter, 1997; Ionica and Coman, 2009) while high losses were reported in intensively managed broiler flocks (Calnek and Witter, 1997).

The predominant complaints of flock owners were uneven growth of their birds and inability to put on weight. Clinical observations often revealed weight loss and ruffled feathers. Gross lesions of emaciation and flabby heart with loss of coronary fats were consistently observed. Other important diagnostic lesions identified were principally neoplastic affecting many organs and tissues but these were predominantly observed in older birds as earlier reported (Fatumbi and Adene, 1984; Cauchy and Coudert, 1986; Frank, 2001).

Clinical manifestations of MD vary with the form of MD virus infection and where the lesions primarily develop in the host (Cauchy and Coudert, 1986; Stephen et al., 2008).

Flock history (age of birds), organ/and or tissue neoplastic gross lesions, peripheral nerve lesions and histopathology findings have been our strong diagnostic evidences of MD in the clinic as reported by others (Fatumbi and Adene, 1984; Cauchy and Coudert, 1986; Calnek and Witter, 1997) as serology or viral antigen detection might not in our situation differentiate between vaccine and field strains. Nervous signs have not been consistent in assisting the MD diagnosis in our environment, thereby posing further difficulty in

differentiating it from avian leukosis. It has earlier been documented that only a small proportion of the MD affected flocks show nervous signs (Cauchy and Coudert, 1986) and that morbidity, mortality rates and other gross manifestations of MD may vary with virus strain, immune status, age and sex of the host (Calnek and Witter, 1997; Frank, 2001).

We have consistently observed initial MD cases presented with uneven growth, emaciated carcasses and enlarged/thickened proventriculus in birds less than 10 weeks old before visceral tumours fully developed. Such birds may not show any additional classical gross lesions suggestive of MD; but were later returned with fully blown MD gross lesions weeks after first presentation. Meanwhile swelling of the proventriculus has been reported to be one of the classical signs of MD (Frank, 2001).

Available research showed that many host, environmental and management factors do modify the expression of MD (Graham, 1976; Cauchy and Coudert, 1986; Stephen, 2008). It is not the same in all birds, and may vary during spread within a flock, also between flocks and from one region to another (Frank, 2001; Stephen et al., 2008). It is therefore difficult to provide an overall field description of Marek's disease (Cauchy, 1986; Stephen et al., 2008). This is why there is an "acute" form which is more severe in symptoms and lesions than the "classical" form which is close to the original description and such descriptions were based on the disease as it occurred in unvaccinated flocks (Frank, 2001; Stephen et al., 2008).

Conclusion

From this report, it is clear that MD is on the increase despite farmer's efforts to prevent and control it through vaccination. Vaccination alone is therefore not completely reliable in preventing MD outbreaks. Intensive research should be carried out to develop a more potent user friendly MD vaccine, while for the time being, high hygienic standards should be maintained at the hatcheries and farms, and poultry breeds that are known for their resistant to MD should be encouraged to be kept by farmers.

ACKNOWLEDGEMENTS

The authors thank the technical staff of the poultry VTH-ABU clinic and histopathology laboratory for making available the required records and processing the tissue slides.

REFERENCES

ABDU PA, MERA UM, SA'IDU L (1992). A study of chicken mortality in

Zaria Nigeria. Proceedings of National Workshop on Livestock and Veterinary Services, Vom, Nigeria pp. 51-55.

Adene DF, Akpavie SO (2004). An overview of Marek's disease and the options in its control. Poultry Health and Production Principles and Practices. Stering Horden Publishers Nig. LTD. Ibadan, Nigeria. pp. 107-127.

Akpavie SO (2005). Major neoplasm's of poultry with special emphasis on the pathology, diagnosis and control of Marek's disease and Leucosis-J. Proceedings Poultry Health and Production Efficiency and Satellite Colloquium on Avian Flu. Workshop on improved disease diagnosis, health nutrition and risk management practices in poultry production efficiency, Zaria. pp. 26-30.

Baker FJ, Silverton RE, Palister CJ (2001). Cellular pathology. Introduction to Medical Laboratory Technology, 7th Edition. Bounty Press Limited, Nigeria. pp. 175-248.

Calnek BW, Witter RL (1997). Marek's disease. In Calnek, B.W, Barnes, J.H. Beard, C.W. LARRY, R.M and SAIF, Y.M. eds. Diseases of Poultry, 9th ed. Iowa State University Press, Iowa. pp. 356-430

Cauchy L, Coudert F (1986). Marek's disease. Rev. Sci. Tech. Int Off des Epiz. 5:1025-1035.

David S (2000). Marek's disease and leukosis. Poultry health and management, 4th edn. Blackwell Publishers, Cowley road Oxford. pp. 235-238.

Dong H, Kumar PM, Tavlarides-Hontz P, Arumugaswami V, Jarosinski K, Osterrieder N, and Parcells MS (2008). The Meq oncoprotein of Marek's disease virus: Mutations in high virulence MDVs and their putative role in changing pathogenesis. Proceedings of the 80TH Northeastern Conference on Avian Diseases. State College, PA, USA 2008, www.ivis.org. pp. 1-2.

Fatumbi OO, Adene DF (1984). A ten year prevalence study of Marek's disease and avian leukosis at Ibadan, Nigeria. Acta Vet. Brno. 55:49-53.

Frank F (2001). Marek's disease: History, actual and future Perspectives. Lohmann Inform. No. 25, pp. 1-5.

Geidam YA, Bukar MM, Ambali AG (2006). Chick quality control, a key to sustainable poultry production Nigeria. Nig. Vet. J. 27:1-6.

Gordon RF (1979). Poultry disease. Bulliere Tindall, London. 2:80.

Graham H (1976). Prevention of mareks disease, a review. Can. Res. 36:696-700.

Ionica FM, Coman NC (2009). An outbreak of Marek's disease in broiler chickens: Epidemiological, clinical and anatomopathological aspects. Luc. Sti. Med. Vet. 1:224-227.

Katherine EA, Andrew FR, Nicholas JS, Katrin G, Stephen WW, Mark EJW (2011). Modelling Marek's Disease Virus infection, parameter estimates for mortality rate and infectiousness. Vet. Res. 2:52-55.

Olabode HOK, Jwander LD, Moses GD, Ighadalo E, Egbaidome SA (2009). Prevalence of Avian Leucosis and MAREK'S disease in Ilorin, Kwara State, Nigeria. Nig. Vet. J. 30:64-68.

Powel PC (1981). Immunity to Marek's Disease. In Rose, M.E. Payne, L.N. and Freeman, B.M. (Eds). Avian Immunology, Poultry Science Symposium number sixteen. pp. 263-283.

Sa'id L, Abdu PA, Tekdek LB, Umoh, JU (2004). Retrospective study of Newcastle disease cases in Zaria, Nigeria. Nig. Vet. J. 27:53-56.

Stephen L, Claire K, Philip H (2008). Marek's disease-a silent enemy. Bri. Free Range Egg Prod. Ass. pp. 1-5.

Payne L (2004). Understanding Marek's disease. Proceedings of the 7th International Marek's Disease Symposium, Catherine's College, Oxford, England pp.1-4.

19

Characterization of bacterial strains associated with sheath rot complex and grain discoloration of rice in North of Iran (Mazandaran province)

Esmaeil Saberi[1], Naser Safaie[1] , Heshmatollah Rahimian[2]

[1]Department of Plant Pathology, Faculty of Agriculture, Tarbiat Modares University, PO Box 14115-336, Tehan, Iran.
[2]Department of Plant Protection, Faculty of Agriculture, Mazandaran Agricultural Sciences, Sari, Iran.

In recent years, similar symptoms to sheath rot complex (SH C) and grain discoloration (GD) has been observed in paddy fields in the north of Iran. To survey the etiology of this complex disease, 207 samples of rice plants showing wide range of sheath and grain discoloration were collected during 2002-2005 at booting and ripening stages from various geographical locations in Mazandaran province. Pathogenic strains were often isolated from samples with the symptom of longitudinal brown to reddish brown necrosis 5 mm wide extending the entire length of flag leaf sheath. Over 800 strains were isolated from collected samples and tested for pathogenicity on rice and/or hypersensitivity on tobacco or pelargonium. Pathogenic strains were isolated from 20.28% samples which contained only 5.3%of total strains. Eighty two strains comprising 72 strains from plants showing symptoms of sheath rot and grain discoloration, five strains from diseased seedling and five standard strains were analyzed for phenotypic studies. Pseudomonas-specific primers were used to confirm identification of the genus.On the basis of phenotypic characters and genus specific primers, the strains belonged to genera *Acidovorax* and *Pseudomonas*. Cluster analysis of 67 biochemical characters grouped 70 selected strains into seven distinct cluster and six groups with one member. The results confirmed that pathogenic strains associated with SH C and GD in Iran (Mazandaran province) belong to *Acidovorax avenae* subsp *avenae* (2.32%), *Pseudomonas putida* (2.32%), *Pseudomonas marginalis* (6.65%), *Pseudomonas syringae* (76.7%) and two unidentified species of *Pseudomonas* (13.95%). Accordingly, the *P. syringae* was revealed that is the major causal agent of SH C and GD in north of Iran.

Key words: *Oryza sativa* L., bacterial flag leaf sheath rot, *Pseudomonas*.

INTRODUCTION

Sheath rot complex and grain discoloration of rice (*Oryza sativa* L.) generally describes a disease which involves brown discoloration or the flag leaf sheath rot and grain discoloration. Both fungi and bacteria are reported to be associated with this disease (Cottyn et al., 1996a, b; Zeigler and Alvarez, 1990). The symptoms apparently caused by these agents are similar, preventing reliable diagnosis only based on symptomology (Duveiller et al.,

1998; Cottyn et al., 1996a; Zeigler and Alvarez, 1990). This disease is widespread in temperate (Joana et al., 2007) and tropical areas (Cottyn et al., 1996a; Jaunet et al., 1996) and more prevalent in areas where low temperature (and humidity) occur during the rice booting and heading stages (Jaunet et al., 1996) and the disease is especially apparent during the rainy season and the intensity of infection varies from mild to severe (Cottyni et al.,1996). Several pathogenic bacteria have been associated to be the cause of this disease in various parts of the world such as Asia, Africa, Latin America, Europe and Australia (Cottyn et al., 1996a; Cottyn et al., 1996b; Rott et al., 1989; Zeigler and Alvarez, 1990; Zeigler et al., 1987; Xie et al., 2012; Adoradaet al., 2013). Bacterial sheath rot of rice caused by *Pseudomonas oryzicola* was first described in 1955 (Ou, 1985) but the name was changed later to *P. syringae pv. syringae* (Ou, 1985; Zeigler et al., 1987). *Pseudomonas fuscovaginae* which causes sheath brown rot was first reported by Tani et al. (Miyajima et al., 1983) in Hokkaido in the north of Japan. Subsequently, this pathogen was reported in Burundi, Colombia, Madagascar, China, Nepal and Philippines (Cottynet al., 1996 a, b; Duveiller et al., 1998; Xie, 2003; Zeigler and Alvarez, 1987). *Burkholderiaglumae* (the causal agent of both seedling and grain rot) and *Acidovorax avenae subsp avenae* (A causal agent of bacterial stripe) have been reported to be associated with this disease (Cottyn et al., 1996a; Zeigler and Alvarez, 1990). In addition to these bacterial pathogens that commonly reported to be involved in sheath rot complex and grain discoloration, bacteria with similar characteristics to *Pseudomonas marginalis, Pseudomonas fluorescens, Pseudomonas corrugata, Pseudomonas aurofacience, P. fluorescensbv.* V and *P. fluorescens* bv. IV have been pathogenic and were able to induce sheath rot of rice (Joana et al., 2007; Cottyn et al., 1996). In Iran, a few surveys were made in paddy fields of Mazandaran province (the major rice growing region) to determine the causal agents of this complex disease. In one study, some fungi including *Sarocladium oryzae, Cochiobolus miyabeanus, Althernaria padwichii* and *Fusarium* spp were confirmed to be associated with this disease (Naeimi et al., 1983). There are two short reports on engagement of *Pseudomonas fuscovaginae* (Rostami et al., 1985) and *P. syringae* (Khoshkadam et al., 2008) in this disease .The present study was designed to investigate the occurrence of pathogenic bacteria involved in the sheath rot complex and grain discoloration in Iran (Mazandaran province).

MATERIALS AND METHODS

Isolation of bacteria from plant material

Diseased flag leaf sheath and grains of rice at booting and ripening stages were collected though out the geographical regions of Mazandaran province during 2002 - 2005. Small segments of sheath from the margin of discoloration tissue were washed under running tap water for 20 min and then macerated in sterile distilled water. Loopful of macerated tissue were streaked onto King's B medium and incubated at 28°C for 1 to 3 days. Three to five predominant colony types and all fluorescent colonies were purified by re-streaking and kept on nutrient agar slants at 4°C for routine work. To isolate the pathogenic bacteria from seeds, about 20 seeds of each sample were washed under tap water for 30 min and the rest of the procedure was as described for sheath samples. All obtained strains were evaluated on rice at both seedling and booting stages and pathogenic strains at these stage as well as some non-patho-

genic fluorescent strains and yellow pigment strains were selected for further studies.

Bacterial strains

A total number of 82 strains comprising72 strains from rice plants showing symptoms of sheath rot and grain discoloration, five strains from diseased seedling and five standard strains obtained from BSBF were studied (Table 1).

Pathogenicity test

Inoculations were made on rice Fajr and Tarom cultivars on both seedling and adult plants. Pathogencity teston seedling was conducted on 20 day old rice. Inoculation was made by injection of a suspension of 10^8cfu/ml of 24 h old bacterial culture in sterile water between leaf sheath located about 5 cm above soil (Duveiller et al., 1998; Rott et al., 1991). At least four seedlings were inoculated and for control two seedlings were injectedby sterile distilled water. The plants were incubated in a moist chamber at 100% relative humidity (RH) at 24°C for 48 h before being moved to greenhouse. The same type of inoculation was performed on the flag leaf sheath on adult plants at booting stage. The inoculated plants were incubated for 48 h at 100% RH at 24°C and transferred to the greenhouse (Duveiller et al., 1998; Zeigler and Alvarez, 1987; Zeigler and Alvarez, 1990). Two plants each with 5 tillers were inoculated and for control, two tillers were injected by sterile distilled water. The plants were evaluated in two day intervals during the first week and then in one- week intervals up to one month.

Biochemical characteristics

The physiological and morphological characteristics of 70 strains (Table 2) were determined by 67phenotypic tests using standard microbiological techniques (Schaad et al., 2001; Mew and Cottyn, 2001).

Genus-specific primers

Genus-specific 16Sr RNA of *Pseudomonas* (sensostrictu) were used for selective detection of strains of *Pseudomonas* (Widmer et al., 1992; Porteouset al., 1985).

DNA extraction

For genomic DNA preparation, the bacteria were grown on Tryptic soy broth (Merck), shaking for 24 - 48 h at 28°C. Cells were harvested by centrifugation (10/000 rpm, 5 min), washed with sterile water and resuspended in sterile distilled water to give an OD 600 of 0.5. Cells were lyzed by NaOH 3% (g/vol) and centrifuged (10/000 rpm, 5 min) to pellet cell debris. Clear supernatant DNA solution was extracted twice with chloroform. Precipitation of nucleic acids was made with sodium acetate (final concentration, 0.3 M) and 2 volume of absolute ethanol (Rahimian, unpublished). The DNA was dissolved in sterile water and the concentration of DNA was determined spectrophotometrically at 260 nm.

PCR amplification

The primers PS - for (20-mer[5'-GGTCTGAGAGGATGATCAGT-3']) and PS - rev (18 -mer [5'-TTAGCTCCACCTCGCGGC-3']) were used for PS – PCR (Widmer et al., 1992). Amplification reactions were performed in 20 μl of a solution containing 1X reaction buffer (

Table 1. Geographical origin, tissue of origin and pathogenicity test on rice forbacterial strains used in this study.

Location	Tissue origin	Rice pathogenicity		Strain number
		Seedling stage	Booting stage	
Amol	Sheath	+	+	1, 2, 18, 19, 43
	Sheath	W	+	3, 34
	Sheath	-	+	101
	Sheath	+	W	31
	Sheath	-	-	102, 103, 104, 105
	Sheath	ND	-	302, 303
	Seed	+	+	16, 17, 109
	Seed	-	-	108, 122
	Seedling	+	W	201
	Seedling	-	-	202
Babol	Sheath	+	+	4, 5, 20, 21
	Sheath	+	-	33
	Seed	-	-	110
	Seedling	-	-	203
Babolsar	Sheath	+	+	9, 10
	Seed	-	-	111
Chalos and Noshah	Sheath	W	+	42
	Sheath	W	-	36
	Seed	+	+	29
Ghaemshah	Sheath	+	+	11, 22, 23, 26
	Sheath	+	-	35
	Seed	+	+	8
	Seed	ND	-	304
Fridonkenar	Sheath	+	+	28
	Seed	-	-	120
	Seedling	+	W	204
Mahmodabad	Sheath	+	+	12, 13. 25, 30
	Sheath	+	-	39, 41
	Seed	+	+	27, 44
	Seed	W	W	32
	Seed	-	-	118
	Seedling	-	-	205
Noor	Sheath	+	+	14
	Sheath	W	W	37
	Seed	+	-	38
	Seed	-	-	40
Sari	Sheath	-	+	106
	Seed	-	+	112
	Seed	-	-	113, 121
Savadkouh	Sheath	-	W	107
	Sheath	+	-	24
	Seed	+	+	6, 7
	Seed	-	W	116, 117
Shiraze	Seed	-	-	123
Tonokabon	Seed	+	+	15
	Seed	-	-	115, 119
	Seed	-	w	114
NO	Seed	-	-	403
Strain standard				
IBSBF(674)	-	ND	ND	401(*P. viridiflava*Typestrine)
IBSBF(676)	-	ND	ND	402 (*P. s.* pv. *panici*)
IBSBF (674)	-	ND	ND	404 (*P. s.* pv. *atrofaciens*)
IBSBF (199)	-	ND	ND	405(*B.andropogonis*,Typestrine)
IBSBF (1890)	-	ND	ND	406 (*B. gladioli*pv. *gladiol*)

W, Weak pathogenesis; ND, not determined; NO, unknown.

Table 2. Phenotypic characteristics of strains associated with sheath rot complex and grain discoloration of rice in Mazandaran province (Iran).

Character [b]	Cluster number [a]												
	A	B	C	D	E	F	G	40	201	29	122	405	406
Gram stain(KoH)	-	-	-	-	-	-	-	-	-	-	-	-	-
Fluorescent pigment	+	+	+	+	+	+	+	+	+	-	+	-	-
Catalase test	+	+	+	+	+	+	+	+	+	+	+	+	+
Levan production from sucrose	-	-	-	-	v	-	-	-	-	-	-	-	-
Oxidase	+	+	+	+	-	-	-	+	+	+	-	-	-
Soft rot of potato	-	-	-	-	-	-	-	-	-	-	-	-	-
Arginine dihydrolase	+	+	+	+	-	-	-	+	+	+	-	+	+
Tobacco hypersensitivity	+	+	+	+	+	+	+	-	+	+	+	+	+
2 -ketogluconate production	+	v	v	+	-	+	v	+	-	-	-	-	-
Nitrate reduction	v	v	v	-	-	-	-	+	-	+	-	-	+
eculin hydrolysis	-	-	-	-	+	+	+	-	-	-	+	-	+
Arbutin hydrolysis	-	-	-	-	+	+	v	-	-	-	+	-	+
Starch hydrolysis	-	-	+	+	-	-	+	-	-	-	-	-	-
Gelatin liquefaction	-	+	+	+	v	-	+	+	+	-	-	-	-
Tween 80 hydrolysis	-	+	+	+	+	+	+	+	+	+	+	-	-
Casein hydrolysis	v	+	+	-	+	v	+	+	-	-	+	-	+
Urease production	+	+	+	+	+	+	+	+	-	+	+	+	-
Indol Production	-	-	-	-	-	-	-	-	-	-	-	-	-
Phosphatase	v	v	-	+	-	v	+	+	-	+	-	+	+
Production of reducing substance from sucrose	-	-	v	-	+	v	v	-	-	-	-	-	-
H2s from cysteine	±	±	v	v	±	±	v	+	±	+	±	±	±
H2s from thiosulfate	±	±	±	+	±	±	v	±	±	+	±	±	±
Growth in NaCl 3%	+	+	+	+	+	+	+	+	+	+	-	-	+
Growth in NaCl 5%	+	+	-	v	v	v	-	+	-	-	-	-	-
Fermentation Test	-	-	-	-	-	-	-	-	-	-	-	-	-
Lecithinase test	v	+	v	+	-	v	v	+	-	-	+	-	+
Voges-proskauer test	-	-	-	-	-	-	-	-	-	-	-	-	-
Methyl red reaction	-	-	-	-	-	-	-	-	-	-	-	-	-
Ice nucleation	-	v	-	-	+	+	+	-	-	-	-	-	-
Growth at 37°C	+	+	v	v	-	+	-	+	+	+	+	-	+
Growth at 40°C	v	-	-	-	-	-	-	+	+	+	-	-	+
Tyrosinase	-	-	v	-	-	v	-	+	-	-	-	-	-
Utilization of betain	+	+	+	+	+	+	v	+	+	+	-	+	+
Acid production from:													
Trehalose	v	v	+	+	-	-	-	+	+	-	-	+	+
Xylose	+	-	+	+	+	+	+	+	-	+	-	+	+
L - Arabinose	-	-	+	+	v	+	-	+	-	-	-	-	+
Mannose	+	+	+	+	+	+	+	+	-	+	-	+	+
L -Rhamnose	-	-	+	+	-	-	v	+	-	+	-	-	-
Acid production from:													
Sucrose	-	v	v	-	+	+	+	-	-	-	-	-	-
Cellobiose	-	-	+	-	-	v	-	+	-	-	-	-	+
Lactose	-	-	+	-	-	-	v	+	-	+	-	-	-
Raffinose	-	-	v	-	-	-	-	-	-	+	-	-	-
Salicin	-	-	+	-	-	-	-	-	-	+	-	-	+
Inuline	-	-	-	-	-	-	-	-	-	-	-	+	+
Inositol	-	v	+	-	+	+	+	+	-	-	-	+	+
Sorbitol	-	-	+	+	+	+	+	+	-	+	-	-	+

Table 2. Contd.

Character [b]	Cluster number[a]												
	A	B	C	D	E	F	G	40	201	29	122	405	406
Mannitol	-	+	+	+	+	+	+	+	-	+	-	+	+
Adonitol	-	v	+	-	v	-	-	-	-	-	-	+	+
Dulcitol	-	-	-	-	-	-	-	-	-	-	-	-	+
L -Arabitol	-	+	+	+	+	+	+	+	-	-	-	+	+
Maltose	-	-	v	v	-	+	v	+	-	-	-	-	-
Etanol	v	-	v	+	-	+	v	+	-	+	-	+	+
Melibiose	v	v	+	+	-	+	+	+	+	+	-	+	+
Erythitol	v	-	-	-	-	-	-	-	+	-	-	-	-
N - Propanol	v	-	v	+	-	+	v	+	-	+	-	-	+
Geraniol	v	-	-	-	-	-	-	-	-	-	-	-	-
Utilization for growth:													
Acetate	+	+	+	+	+	v	+	+	+	+	+	+	+
Malonate	+	v	+	+	+	+	+	+	-	+	-	+	+
maleate	+	+	+	+	+	+	+	+	+	+	-	+	-
Citrate	+	+	+	+	+	+	+	+	+	+	+	+	+
D-tartrate	-	-	-	-	+	-	-	-	-	+	-	-	-
L-tartrate	v	v	-	-	-	-	-	-	-	-	-	-	+
Benzoate	v	-	-	-	-	-	-	-	-	-	-	+	-
Arginine	+	+	+	+	v	+	v	+	+	-	-	-	+
β-Alanine	+	+	+	+	-	-	-	+	+	+	-	+	+
Quinat	+	+	+	+	+	+	+	+	+	+	+	+	+
Oxalate	v	v	-	-	-	-	-	-	+	-	-	+	+

[a],Cluster number from numerical analysis of phenotypic data; +, 80% or more strains positive; v, between 21 - 79% strains positive (numbers in parentheses are percentages of strains that tested positive); -, 80% or more strains negative.

500 mMKCl, Tris-Hcl (PH = 8.4), 200 mM Amoniom sulfate) 20pmol of the two opposing primers (Cinna Gen of Iran), 200 μM of each dexoynucleoside triphosphate (Cinna Gen of Iran), 1.5 mM $MgCl_2$ (Cinna Gen of Iran), 1 μl (/ ng) of template DNA and 2.5 U of *Taq* DNA polymerase (Cinna Gen of Iran). Amplification were performed with a DNA thermocycler (Eppendorf AG, Germany) with an initial denaturation at 94°C for 4 min, followed by 30 cycles of denaturation at 94°C for 1 min, annealing at 68°C for 1 min, extension at 72°C for 1 min and a final extension at 72°C for 7 min (Widmer et al., 1992; Porteous et al., 1985).

Data analysis

The clustering of the strains was performed by the unweighted pair group method using arithmetic average (UPGMA) and simple matching coefficient using MVSP 32 software.

RESULTS

The collected samples showed a wide range of sheath and grain symptoms. Pathogenic strains were often isolated from samples with symptoms that include longitudinal brown to reddish brown necrosis 5 mm wide extending the entire length of flag leaf sheath. Panicles

emerging from these samples often were affected so that 1/3 to 1/4 of panicles failed to emerge from the boot. These panicles produced discolored brown to reddish brown and poorly filled grain. These symptoms mostly were observed on samples from high lands (Savadkouh and Kiasar). These symptoms were distinct from sheath and seed discoloration on the samples from other regions. Sheath discoloration symptoms in coastal low lands and the plains especially in Mahmoodabad where no pathogenic strain was isolated, were different from above-mentioned typical symptoms. The latter samples showed a dark brown to black sheath discoloration which mostly extended to stem.

Over 800 bacterial strains were isolated from 207 samples collected from 12 cities (Figure 1) and tested for HR and/or pathogencity on rice plant (Table 1). Distribution and isolation frequency of the 64 strains, which were grouped into clusters (Figure 2) are presented in Table 3. No pathogenic strain was isolated from eight samples collected from Ramsar. A wide range of differences in morphology and color of colonies were observed on isolation plates. Yellow and fluorescent colonies, which were non-pathogenic (with negative HR), were predominant. These colonies grow very rapidly and

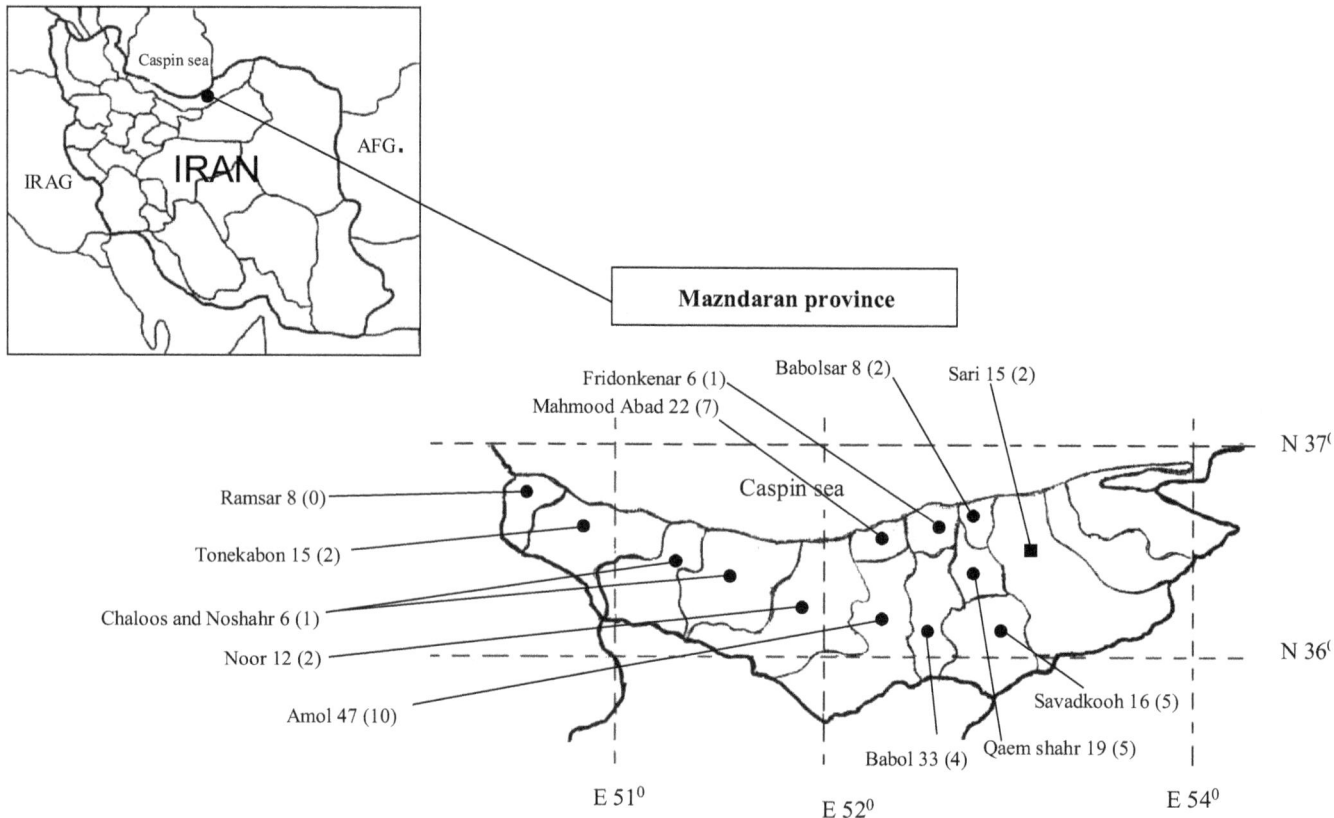

Figure 1. Sampling sites and respective number of Healthy and infectedsamples collected from the major growing region of the Mazandaran province over a period of 4 years (2002 - 2005).

interfere with the isolation procedure. Pathogenic and/or fluorescent HR positive strains were isolated from samples with typical symptoms of bacterial sheath rot and these colonies were predominant on the plate. These strains showed HR in test plant less than 12h and considering higher growth rate, colony morphology, rapid and intensive fluorescent pigment production were different from saprophytic fluorescent bacteria. These isolates were recovered from both seeds and sheaths.

Pathogenicity test

All pathogenic strains on seedling after 2 - 3 days of inoculation showed, elongated water soaked blotches on the leaf sheaths. These blotches developed into a brownish black necrosis that eventually spread the full length of the sheaths. Plants inoculated at booting stage showed progressive water - soaking in the inoculated region after 2 - 4 days. The lesions became necrotic and progressed up and down the inoculated region, induced the necrotic stripe on the sheath that eventually spread the full length of the sheaths. Panicles emerged from these plants were poor and produced brown to reddish brown grains that often were sterile. In some cases, the symptoms were restricted to the inoculation point as gray to dark brown spots. These strains and those which

caused only discoloration on seeds were considered as weak pathogen. For most of the strains, the symptom severity was less in greenhouse than in the field. Although the observed symptoms were generally the same, the severity of the produced lesions greatly varied among the strains.

Biochemical characteristics

The tested strains were divided into two groups according to fluorescent pigmentation on king's B medium. The non-fluorescent group only contained strain 29 while the other strains belonged to the fluorescent group. The 70 tested strains were grouped into seven distinct clusters and six groups with one member based on the biochemical characteristics (Table 2, Figure 2). Strains of cluster A and strains 40 and 122 were not pathogenic on rice. Cluster A contained seven strains with characteristics similar to *Pseudomonas putida*. Cluster B included strains that in 84% similarity were grouped into the distinct B1, B2 and B3 sub-clusters. In B1 and B2 subclusters, only strain 14 caused grain discoloration and other strains were not pathogenic on rice. The strains in B2 sub-cluster were similar to *P. putida*al though they showed differences in gelatin hydrolysis, lecithinase and tobacco hypersensitivity. B2 sub-cluster (strains 104, 108

UPGMA

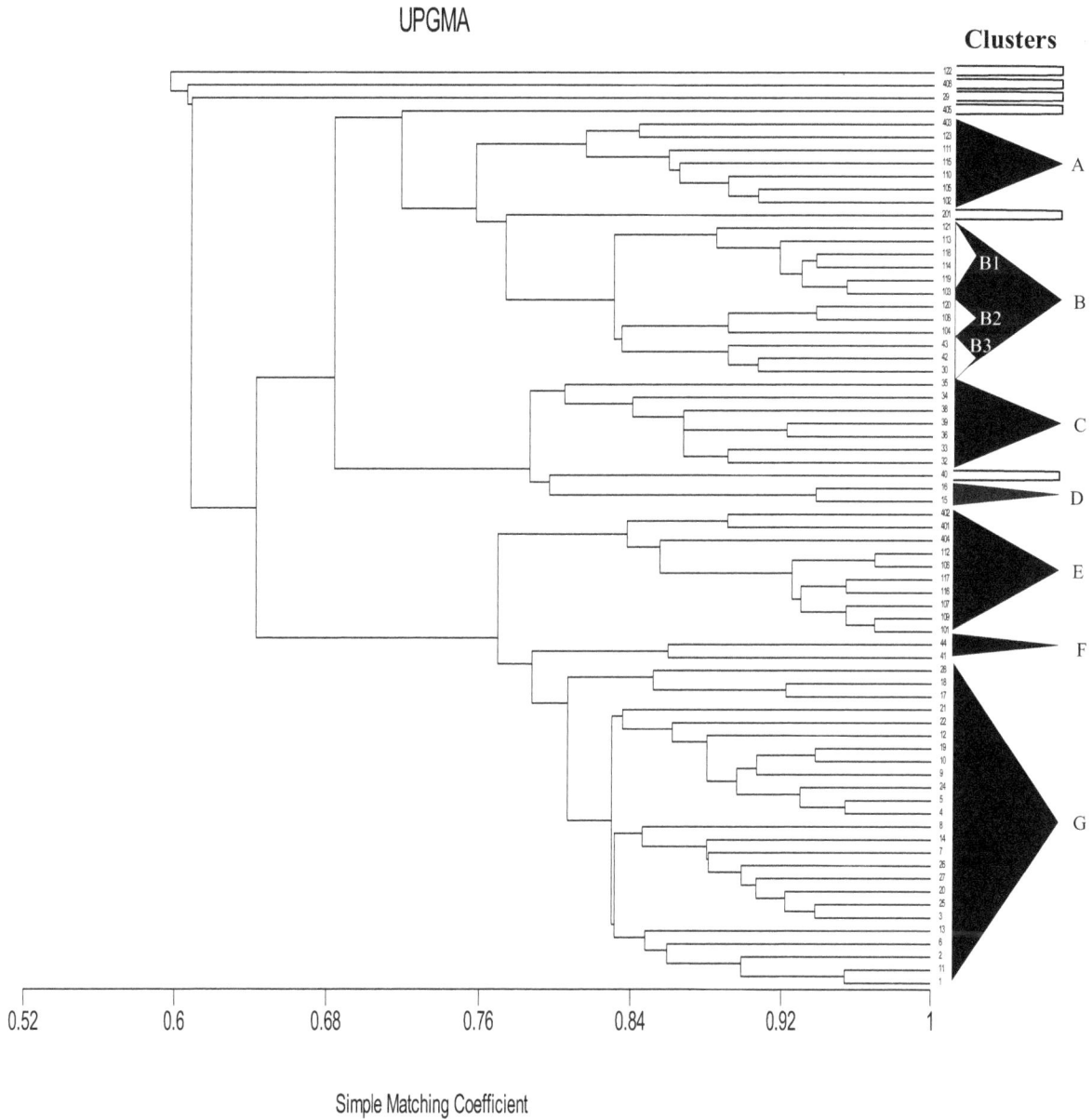

Figure 2. Cluster analysis of phenotypic data of bacterial strains isolated from rice samples with sheath rot and grain discoloration collected in Mazandaran province.

and 120) contained strains similar to *P. fluorescens* bv III. They differed from B1 cluster in acid production from trehalose, sucrose, inositol and adonitol. B3 sub-cluster contained the strains (30, 42 and 43) that caused disease on rice at both seedling and booting stages. These strains differentiated from all pathogenic fluorescent pseudomonades associated with sheath rot and grain discoloration. They were similar to *P. fuscovaginae* for soft rot of potato, 2-ketogluconate production and acid production from inositol and similar to *P. marginalis* for nitrate reduction, lecithinase and growth at 37°C. They were variable for acid production from sorbitol, trehalose, sucrose and adonitol. Cluster C included seven strains

that all were pathogenic on rice seedling but only strains 32 and 34 caused disease at booting stage. Phenotypic characters of this cluster were variable which differentiated it from all common pathogenic fluorescent pseudomonads associated with the sheath rot (Garden et al., 2002; Schaad et al., 2001; Zeigler and Alvarez, 1990). Phenotypiccharacters of strain 35 which is grouped relatively distant from this cluster were similar to *P. marginalis*.

Cluster D contained two strains (15 and 16) that caused severe disease on rice in both stages. These strains had similar characteristics to *P. fuscovaginae* in comparison with other pathogenic *Pseudomonas*. When the eight

Table 3. Origin and isolation frequency of (pathogenic) strains associated with sheath rot complex and grain discoloration of rice in Mazandaran province (Iran).

Origin	Sample (total)[a]	Pathogens[b]	Cluster[c]						
			A	B	C	D	E	F	G
Amol	10 (47)	11	2	4	1	1	2	-	6
Babol	4 (33)	4	1	-	1	-	-	-	4
Babolsar	2 (8)	2	1	-	-	-	-	-	2
Chalos and Noshah	1 (6)	1	-	1	1	-	-	-	-
Fridonkenar	1 (6)	1	-	1	-	-	-	-	1
Ghaemshah	5(19)	5	-	-	1	-	-	-	5
Mahmodabad	7 (22)	7	-	2	2	-	-	2	4
Noor	2 (12)	2	-	-	-	-	-	-	1
Ramsar	0 (8)	0	-	-	-	-	-	-	-
Sari	2 (15)	2	-	2	-	-	2	-	-
Savadkoh	5 (16)	5	-	-	-	-	3	-	3
Tnokabon	3 (15)	3	1	2	-	1	-	-	-

a, Number of samples from which the pathogenic strains were isolated and (total sample); b, number of pathogenic strains; c, number of strains grouped according to numerical analysis of phenotypic data (one member groups are not listed 7 in Table).

biochemical tests useful in identifying *P. fuscovaginae* (Rott et al., 1991) were used for comparison, these strains differed from *P. fuscovaginae* for production of 2-ketogluconate and acid from sorbitol. Moreover, comparison with more phenotypic characteristics (55 tests) showed they were differentiated from *P. fuscovaginae* (Miyajimaet al., 1983). These strains differed from *P. fuscovaginae* for urease test, production of H_2S from thiosulfate, growth at 3% NaCl, lecithinase, acid production from xylose, melibiose and L-arabitol. They were variable for growth at 37°C, acid production from maltose and N-propanol.Strain 40 had similar characteristicstoP. *Fluorescens* bv. I. This strain was not pathogenic on rice. Strain 201 which was isolated from seedling with blight had similar characteristics to *P. fuscovaginae*. In comparison to eight biochemical tests (Rottet al., 1991), only this strain had complete similarity to *P. fuscovaginae*al though more phenotypic characters differentiated it from *P. fuscovaginae*.

Reference strains 405 (*B. andropogonis*) and 406 (*B. gladioli pv. gladioli*) were grouped into separate clusters with one member. Strain 29 was pathogenic on rice in both seedling and booting stages. This strain was phenotypically similar to *A. avenae* subsp. *avenae*. Strain 29 produced the brown stripe on the leaf midrib of inoculated seedling and was the only pathogen recognizable by a typical symptoms expression in the pathogenicity tests on seedling. It was phenotypically distinct from fluorescent strains as well and grouped into different one member group. Its phenotypic characteristics were similar to those reported by Schaad et al. (2001) for *A. avenae* subsp. *avenae* except for utilization ofbetain, growth on arabinose and starch hydrolysis (Table 2).

Cluster analysis of phenotypic data grouped all oxidase and arginine dehydrolase negative strains except for strain 122 into E, F and G clusters. According to their phenotypic characteristics, all of these strains were identified as *P. syringae* (Table 4). The E cluster contained seven rice strains and the standard strains *P. s.* pv. *Atrofaciens P. s.* pv. *panici* and *P. viridiflava* at 80% similarity. All rice strain of E cluster showed weak to strong pathogenicity on rice at booting stage. Only strain 109 was pathogenic on rice seedling (Table 1). All strains of this cluster showed weak levan production in comparison with *P. s.* pv. *atrofaciens* standard strain. These strains were distinct from F and G clusters in levan production, alkaline production from D-tartrat. F cluster contained only two (41 and 44) strains of which only strain 44 was pathogenic on rice sheath. Phenotypic characteristics of this cluster showed its similarity to *P.syringae*.

However, some characteristics such as growth at 37°C, cellobiose utilization and 2-ketogluconate production differed from *P. syringae* and strains in E and G clusters. ClusterG contained 25 strains with characteristics similarto *P. syringae*, however, they were variable in some important tests including, arbutin hydrolysis and betain. Also, they were variable in 2-ketogluconate production, production of reducing substances from sucrose, H_2S production from cysteine and thiosulfate, lecithinase, tyrosinase, acid production from maltose, rhamnose, lactose, salicin, ethanol, n - propanol, and alkalines production from arginine.

Genus-specific primers

Ps-for/Ps-rev primers amplified a DNA fragment of 990 kb in all strains except for strain 29. This band was not amplified in negative controls including *B. andropogonis*

Table 4. Comparison of the phenotypic characteristics of *P. syringae* strains associated with sheath rot and grain discoloration of rice in Iran (Mazandaran province) and published characters of *P. syringae* described by Schaad et al. (2001).

Character	*P. syringae*[c]	Cluster number [a]		
		E (7[b])	F (2)	G (25)
Fluorescent pigment	+	+	+	+
Levan production	D	V (5) b	-	- (5)
Oxidase	-	-	-	-
Arginine dehydrolase	-	-	-	- (1)
Soft rot of potato	-	-	-	-
Tobacco hypersensitivity	+	+	+	+
Growth at 37°C	-	-	-	- (2)
Nitrate reduction	-	-	-	-
Gelatin liquefaction	D	V (5)	-	+ (20)
Utilization for growth:				
Mannitol	D	+	+	+
Geraniol	-	-	-	-
Sorbitol	+	+	+	+
Cellobiose	-	-	V (1)	-
benzoat	-	-	-	-
Sucrose	+	+	+	+
D -Tartrat	- (D)	+	-	- (4)
L -Tartrat	- (D)	-	-	-
Trehalose	-	-	-	-
Ice nucleation	+	+	+	+
L -Rhamnos	-	-	-	V (15)

[a], Cluster number from numerical analysis of phenotypic data; [b], number of strains grouped according to numerical analysis of phenotypic data; [c], published characters of *P. syringae* described by Schaad et al. (2001); [D], some pathovars are positive; +, 80% or more strains positive; V, between 21 - 79% strains positive (numbers in parentheses are percentages of strains that tested positive); -, 80% or more strains negative.

and *B. gladioli* pv. *gladioli* (Figure 3). Accordingly all strains except the non-fluorescent pathogenic strain 29 which was confirmed to be *Acidovorax*, approved to be Pseudomonas.

DISCUSSION

Eight hundred bacterial stains were collected from rice plants with the sheath rot complex and grain discoloration, of which only 5.3% (43 strains) were pathogenic and majority of this bacteria were saprophytic. The earlier studies (Miyajima et al., 1983; Cottyn et al., 1996a) also reported that less than 15% of their collected strains were pathogenic. These pathogenic strains were isolated from 42 out of 207 collected samples (20.28% of samples) which confirmed the importance of bacterial agents in sheath rot complex and grain discoloration as previously reported by other researchers (Cottyn et al., 1996a, b; Adorada et al., 2013).

In pathogenicity tests, the pathogenic strains showed similar symptoms, lacked any differential symptoms for etiological purpose. Nevertheless, the disease severity was varied among the strains and they were mainly less severe in greenhouse experiments. The large proportion of samples which showed disease symptoms may be caused by abiotic agents or isolation problem. Based on phenotypic characteristics and specific primers, our strains belong to the genera including *Acidovorax* and *Pseudomonas*. The only strain of *Acidovorax* (Strain 29) identified as *A. avenae* subsp. *Avenae* which is seed born and has been detected in seed lots from many countries (Cottyn et al., 1996a; Webstar and Gunnell, 1992) and identified as a common agent associated with sheath rot complex disease (Cottyn et al., 1996a,b; Zeigler and Alvarez, 1990). Strain 29 was isolated from seeds showing symptoms and the role of this strain in sheath rot complex disease remains unclear.

Based on polyphase taxonomic studies, *Pseudomonas* have been revealed to comprise rRNA group I species, it includes the strain *P. aruginosae* and other fluorescent Pseudomonads such as *P.fluorescens*, *P. putida*, *P. syringe* and *P. fuscovaginae* (Anzaiet al., 2000; Kersters et al., 1996; Porteous et al., 1985) which are the principal bacteria known to infect rice grain and sheaths (Zeigler et al., 1987; Cottyn et al., 1996a; Xie et al., 2012). Based on

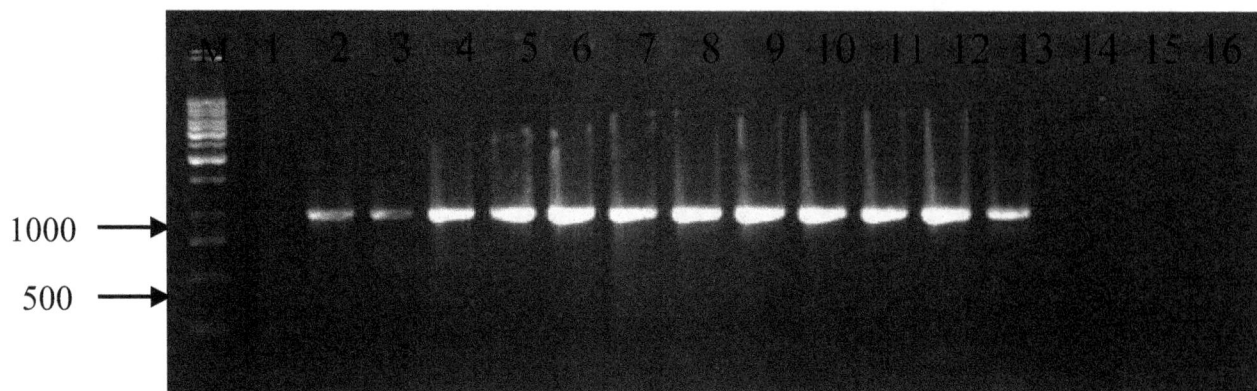

Figure 3. Agarose gel electrophoresis of *Pseudomonas* selective studied strains with *Pseudomonas* specific primers (PS - for , PS - rev) .M, Ladder 1 Kb; 1, negative; 2, *P.s.*pv. *syringe*; 3, *P. fuscovaginae*; 4, *P. marginalis*; 5, 101; 6, 102; 7, 103; 8, 104; 9, 105; 10, 106; 11, 107; 12, 108; 13, 109; 14, *B. andropogonios*; 15, *B. gladioli* pv. *gladioli*; 16, *Acidovoraxavenae*.

phenotypic characteristics, the tested strains belonging to this genus, could be clearly separated into two groups; those positive for arginine dihydrolase and oxidase and those negative for arginine dihydrolase and oxidase, identified as *P. syringae*. According to the resulting dendogram of phenotypic characteristics at 80% similarity level, strains of the first group (positive for arginine dihydrolase and oxidase) were grouped in clusters A to D and the single-member groups 201 and 40. Pathogenic strains from this group belong to *P. putida*(strain 114), *P. marginalis* (strain 32 and 34) and the unidentified species of *Pseudomonas* include strain 201, clusters D and subcluster B3, respectively. Strains of sub-cluster B3 display intermediate characteristics of *P. marginalis* and *P. fuscovaginae* which were different from those of known arginine dehydrolase-oxidase strains (Schaadet al., 2001; Zeigler and Alvarez, 1990; Cottynet al., 1996 a,b; Webster and Gunnell, 1992).Strains 32 and 34 have phenotypic characteristics similar to *P. marginalis*. Phenotypic characteristics of Strains of cluster D (strains 15,16) and strain 201 did not fit with any of described species of phytopathogenic flouresent *Pseudomonas* (Schaadet al., 2001; Zeigler and Alvarez, 1990; Cottynet al., 1996 a, b; Webster and Gunnell, 1992), however they show more similarity to *P. fuscovaginae* than the other pathogenic strains. However, regarding the difference in eight biochemical characteristics (Rottet al., 1991) and 55 phenotypic characteristics (Miyajimaet al., 1983) these strains are considered to be distinct from *P. fuscovaginae*.

Previous studies on samples from Brundi (Duveiller et al., 1990; Duveiller et al., 1998 b) Colombia (Zeigler and Alvarez, 1987) and Madagascar (Rott et al., 1989) have demonstrated strains with similar characteristics to these pathogenic strains. The determination of the exact taxonomic position of these strains needs complementary studies especially investigations at molecular level.

The second group of *Pseudomonas* strains includes the strains which were negative for arginine dehydrolase-

oxidase. As the major bacterial agents associated with sheath rot complex disease in Iran, these strains are grouped in E, F and G clusters. Analysis of phenotypic characteristics explicitly revealed that these strains belong to *P. syringae*, which has been also associated with sheath rot complex disease in Hungary (Ou, 1985), Australia, Chile and several Asian countries (Ou, 1985; Cottynet al., 1996a; Rottet al., 1989; Webstar and Gunnell, 1992). E, F, and G clusters differed from each other for characteristics which are not taxonomically important. Besides, there was no difference in their geographical origin or tissue of isolation.These clusters were considered as taxonomically similar to *P. syringae*. Since phenotypic characteristics are not sufficient for clustering the strains at infraspecific level (pathovar), (Manceau and Horvais, 1997; Cottyn et al., 1996; Joana et al, 2007) distinction of strains through comparison with the common rice disease associated *P. syringae* pathovars was not possible. Taxonomic identification of these strains at subspecies level (pathovar) requires complementary molecular studies such as comparison of electroforetic patterns of total protein, analysis of ITS regions (Manceau and Horvais, 1997; Olczak-Woltman et al., 1983) and DNA fingerprinting (Jaunetet al., 1995; Little et al., 1996; Louwset al., 1994), which are now in progress.

The fluorescent non-pathogenic species such as *P. putida* and *P. fluorescens* were commonly recovered from the samples with disease symptoms and determination of their possible role in development of disease syndrome needs further study. Strains similar to *P.s.* pv. *panici* (Elliott) Young et al. 1987, causing bacterial brown stripe of rice seedling, *B. glumae* causing grain rot (Mew and Cottyn, 1997; Ou, 1985), panicule blight (Yuan et al., 2004) and as common bacteria associated with sheath rot complex (Zeigler and Alvarez, 1990; Cottyni et al., 1996), *B. gladioli* causing leaf - sheath browning of rice (Hiroyuki et al., 2006) were not detected in the collected samples.

ACKNOWLEDGMENTS

We thank the contribution of all individuals in laboratories who helped us through this research.

REFERENCES

Anzai Y, Kim H, Park Y, Wakabayashi H, Oyaizu H (2000). Phylogenetic affiliation of the Pseudomonas based on the 16S rRNA sequence. Int. J. Syst. Evol. Microbiol.50: 1563- 589.

Adorada DL, Stodart BJ, Tpoi RP, Costa SS, Ash GJ (2013). Bacteria associated with sheath browning and grain discoloration of rice in East Timor and implications for Australia's biosecurity. Australasian Plant Dis. Notes, DOI 10.1007/s13314-013-0092-z

Cottyn B, Ceres M T, Van outryve M F, Baroga J, Mew T W (1996). Bacterial diseases of rice. I. Pathogenic bacteria associated with sheath rot complex and grain discoloration of rice in the Philippines. Plant Dis. 80: 429-437.

Cottyn B, Van outryve M F, Cerez M T, Decleene M, Swings J, Mew TW (1996). Baeterial diseases of rice. II. Characterization of pathogenic bacteria associated with sheath rot complex and grain discoloration of rice in the Philippines Plant Dis. 80: 438- 445.

Duveiller E, Maraite H (1990). Bacterial sheath rot of wheat caused by Pseudomonas fuscovaginae in the highlands of Mexico. Plant Dis. 74: 932-935.

Duveiller E, Miyajima K., Snacken F., Autrique A, Maraite H (1998). Characterization of Pseudomonas fuscovaginae and differentiation from other fluorescent Pseudomonas occurring on rice in Brundi.Phytopathology 122: 97-107.

Duveiller E, Snacken F, Maraite H (1989). First detection of Pseudomonas fuscovaginae on maize and sorghum in Brundi.Plant Dis. 73: 514-517.

Garden L, Bella P, Meyer J M, Chisten P R, Achouak W, Samson R (2002).Pseudomonas salomonii sp. Nov., pathogenic on garic and Pseudomonas palleroniana sp. nov.isolated from rice. Int. J. Syst. Evo. Microbiol. 52: 2065-2074.

Hiroyuki U, Naruto F, Kazuhirol (2006). Burkholderia gladioli associated with symptoms of bacterial grain rot and leaf - sheath browing of rice plants. J. Gen. Plant. Pathol. 72: 98 - 103.

Jaunet T (1996). Pathogenicity of process of Pseudomonas fuscovaginae the causal agent of sheath brown rot of rice.J. Phytopatho. 144: 423-430.

Jaunet T, Laguerre G, Lemanceau P, Frutos R, Notteghem JL (1995). Diversity of Pseudomonas fuscovaginae and other fluorescent pseudomonads isolated from diseased rice .Phytopatholo. 85: 1534-1541.

Joana G, Steven V, Roberts J (2007). Discrimination of Pseudomonas syringae isolates from sweet and wild cherry using rep - PCR. Plan.t Patho. 117: 383 - 392.

Kersters K, Wolfgang L, Vancanneyt M, De Vos P, Gillis M, Schleifer K H (1996). Recent changes in the classification of the pseudomonads: an overview. Syst. Appl. Microbio. 19: 465-477.

Khoshkadam M, Kazempor MN, Ebadi AA, MossanegadM, Pedramfar H, 2008. Identification of causal agent of foot and sheath rot of rice in the field of Guilan province of Iran. Agricul.Trua.41: 17-20.

Little L E, Bostock R M, Kirkpatrick B C (1998). Genetic characterization of Pseudomonas syringaepv.syringae strains from stone fruits in California. Appli.Enviro.Microbio. 64: 3818-3823.

Louws FJ, Fulbright DW, Stephens CT, de Bruijn FJ (1994). Specific genomic fingerprints of phytopathogenicXanthomonasand Pseudomonas pathovars and strains generated with repetitive sequences and PCR. Appli.Enviro.Microbio.60: 2286-295.

Manceau C, Horvais A (1997). Assessment of genetic diversity among strains of Pseudomonas syringaeby PCR-restriction fragment length polymorphismanalysis of rRNA operons with special emphasison P. syringaepv. tomatoAppli. Enviro.Microbio.63: 498-505.

Mew T W, Cottyn B (ed) (2001). Seed Health and Seed- Associated Microorganisms for Rice Diseasse Management. Limited Proceedings. IRRI, No 6.

Miyajima K, Tanii A, Akita T (1983).Pseudomonas fuscovaginae sp. nov., nom. rev. Int. J. Syst. Bacteriol. 33: 656-657.

Naeimi S, Okhovvat SM, Hehjaroude GA, Khosravi V (2003). Sheath rot of rice in Iran.Common.Agric. Appl Bio. Sci. 68: 681-684.

Olczak-Woltman H, Masny A, Bartoszeweski G, Ptucienniczak A, Niemirowicz-Szczytt K (2007). Genetic diversity of Pseudomonas syringaepv. lachymans strains isolated from cucumber leaves collected in Poland. Plant patho. 56: 373-382.

Ou SH. 1985. Rice diseases.Second Edition. Commonwealth Mycological Institute, Kew, Surrey, England ,380 p.

Porteous Arlene L, Widmer F, Seidler RJ (2002). Multiple enzyme restriction fragment length polymorphism analysis for high resolution distinction of pseudomonassensustrito 16S rRNA genes. J. Microbio. Metho. 51: 337-348.

Rostam M, Rahimian H Gasemi A (2005). Identification of pseudomonas fuscovaginaeas the bacterial sheath brown rot of rice in the North of Iran.Iran J. Plant.Pathol.41: 57-58.

Rott P, Honegger J, Notteghem J L, Ranomenjanahary S (1991). Identification of Pseudomonas fuscovaginae with biochemical, serological and pathogenicity tests.Plant Dis. 75: 843-846.

Rott P, Nottheghem JL, Frossard P (1989). Identification and characterization of Pseudomonas fuscovaginae, the causal agent of bacterial sheath brown rot of rice, from Madagascar and other countries Plant Dis. 73: 133-137.

Schaad NW, Jones JB, Chun W (2001). Laboratory guide for identification of plant pathogenic bacteria.3rd edition.APS St Paul Minnesota USA, 245 Pp.

Webster RK, Gunnell PS (ed) (1992). Compendium of Rice Diseases.APS Press, St. Paul, Minnesota, USA, 62p.

Widmer F, Seidler RJ, Gillevet PM, Watrud LS, DiGiovanni GD (1998). A highly selective PCR protocol for detecting 16S rRNA genes of the genus Pseudomonas (sensustricto) in environmental samples.Appl.Enviro.Microbio. 64: 2545-2553.

Xie G (2003). First report of sheath brown rot of rice in China and characterization of causal organism by phenotypic tests and biologIRRN 28: 50-53.

Xie G, Cui Z, Tao Z, Qiu H, Liu H, Ibrahim M, Zhu B, Jin G, Sun G, Almoneafy A, and Lia B (2012). Genome Sequence of the Rice Pathogen Pseudomonas fuscovaginae CB98818. J. Bacteriol. 194(19):5479.

Yuan XL (2004).Identification of bacterial pathogens causing panicle blight of rice in Louisiana. M.S.thesis, University of Louisiana state, USA . 102p. Available: on line: etdu. lsu.Edu/ docs/available/

Zeigler GS, Alvarez E (1990). Characteristics of Pseudomonas spp. Causing grain discoloration and sheath rot of rice and associated Pseudomonas epiphytes. Plant Dis. 74: 917-922.

Zeigler RS, Alvarez E (1987).Bacterial sheath brown rot of rice caused by Pseudomonas fuscovaginae in Latin America.Plant Dis. 71: 592-597.

Zeigler RS, Aricapa G, Hoyos E (1987) Distribution of fluorescent Pseudomonas spp. Causing grain and sheath discoloration of rice in Latin America. Plant Dis. 71: 894-900.

20

Brucellosis: Prevalence and retrospective evaluation of risk factors in western cities of Tehran province, Iran

Abas Bahador[1], Noormohamad Mansoori[1], Davood Esmaeili[2]*, Reza Amini Sabri[2]

[1]Department of Medical Microbiology, Faculty of Medicine, Tehran University of Medical Sciences, Tehran, Iran.
[2]Bacteriology Department, Applied Microbiology Research Center, Baqiyatallah University.

Brucellosis is the commonest zoonotic disease that is spread worldwide. In this study, we aimed to compare 40 brucellosis cases with other big series in the literature in view of epidemiological and laboratory findings. A total of 40 brucellosis cases referred to Shahriar Health Center over a 3-year period were included in our study and patient files were reviewed for history and laboratory findings. Of the 40 patients, 19 (47.5%) were females and 21 (52.5%) were males. The mean age of patients was 40 ± 32 years, and 22.5% of cases were aged 24 to 32 years. 19 cases (47.5%) had a history of raising livestock and 52.5% of the cases were found to have no occupational risk for brucellosis. All cases had a history of raw milk and dairy products consumption. The standard tube agglutination (STA) test, Coombs STA test and 2-mercaptoethanol (2ME) test was positive in all cases (100%). Brucellosis is a major health problem in Iran and the world. Pasteurization of milk and dairy products and education on eating habits must be pursued for eradication of human brucellosis; thus, detection of precise brucella and report to WHO is essential.

Key words: Brucellosis, Iran, risk factor.

INTRODUCTION

Brucellosis is the commonest zoonotic disease with worldwide expansion, representing a serious public health problem in many countries, especially those around the Mediterranean Sea, Middle East and South America (Karagiannis and Mellou, 2008; Al-Majali and Shorman, 2009). In these areas, poor diagnosis and treatment may result in serious complications (Boschiroli et al., 2001). Most common symptoms of brucellosis include undulant fever, weakness, night sweats with peculiar odor and chills (Acha and Szyfres 2003).

Each year, more than 500,000 new cases are reported, and this figure underestimates the magnitude of the problem. It is also a disease of considerable economic and social importance (Agasthya et al., 2011). This disease causes high clinical morbidity and various clinical manifestations in humans, and any organ can be affected include encephalitis, meningitis, spondylitis, orchitis, prostatitis, arthritis and endocarditic (Skendros and Pappas, 2011).

Although brucellosis in domestic animals has been controlled in most developed countries, it remains an important public and animal health problem in the developing countries. Brucellosis is endemic in Iran (Al-Majali and Shorman, 2009).

Brucellosis is transmittable disease from animals to humans through consumption of contaminated Dairy products and infected meat from domestic livestock or through the aerosol route and through direct contact with infected animals (Buzgan and Karahocagil, 2010).

The incubation period of brucellosis normally is 1 to 3 weeks, but it can be more to several months before showing signs of infection (Seleem and Boyle, 2010). The diagnosis of brucellosis must be confirmed by laboratory tests. Accurate and fast diagnosis of human brucellosis is very important as delay or misdiagnosis usually results in

---

*Corresponding author. E-mail: esm114@gmail.com.

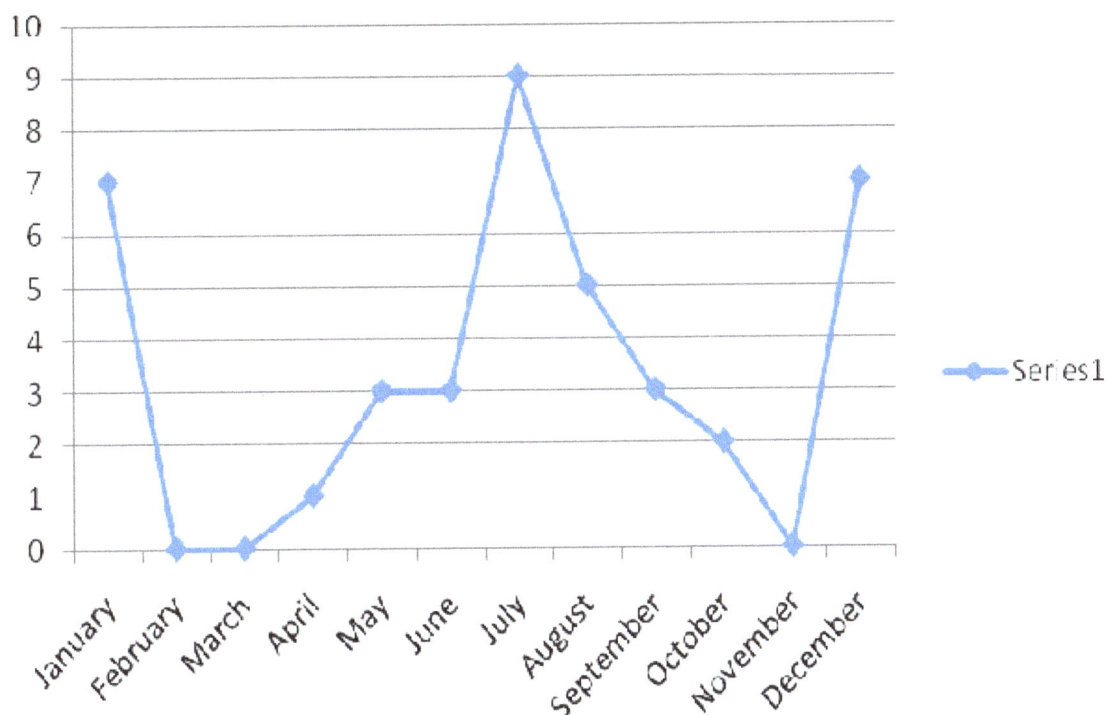

Figure 1. Seasonal distribution of patients over the study period.

treatment failures, chronic courses, focal complications, relapses and increase of mortality rate (Seleem and Boyle, 2010).

Serological tests for antibody detection that measure the ability of serum to agglutinate a standardized amount of killed *Brucella abortus* reflect the presence of antibody against O-side chain. Brucella specific IgM appeared at the end of the first week of the disease followed by IgG. This test is still the most common and useful method for the laboratory diagnosis of brucellosis (Seleem and Boyle, 2010).

Epidemiological studies on brucellosis have suggested that adults have more commonly become seropositive than younger people (Al-Majali and Shorman, 2009).

The objectives of this study were to investigate the prevalence of brucellosis in patients with symptoms and investigate risk factors associated with brucellosis in western cities of Tehran province.

MATERIALS AND METHODS

A total of 40 brucellosis cases referred to Shahriar Health Center over a 3-year period from March 2008 to March 2012 were included in our study. The patients' files were studied retrospective for their history and laboratory findings.

Brucellosis was diagnosed on by a compatible clinical signs, such as undulant fever, night sweats, chills and weakness, supported by the detection of specific antibodies against brucella, at significant titers.

Significant titers were ≥1/80 in the standard tube agglutination test (STA) and in the presence of 2-mercaptoethanol (2ME) agglutination ≥1/20 (Buzgan and Karahocagil, 2010; Sofian and Aghakhani, 2008). *Brucella abortus* s19 (Pasteur Institue, Iran) used for the STA.

RESULTS

The majority of brucellosis cases occurred during January, July and December (Figure 1). Of the 40 patients that were studied, 19 (47.5%) were females and 21 (52.5%) were males. The mean patient age was 40 ± 32 years, ranging from 8 to 72 years; 6 (15%) were aged 8 to 16 years, while 8 (20%) were aged 16 to 24 years, 9 (22.5%) were aged 24 to 32 years, 7 (17.5%) were aged 32 to 40 years, 4 (10%) were aged 40 to 48 years, 3 (7.5%) were aged 48 to 56 years, 2 (5%) were aged 56 to 64 years, and 1 (2.5%) were over 64 years of age.

In studied patients, 19 (47.5%) had a history of raising livestock (4 housewives, 4 self-employed persons, 3 students, 3 workers, 2 farmers, 1 butcher, 1 goat sales and 1 unemployed (Figure 2).

In this study, 78.9% of cases were found to have no occupational risk for brucellosis. All cases had a history of raw milk and dairy products consumption. Education level, ranged from illiterate to graduate (Figure 3). The STA test was positive in all (100%) cases, with titers

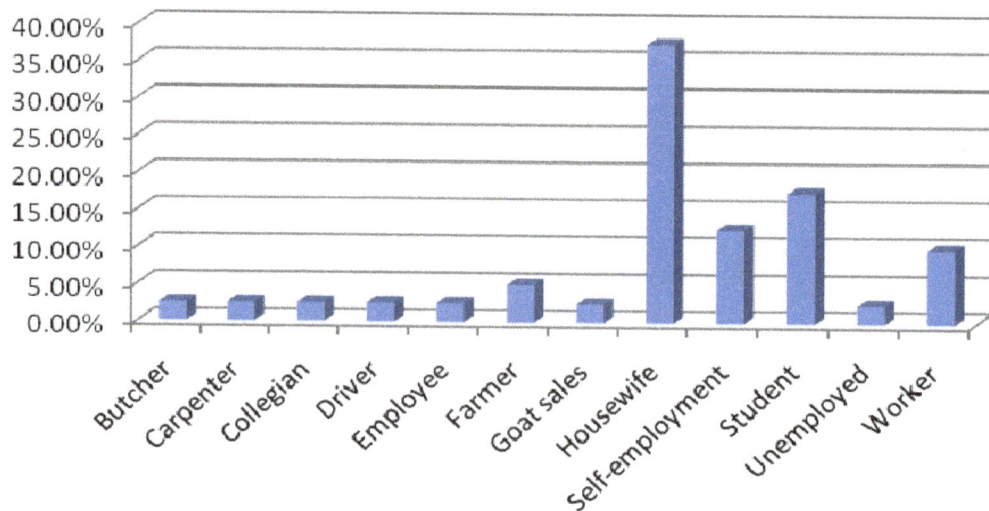

Figure 2. Occupational characteristics of cases.

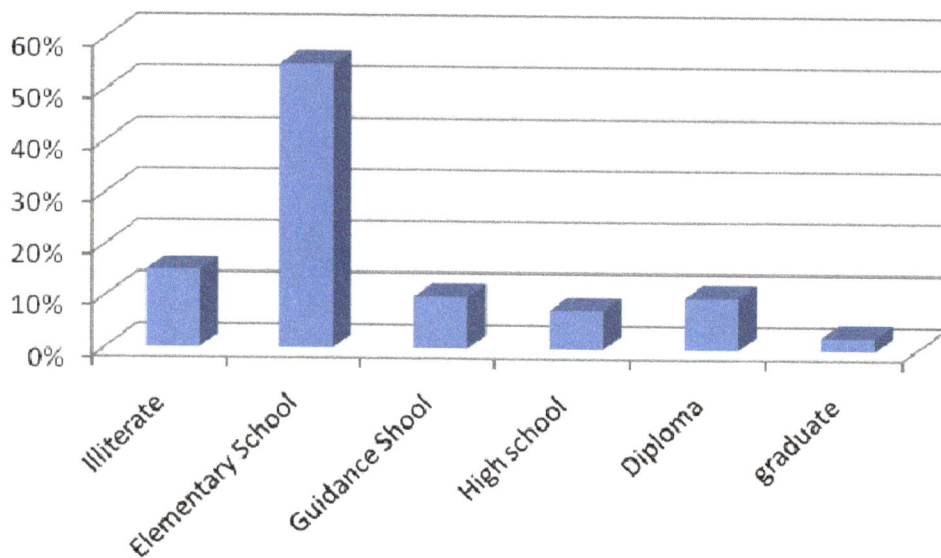

Figure 3. Educational characteristics of case.

ranging from 1/80 to 1/20480.

DISCUSSION

Brucellosis is the commonest zoonotic bacterial disease in the world; transmissions to human were reported by direct and indirect routes, affecting more than 500,000 people each year. Alballa reported that consump-tion of dairy products and direct contact with domestic animals was the main risk factors for this disease (Alballa, 1995). In a study in Iran, the incidence of brucellosis in Tehran province was 17.5/100 000 (data published in Iranian journals).

Brucellosis is more prevalent in age 15 to 35 years groups, in endemic countries (Young et al., 2005; Dog̈anay and Mese Alp, 2008). In this study, 75% of patients were between 8 to 40 years of age. This research indicates that unhealthy dairy products con-sumption have important role in spreading of pollution.

In a study in Iran by Hasanjani Roushan et al., 469 cases were studied and mean age of 36.97 ± 15 years was reported in a population which includes 56.9% male and43.1% female (Hasanjani et al., 2004) (Table 1).

Table 1. Comparison of gender and age of patients in various studies

Author [ref.]	Year	Country	No. of cases	Male (%)	Female (%)	Mean age (years)
Sofian and Aghakhani	2008	Iran	150	55.3	44.7	33.37 ± 21.3
Buzgan and Karahocagil	2010	Turkey	1028	47.6	52.4	33.7 ± 16.34
Earhart and Vafakolov	2009	Uzbekistan	144	83.3	16.7	42 ± 32
Mantur et al.	2006	India	495	78.8	21.2	31
This study	2012	Iran	40	52.5	47.5	40 ± 32

In endemic countries, the primary transmission route of brucellosis is by the ingestion of unpasteurized milk and dairy products, but in developed ountries, occupational exposure is important (Young et al., 2005). History of raw dairy product consumption was present in all cases in our study. The consumption of raw dairy products in other studies has been reported as occurring in 23.6% of cases in Spain (Colmenero et al., 1996), 69% in Kuwait (Mousa et al., 1988), 34.7% in the Balkan Peninsula (Bosilkovski et al., 2007) and 22.4% in Iran (Hasanjani et al., 2004).

In developed countries, most of the brucellosis cases occur due to occupational exposure. High-risk occupations for the disease are the raising of livestock, butchery, farming, and veterinary medicine, laboratory transmission has also been reported (Dog¨anay and Mese Alp, 2008).

History of livestock raising in countries has been reported in 11.1% from Kuwait (Mousa et al., 1988), 9% also from Kuwait (Lulu et al., 1988), 11.3% from Iran (Hasanjani et al., 2004) and 20% from Greece (Andriopoulos et al., 2007). In this study, 19 cases (47.5%) have history of livestock raising. However in Iran, most of the families, particularly housewives, deal with the raising of livestock, which increases the contact rate in this study. Occupational contact was found in 4 patients in our study (10%; 2 farmers, 1 butcher and 1 goat seller). No contact history was identified in 52.5% of patients. In our study, all patients had risk factor for brucellosis. Risk factors in other studies have been found for between 10.9 and 28.7% of cases (Gur et al., 2003; Tasova et al., 1998) and even higher rates 41.8 and 56.7% have been reported (Yu¨ ce et al., 2006; Hasanjani et al., 2004). Our data showed that occupation levels were not important risk factors for brucellosis, but educational levels were important risk factors.

Serology is the preferred method for the diagnosis of brucellosis when bacterial isolation is not possible, and serologic testing is widely used in the diagnosis of brucellosis (Gotuzzo and Celillo, 1992). STA test positivity was reported in 95 and 87% (Akdeniz et al., 1998; Tasova et al., 1998).

The STA test was positive in 100% of our study population. Coombs STA test is preferred when the STA test is found negative. In endemic regions for brucellosis such as Iran, serological test results should only be interpreted as significant in the presence of clinical findings compatible with brucellosis.

Findings of this study suggest that brucellosis is an important health problem in Iran. The disease has a significant morbidity and mortality. Additionally, since the disease primarily affects persons in their productive age, it causes important work-power losses. Pasteurization of milk and dairy products and education for eating habits must be pursued for eradication of human brucellosis. Eradication of the disease in humans can be achieved by the control of the disease in animals; this necessitates a multidisciplinary approach involving both humans and animals. On the other hand, development of detection methods and prevention of brucellosis is important.

REFERENCES

Acha NP, Szyfres B (2003). Zoonoses and Communicable Diseases Common to Man and Animals. PAHO 3:132-137.

Agasthya AS, Isloor S (2012). Seroprevalence study of human brucellosis by conventional tests and indigenous indirect enzyme-linked immunosorbent assay. J. Sci. Work 201: 86-91.

Akdeniz H, Irmak H, Demiroz AP (1998). Evaluation of brucellosis cases in Van region of Eastern Anatolia: a-3 year experience. Nagoya Med. J. 42:101-110.

Alballa SR (1995). Epidemiology of human brucellosis in southern Saudi Arabia. J. Trop. Med. Hyg. 98:185-189.

Al-Majali AM, Shorman M (2009). Childhood brucellosis in Jordan: prevalence and analysis of risk factors. Int. J. Infect. Dis. 13:196-200.

Andriopoulos P, Tsironi M, Deftereos S, Aessopos A, Assimakopoulos G (2007). Acute brucellosis: presentation, diagnosis and treatment of 144 cases. Int. J. Infect. Dis.11:52-7.

Boschiroli ML, Foulongne V (2001).Brucellosis: a worldwide zoonosis. Curr. Opin. Microbiol. 4: 58-64.

Bosilkovski M, Krteva L, Dimzova M, Kondova I (2007). Brucellosis in 418 patients from the Balkan Peninsula: exposure-related differences in clinical manifestations, laboratory test results, and therapy outcome. Int. J. Infect. Dis. 11: 342-347.

Buzgan T, Karahocagil MK (2010). Clinical manifestations and complications in 1028 cases of brucellosis: a retrospective evaluation and review of the literature. Int. J. Infect. Dis. 14:469-478.

Colmenero JD, Reguera JM, Martos F, Sanchez De Mora D, Delgado M, Causse M (1996). Complications associated with Brucella melitensis infection: a study of 530 cases. Medicine 75:195-211.

Dog¨anay M, Mese Alp E (2008). Infeksiyon hastalıkları ve mikrobiyolojisi. Nobel Tıp Kitabevleri 3: 897-909.

Earhart K, Vafakolov S (2009). Risk factors for brucellosis in Samarqand Oblast, Uzbekistan. Int. J. Infect. Dis. 13:749-753.

Gotuzzo E, Celillo E (1992). Brucella. Philadelphia. Harcourt Brace Jovanovich Inc. 24:1513-1518.

Gür A, Geyik MF, Dikici B, Nas K, Cevik R, Sarac J (2003). Complications of brucellosis in different age groups: a study of 283 cases in Southeastern Anatolia of Turkey. Yonsei Med. J. 44:33-44.

Hasanjani Roushan MR, Mohrez M, Smailnejad Gangi SM, Solemani Amiri MJ, Hajiahmadi M (2004). Epidemiological features and clinical manifestations in 469 adult patients with brucellosis in Babol, Northern Iran. Epidemiol Infect. 132:1109-1114.

Karagiannis I, Mellou K (2008). Outbreak investigation of brucellosis in Thassos, Greece, 2008. Euro Surveill. 17: 34-37.

Lulu AR, Araj GF, Khateeb MI, Mustafa MY, Yusuf AR, Fenech FF (1988). Human brucellosis in Kuwait: a prospective study of 400 cases. Q. J. Med. 66:39-54.

Manture BG,Amarnath SK,Shinde RS(2006). Review of Clinical and laboratory features of human Brucellosis. Indian Journal of Medical Microbiology.25(3):188-202

Mousa AR, Elhag KM, Khagali M, Marafie AA (1988). A nature of human brucellosis in Kuwait: study of 379 cases. Rev. Infect. Dis. 10:211-217.

Seleem MN, Boyle SM (2010). Brucellosis: a re-emerging zoonosis. Vet Microbiol. 140: 392-398.

Skendros P, Pappas G (2011). Cell-mediated immunity in human brucellosis. Microbes Infect. 13:134-42.

Sofian M, Aghakhani A (2008). Risk factors for human brucellosis in Iran: a case-control study. Int J Infect Dis. 12: 157-61.

Tasova Y, Saltoğlu N, Yılmaz G,Inal S, Aksu HS (1988). Bruselloz: 238 eris, kin olgunun klinik, laboratuvar ve tedavi ö zelliklerinin değerlendirilmesi. Turk. J Infect. 12:307-312.

Young EJ, Mandell GL, Bennett JE, Dolin R (2005). Principles and practice of infectious diseases. Churchill Livingstone 2: 2669-2672.

Yüce A, Alp-Cavus S, Yapar N, Cakır N (2006). Bruselloz: 55 olgunun değerlendirilmesi. Klimik Derg. 19:13-17.

21

Effect of high temperature on viability of *Lactobacillus casei* and analysis of secreted and GroEL proteins profiles

Najla Haddaji, Boubaker Krifi, Rihab Lagha, Sadok Khouadja and Amina Bakhrouf

Laboratoire d'Analyse, Traitement et Valorisation des Polluants de l'Environnement et des Produits, Faculté de Pharmacie Rue Avicenne, Monastir 5000, Tunisie,

The bacterial heat shock response is characterized by the elevated expression of a number of chaperone complexes including the GroEL and the rate change of synthesis of certain proteins (total and secreted). In this work, after incubation at 45°C, total and secreted proteins profiles of stressed bacteria were found to be altered when analyzed by sodium dodecyl sulfate-polyacrylamide gel electrophoresis (SDS-PAGE). In addition, the level expression of GroEL was evaluated with Western blot. Our results show a marked increase in both GroEL expression and protein synthesis at 45°C. These modifications were manifested by the appearance and/or disappearance of bands as well as in the level of expression of certain proteins.

Key words: *Lactobacillus casei*, heat shock, GroEL, secreted and total proteins, sodium dodecyl sulfate-polyacrylamide gel electrophoresis (SDS-PAGE), Western blotting.

INTRODUCTION

The viability of probiotic in foods depends on various factors during processing and storage. Heat is used in the process of a lot of foods, so the survival of probiotics in thermal processing is the main obstacles to food manufacturers (Mansouripour et al., 2013). Thus, paying special attention to the effect of heat shock on the survival of probiotics is necessary. For an organism to grow at high temperatures, especially as high as those of the hyperthermophiles discussed here, all cellular components, including proteins, nucleic acids and lipids, must be heat protected. Desmond et al. (2004) showed that the mechanisms involved in thermoprotection of *Lactobacillus paracasei NFBC338* are probably controlled at the protein synthesis level. It has been previously reported that stress tolerance in *Lactobacillus delbrueckii ssp. bulgaricus* induced by a moderate heat shock was dependent on protein synthesis (Gouesbet et al., 2002) and Whitaker and Batt (1991) demonstrated the enhanced synthesis of a number of heat shock proteins, including GroEL and DnaK in a heat-adapted culture of *Lactococcus lactis* ssp. *lactis*. In addition, destabilization of macromolecules as ribosomes and RNA, and alterations of membrane fluidity were also described (Earnshaw et al., 1995; Teixera et al., 1997; Hansen et al., 2001). The heat-shock response has been studied notably in *Escherichia coli* (Gram negative) and *Bacillus subtilis* (Gram positive). Physiological studies demonstrated that lactobacilli elicit heat-shock responses similar to that of other Gram positive bacteria: *L. lactis* (Whitaker and Batt, 1991; Auffray et al., 1992; Kilstrup et al., 1997), *Leuconostoc mesenteroides* (Salotra et al., 1995), *Enterococcus faecalis* (Flahaut et al., 1996), *Oenococcus oeni* (Guzzo et al., 1997) and *L. bulgaricus* (Gouesbet et al., 2002). In order to understand the mechanisms of stress tolerance of lactobacilli, numerous studies have examined the physiological and genetic adaptations of these organisms during growth and survival in diverse environmental stresses (Corcoran et al., 2008; Spano and Massa, 2006; Van de Guchte et al., 2002) and postgenomic approaches have accelerated the understanding of the global (genome-wide) stress responses in lactobacilli to acid, lactate, oxidative, bile and heat stresses (Broadbent et al., 1998; Stevens, 2008; Serrano et al., 2007; Bron et al., 2006; Pieterse et al., 2005; De Angelis et al., 2004). These studies have shown that lactobacilli respond rapidly to their environment by modulating expression levels of genes involved in different cellular processes, including stress response pathways, cell division, transport and cell envelope composition. However, the proteins synthesized by the stressed cells during the stationary phase were necessary for maintaining of the viability during prolonged stress (Kolter et al., 1993). These proteins enable the cell to neutralize stress to adapt or repair damages caused by stress (Hecker et al., 1996). Examination of lactobacilli heat shock responses using bi-dimensional electrophoresis revealed variable numbers of induced proteins: 34 in *E. faecalis* (Flahaut et al., 1996), 17 in *L. lactis* (Kilstrup et al., 1997) and 40 in *Streptococcus mutans* (Svensater et al., 2000), among them, only six proteins were specifically up regulated by heat (Svensater et al., 2000). In *L. lactis*, the 12 proteins were induced by NaCl and detected on bi-dimensional electrophoresis, all belong to the heat shock stimulon (Kilstrup et al., 1997). This striking overlap between heat-shock and osmotic-stress responses may also exist in *S. mutans* as 21 heat-inducible proteins also belong to the osmotic stress response. Among the heat shock proteins, well conserved chaperones (DnaK, DnaJ, GrpE, GroES and GroEL) and proteases (Clp, HtrA and FtsH) have often been identified.

Some of the most intensively investigated heat shock proteins include the molecular chaperones GroEL and GroES, which are highly conserved in *E. coli* and eukaryotic cells (Gupta, 1995). The GroEL and GroES chaperones (also known as Hsp60 and Hsp10 chaperones) have been recognized as heat shock proteins in many bacteria, including *E. coli*, *B. subtilis* (Hecker et al., 1996), *Agrobacterium tumefaciens* (Segal et al., 1996), *Streptomyces lividans* (De Leon et al., 1997), *L. lactis* (Kilstrup et al., 1997), *Lactobacillus helveticus* (Broadbent et al., 1998), *Pseudomonas aeruginosa* (Fujita et al., 1998) and *Lactobacillus*

johnsonii VPI 11088 (Walker et al., 1999).

In the present work, we focused on the effect of heat stress on GroEL expression of two strains of *Lactobacillus* and their influence on secreted and total proteins profiles of stressed cells. The cells were analyzed by sodium dodecyl sulphate (SDS)-polyacrylamide gel electrophoresis (PAGE).

MATERIALS AND METHODS

Bacterial strain and growth conditions

Two strains of *Lactobacillus casei* were used in this study: S1, *L. casei* (ATCC 393); S2, *L. casei*, a potential probiotic, obtained from the laboratory collection; this strain was identified by sequencing the 16S rRNA gene (BankIt1773923 BL2 KP123430). All strains were stored on De Man, Rogosa, Sharpe (MRS; LAB M, Bury, UK) broth added with 33% of sterile glycerol at -20°C. Working cultures were grown at 37°C in modified MRS broth at pH of 6.4 for 24 h, which contained 0.5% maltose, 1% peptone, 0.5% yeast extract, 1 ml Tween 80, 0.2% K_2HPO_4, 0.2% ammonium citrate, 0.02% $MgSO_4$ and 0.005% $MnSO_4$. The pH of the modified MRS was 6.4 and, unless otherwise stated, it was maintained constant by the on-line addition of 1 M NaOH. In this study, the cells were grown at 37°C on MRS for 24 h.

Stress conditions

Cells were recovered by centrifugation and re-suspended to an optical density at 600 nm of 0.5 immediately before heat shock treatment at 45, 55, and 65°C. One culture (1 ml) was maintained at 37°C as a control, and three cultures were shifted to 45, 55 and 65°C. After 10 min, heat shock was stopped by placing samples on ice for 5 min.

Determination of the lethal temperature of *Lactobacillus*

100 μl of a bacterial suspension was inoculated with an optical density at 595 nm of about 0.6 prepared from a bacterial culture in MRS incubated overnight. These bacterial suspensions were treated at temperatures ranging from 45 to 95°C with an interval of 10°C for 10 min. All suspensions were prepared at the same time; one sample of each is removed after every increase of 10°C. All experiments were performed in triplicate.

Enumeration of cells

To determine the number of cultivable cells, decimal dilutions were performed in series in sterile saline solution. A volume of 0.1 ml of each dilution was then spread on the surface of three plates of MRS at 37°C. After 24-48 h of incubation, the enumeration of CFU/g of the appropriate dilution were made. All experiments were performed in triplicate.

After incubation and enumeration, we observed temperatures for which a bacterial growth is possible. Thus, we performed the heat shock at three temperatures: 45, 55 and 65°C.

Total protein extraction

Total proteins of two strains of *Lactobacilli* were prepared according to the method described previously (Sabri et al., 2000). Cultures of 24 h at 37°C were centrifuged (7000 ×g for 10 min at 4°C) and the cells were then washed three times in 40 ml of sterile saline solution

(0.9% NaCl) and resuspended in 5 ml of sterile saline solution. Cells were disrupted by lysosyme solution (10 mg/ml) and incubated at 37°C for 30 min. 30 µl of loading buffer was added to the solution and the pellet was incubated in a water bath at 100°C for 5 min. Supernatant was collected and further centrifuged at 100,000 ×g for 40 min at 4°C. The concentration of the total proteins in the final preparation was determined using the Bradford assay (Bradford, 1976).

Secreted protein extraction

Cultures of 24 h at 37°C were centrifuged (7000 ×g for 20 min) and subsequently filtered through a 0.45-µm pore-size filter. Proteins from the cell-free culture supernatants were then precipitated by addition of 10% (vol/vol) trichloroacetic acid and recovered by centrifugation at 7000 ×g for 20 min. Pellets were resuspended in 4 ml of phosphate-buffered saline (PBS), and protein were precipitated by addition of 20 ml of cold acetone. After centrifugation at 7000 ×g for 20 min, the pellets were washed once with cold acetone, air-dried and re-suspended in 25 µl of PBS. The concentration of the secreted proteins in the final preparation was determined using the Bradford assay (Bradford, 1976).

Sample preparation and SDS-PAGE

Protein pellet were denatured with sample containing β-mercaptoethanol (Sigma, Chemical Co., St Louis, MO, USA). These protein samples were subjected to SDS-PAGE analysis using a Mini Protean 3 gel Unit (BIORAD; Richmond, CA) on 15% (w/v) polyacrylamide gels (Laemmli, 1970). Wide range protein markers (from 212 to 6.5 kDa) were used as molecular weight standards (High-Range Rainbow; Amersham, Little Chalfont, Buckinghamshire, UK). Proteins were visualized by Biosafe colloidal Coomassie blue (Sigma, Chemical Co., St Louis, MO, USA) according to the manufacturer's instructions.

Western blotting

20 µg of total proteins were analyzed by SDS PAGE as reported above and blotted onto a nitrocellulose membrane (Hybond ECL, Amersham), using the Mini Trans-Blot equipment (BioRad Inc.), at 90 mA, 4°C, for 16 h in transfert Buffer. The membrane was incubated in TBST buffer (20 mM Tris base, 137 mM NaCl, 0.15% Tween 20) + nonfat dry milk 4% for 1 h at room temperature, with orbital shaking, then incubated in rabbit polyclonal anti-GroEL antibodies (Abcam) diluted 1:10000 in blocking buffer, at room temperature for 1 h. After three washes with TBST, the membrane was incubated with a 1:10000 dilution of the secondary antibody conjugated with horseradish peroxidase (GE Healthcare), in blocking buffer at room temperature for 1 h. After three washes with TBST, the bound antibodies were revealed with revelation solution (tris HCl pH 9, 0.1 M MgCl₂, 1 M Nacl, NBT, BCIP).

Statistical analysis

The statistical analysis was performed on SPSS v.17.0 statistics software. The statistical differences and significance were assessed by ANOVA test. $P < 0.05$ was considered significant.

RESULTS

Effect of temperature on survival of *Lactobacillus*

The survival of *L. casei* at different temperatures was

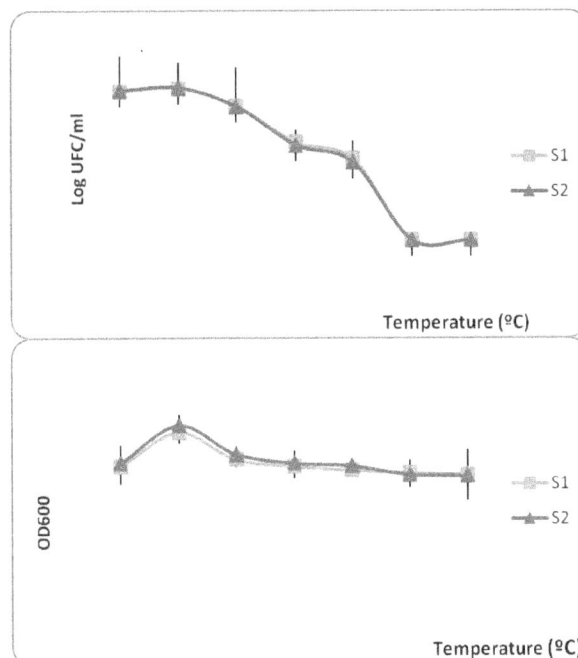

Figure 1. Effect of heat stress on the survival of *L. casei* ATCC 393 (S1) and *L. casei* BL2 KP123430(S2) after 48 h of incubation at pH 7.

investigated. Our results show that the two strains used remained cultivable up to 65°C. The viability was studied for an initial population of about 10^8 CFU/g. It was shown that both strains can survive at 75°C for 10 min (Figure 1). Remarkably, at 45°C, we observed an increase of the optical density of the culture (Figure 1).

Analysis of total proteins

Total protein of *Lactobacilli* strains before and after heat shock were analyzed by SDS-PAGE. They displayed different profiles before and after exposure of heat shock. Before heat stress, the strains S1 and S2 of *L. casei* had the same total protein profile. Five clear bands were detected in each profile at 27, 35, 38 43, 90 kDa (Figure 2). After heat stress, some modifications were observed in total protein profile in both strains. However, at 45 and 55°C, the expression of band corresponding to protein of 35 kDa was significantly decreased. Interestingly, bands corresponding to molecular weights of 38 and 43 kDa were significantly slightly increased at 55°C (Figure 2). At 65°C, the protein synthesis was completely stopped; all bands were not visible after heat shock to this temperature for 10 min.

Analysis of secreted proteins profiles

Secreted protein of Lactobacilli strains displayed different

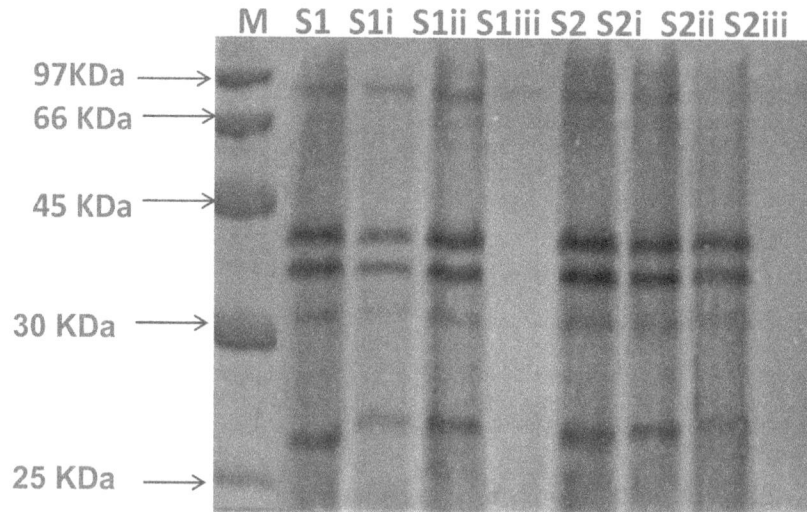

Figure 2. Total protein profile of two strains of *L. casei* subjected to heat stress: M: High-Range Rainbow (Amersham, Little Chalfont, Buckinghamshire, UK), S1: *L. casei ATCC339*; S2: *L. casei* BL2. Si, ii and iii: strains subject respectively, at temperatures of 45, 55 and 65°C.

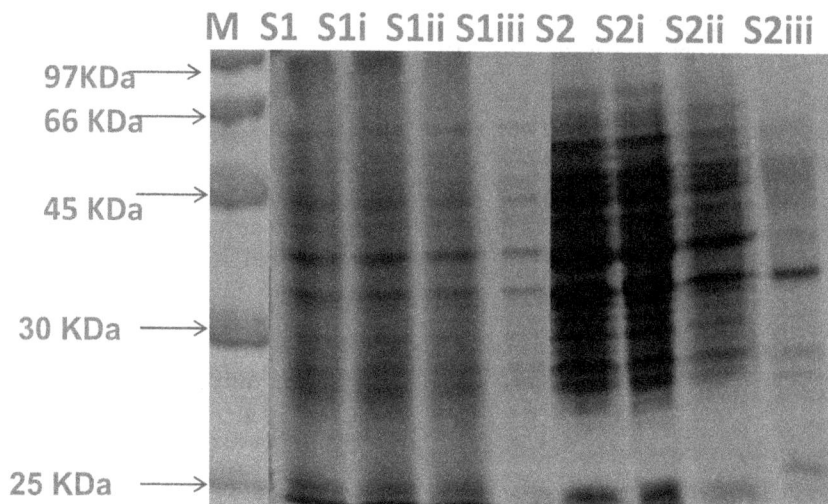

Figure 3. Secreted protein profile of two strains of *L. casei* subjected to heat stress: M: High-Range Rainbow (Amersham, Little Chalfont, Buckinghamshire, UK), S1: *L. casei ATCC339*; S2: *L. casei* BL2. Si, ii and iii: strains subject respectively, at temperatures of 45, 55 and 65°C.

profiles before and after exposure to heat shock (Figure 3). *Lactobacillus* presents five or six major proteins in their profiles. After heat stress, both strains have a different extracellular protein profile, whereas, bands corresponding to protein of molecular weight of 65 kDa were not visible at 55 and 65°C for S2. In comparison, the intensity of this band for S1 was increased in intensity after heat stress. In addition, we have observed the level intense of one band corresponding to molecular weight of 35 kDa for S2. Interestingly, bands corresponding to

molecular weights of 25 and 27 kDa approximately were slightly increased at 55 and 65°C.

GroEL expression

As shown in the Figure 4, the expression of protein was greater at 45°C (column S1i and S2i) and the higher levels were found in both strains at 55°C. However, the proteins were not visible at 65°C (Figure 4).

Figure 4. Western blot analysis of GroEL present in total protein extracts *L. casei* detected with antibodies raised against GroEL., M: high-range rainbow (Amersham, Little Chalfont, Buckinghamshire, UK), S1: *L. casei ATCC339*; S2: *L. casei* BL2. Si, ii and iii: strains subject, respectively, at temperatures of 45, 55 and 65°C.

DISCUSSION

The results show that Gram positive bacteria, such as *Lactobacilli*, are able to survive heat shock up to 75°C. In addition, we found that *L. casei* cells remained cultivable at temperature of 65°C, which demonstrates that bacteria are able to withstand such adverse environments (Van de Guchte et al., 2002). It has been shown that organisms respond to environmental stress by modifying the rate of synthesis of certain proteins (Hennequin et al., 2001). In our study, we have shown that the heat shock induced alterations of secreted and total proteins and induced elevated GroEL expression. The marked increase on both GroEL expression and protein synthesis may suggest that *Lactobacillus* species are able to grow and survive under suboptimal conditions during food and beverage fermentations. Desmond et al. (2004) revealed by 2D-PAGE that the chaperone protein GroEL was among the most strongly expressed proteins in the cell under heat adaptation conditions; indeed, densitometry analyses indicated an approximately 50-fold increase in cells that were pre-adapted to heat shock. Kim et al. (2001) demonstrated that the transcriptional activity of both chaperones is increased dramatically in response to heat shock and was increased to a lesser extent by ethanol stress. Other studies have shown that culture of log-phase *Lactobacillus* cultures were subjected to heat stress, a 15-fold increase in GroEL synthesis was observed when compared with only 1.5-fold increase in protein synthesis in stationary phase cultures (Prasad et al., 2003).

These chaperones play a key role in the maturation of synthesized proteins and are pivotal in the degradation or refolding of denatured proteins (Rechinger et al., 2000). Thus, these heat-shock proteins can also be induced by multiple stresses, such as acid, heat, bile salts, high pressure stress and so on (De Angelis et al., 2004). Several studies have been done in the elucidation of the major chaperones belonging to the Hsp60 (GroEL) and Hsp70 (DnaK) families because protein folding has been recognized as one of the central problems in biology (Narberhaus, 2002). The study of GroESL-overproducing *L. lactis* and *L. paracasei NFBC 338* demonstrated that technologically sensitive cultures can be potentially manipulated to become more robust for survival under harsh conditions, such as food product development and gastrointestinal transit (Desmond et al., 2004). The main function of these HSPs appeared to be the prevention of

the accumulation of unfolded protein intermediates during periods of stress (Veinger et al., 1998). Although, the HSPs are always constitutively expressed, the rate of synthesis is significantly enhanced under stress (Ang et al., 1991).

The production of the HSPs is the origin of the beneficial effect of heat stress in the conservation of *Lactococcus*. Indeed, the physiological and biochemical riposte of the unicellular organisms to all occasional thermal elevation is universal. Some works have been done on *L. lactis* (Boutibonnes et al., 1995) which shows that the temporary exhibition of the cells to a temperature of 40-45°C provoked a disturbance followed by an adjustment of the metabolism. Thus, production of the majority of the usual proteins in stalled synthesis of HSP is unregulated. It has been demonstrated for *E. coli*, that a thermal treatment of 8 h has a beneficial effect on the cryodessiccation and the lyophilization (Joe et al., 2000).

In conclusion, our data suggest that overexpression of stress-induced proteins has the potential to improve the performance of lactobacilli strains. In particular, the GroES/EL chaperone complex can be exploited to prepare *Lactobacillus* for industrial processes. Indeed, the innate probiotic characteristics of the strain, such as adherence to the host cell wall and acid tolerance during gastric transit, may also be improved by the overexpression of GroESL.

Conflict of interests

The authors did not declare any conflict of interest.

ACKNOWLEDGEMENTS

The authors are grateful to Pr. Gaspar Perez Martinez [Laboratory of Lactic Acid Bacteria and Probiotics of the Institute of Agro-chemistry and Food Technology (Spanish National Research Council] for sequencing the 16S rRNA gene of *L. casei*.

REFERENCES

Ang D, Liberek K, Skowyra D, Zylicz M, Georgopoulos C (1991). Biological role and regulation of the universally conserved heat shock proteins. J. Biol. Chem. 266:24233-24236.
Auffray Y, Gansel X, Thammavongs B, Boutibonnes P (1992). Heat-

shock induced protein synthesis in Lactococcus lactis subsp. lactis. Curr. Microbiol. 24:281-284. http://dx.doi.org/10.1007/BF01577333

Boutibonnes P, Bison V, Thammavongs B, Hartke A, Panoff J, Benachour A, Auffray Y (1995). Induction of thermo tolerance by chemical agents in Lactococcus lactis subsp. lactis IL1408. Int. J. Food Microb. 25:83-94. http://dx.doi.org/10.1016/0168-1605(94)00149-Z

Bradford MM (1976). A Rapid and sensitive method for the quantitation of microgram quantities of protein utilizing the principle of protein-dye binding. Anal. Biochem. 72: 248-254. http://dx.doi.org/10.1016/0003-2697(76)90527-3

Broadbent JR, Orberg CJ, Wei L (1998). Characterization of the Lactobacillus helveticus groELS operon. Res. Microbiol. 149:247-253. http://dx.doi.org/10.1016/S0923-2508(98)80300-8

Bron PA, Molenaar D, de Vos WM, Kleerebezem M (2006). DNA micro-array-based identification of bile-responsive genes in Lactobacillus plantarum. J. Appl. Microbiol. 100:728-738. http://dx.doi.org/10.1111/j.1365-2672.2006.02891.x

Corcoran BM, Stanton C, Fitzgerald G, Ross RP (2008). Life under stress: the probiotic stress response and how it may be manipulated. Curr. Pharm. Des. 14:1382-1399. http://dx.doi.org/10.2174/138161208784480225

De Angelis M, Di cagno R, Huet C, Crecchio C, Fox PF, Gobetti M (2004): Heat shock response in Lactobacillus plantarum. Appl. Environ. Microbiol. 70:1336-1346. http://dx.doi.org/10.1128/AEM.70.3.1336-1346.2004

De Leon P, Marco S, Isiegas C, Marina A, Carrascosa JL, Mellado RP (1997). Streptomyces lividans groES, groEL1 and groEL2 genes. Microbiology 143:3563-3571. http://dx.doi.org/10.1099/00221287-143-11-3563

Desmond C, Fitzgerald GF, Stanton C, Ross RP (2004). Improved stress tolerance of GroESL-overproducing Lactococcus lactis and probiotic Lactobacillus paracasei NFBC 338. Appl. Environ. Microbiol. 70:5929-5936. http://dx.doi.org/10.1128/AEM.70.10.5929-5936.2004

Earnshaw RG (1995). Understanding physical inactivation processes: combined preservation opportunities using heat, ultrasound and pressure. Int. J. Food Microbiol. 28:197-219. http://dx.doi.org/10.1016/0168-1605(95)00057-7

Flahaut S, Hartke A, Giard JC, Benachour A, Boutibonnes P, Auffray Y (1996). Relationship between stress response toward bile salts, acid and heat treatment in Enterococcus faecalis. FEMS Microbiol. Lett. 138:49-54. http://dx.doi.org/10.1111/j.1574-6968.1996.tb08133.x

Fujita M, Amemura A, Aramaki H (1998). Transcription of the groESL operon in Pseudomonas aeruginosa PAO1. FEMS Microbiol. Lett. 163:237-242. http://dx.doi.org/10.1111/j.1574-6968.1998.tb13051.x

Gouesbet G, Jan G, Boyaval P (2002). Two-dimensional electrophoresis study of Lactobacillus delbrueckii subsp. Bulgaricus thermotolerance. Appl. Environ. Microbiol. 68:1055-1063. http://dx.doi.org/10.1128/AEM.68.3.1055-1063.2002

Gupta RS (1995). Evolution of the chaperon in families (Hsp60, Hsp10 and Tcp-1) of proteins and origin of eukaryotic cells. Mol. Microbiol. 15:1-11. http://dx.doi.org/10.1111/j.1365-2958.1995.tb02216.x

Guzzo J, Delmas F, Pierre F, Jobin MP, Samyn B, Van Beeumen J, Cavin JF, Divies C (1997). A small heat shock protein from Leuconostoc oenos induced by multiple stresses and during stationary growth phase. Lett. Appl. Microbiol. 24:393-396. http://dx.doi.org/10.1046/j.1472-765X.1997.00042.x

Hansen MC, Nielsen AK, Molin S, Hammer K, Kilstrup M, Palmer JrRJ, Udsen C, White DC (2001). Changes in rRNA levels during stress invalidates results from mRNA blotting: fluorescence in situ rRNA hybridization permits renormalization for estimation of cellular mRNA levels. J. Bacteriol. 183:4747-4751. http://dx.doi.org/10.1128/JB.183.16.4747-4751.2001

Hecker M, Schumann,W, Volker U (1996). Heat-shock and general response in Bacillus subtilis. Mol. Microbiol. 19:417-428. http://dx.doi.org/10.1046/j.1365-2958.1996.396932.x

Hennequin C, Porcheray F, Waligora-Dupriet AJ, Collignon A, Barc M, Bourlioux P, Karjalainen T (2001). GroEL (Hsp60) of Clostridium difficile is involved in cell adherence. Microbiol. 147:87-96.

Joe MK, Parck SM, Lee YS, Hwang DC, Hong CB (2000). High temperature stress resistance of E. Coli induced by a tobacco class I low molecular weight heat-shock proteins. Moll. Cells 10:519-524.

http://dx.doi.org/10.1007/s10059-000-0519-1

Kilstrup M, Jakobsen S, Hammer K, Vogensen FK (1997). Induction of heat shock proteins Dnak, GroEL, and GroES by salt stress in Lactococcus lactis. Appl. Environ. Microbiol. 63:1826-1837.

Kim WS, Perl L, Park JH, Tandianus JE, Dunn NW (2001). Assessment of stress response of the probiotic Lactobacillus acidophilus. Curr. Microbiol. 43:346-350. http://dx.doi.org/10.1007/s002840010314

Kolter R, Siegele DA, Tormo A (1993).The stationary phase of the bacterial life cycle. Annu. Rev. Microbiol. 47:855-874. http://dx.doi.org/10.1146/annurev.mi.47.100193.004231

Mansouripour S, Esfaudiari Z, Natelghi L (2013). The effect of heat process on the survival and increased viability of probiotic by microcapsulation : a review. Ann. Biolog. Res. 4:83-87.

Narberhaus F (2002). Alpha-crystallin-type heat shock proteins: socializing minichaperones in the context of a multichaperone network. Microbiol. Mol. Biol. Rev. (66): 64-93. http://dx.doi.org/10.1128/MMBR.66.1.64-93.2002

Pieterse B, Leer RJ, Schuren FH, Van der Werf MJ (2005). Unravelling the multiple effects of lactic acid stress on Lactobacillus plantarum by transcription profiling. Microbiology 151:3881-3894. http://dx.doi.org/10.1099/mic.0.28304-0

Prasad J, McJarrow P, Gopal P (2003). Heat and osmotic stress responses of probiotic Lactobacillus rhamnosus HN001 (DR20) in relation to viability after drying. Appl. Environ. Microbiol. 69:917-925. http://dx.doi.org/10.1128/AEM.69.2.917-925.2003

Rechinger KB, Siegumfeldt H, Svendsen I, Jakobsen M (2000). "Early" protein synthesis of Lactobacillus delbrueckii ssp. bulgaricus in milk revealed by [35S]methionine labeling and two-dimensional gel electrophoresis. Electrophoresis 21:2660-2669. http://dx.doi.org/10.1002/1522-2683(20000701)21:13<2660::AID-ELPS2660>3.0.CO;2-7

Sabri MY, Zamri-Saad M, Mutalib AR, Israf DA, Muniandy N (2000). Efficacy of an outer membrane protein of Pasteurella haemolytica A2, A7 or A9-enriched vaccine against intratracheal challenge exposure in sheep. Vet. Microbiol. 73: 13-23. http://dx.doi.org/10.1016/S0378-1135(99)00205-9

Salotra P, Singh DK, Seal KP, Krishna N, Jaffe H, Bhatnagar R (1995). Expression of DnaK and GroEL homologs in Leuconostoc esenteroides in response to heat shock, cold shock or chemical stress. FEMS Microbiol. Lett. 131:57-62. http://dx.doi.org/10.1111/j.1574-6968.1995.tb07754.x

Segal G, Ron EZ (1996). Regulation and organization of the groE and dnaK operons in eubacteria. FEMS Microbiol. Lett. 138:1-10. http://dx.doi.org/10.1111/j.1574-6968.1996.tb08126.x

Serrano LM, Molenaar D, Wels M, Teusink B, De Vos WM, Smid EJ (2007). Thioredoxin reductase is a key factor in the oxidative stress response of Lactobacillus plantarum WCFS1. Microb. Cell Fact. 10:6-29. http://dx.doi.org/10.1186/1475-2859-6-29

Spano G, Massa S (2006). Environmental stress response in wine lactic acid bacteria: beyond Bacillus subtilis. Crit. Rev. Microbiol. 32:77-86. http://dx.doi.org/10.1080/10408410600709800

Stevens MJA (2008). Transcriptome response of Lactobacillus plantarum to global regulator deficiency, stress and other environmental conditions. The- sis. Wageningen University, Wageningen, The Netherlands.

Svensater G, Sjogreen B, Hamilton IR (2000). Multiple stress responses in Streptococcus mutans and the induction of general and stress-specific proteins. Microbiology 146: 107-117.

Teixera P, Castro H, Mohacsi-Farkas C, Kirby R (1997). Identification of sites of injury in Lactobacillus bulgaricus during heat stress. J. Appl. Microbiol. 83: 219-226. http://dx.doi.org/10.1046/j.1365-2672.1997.00221.x

Van de Guchte M, Serror P, Chervaux C, Smokvina T, Ehrlich SD, Maguin E (2002). Stress responses in lactic acid bacteria. Antonie Leeuwenhoek 82:187-216. http://dx.doi.org/10.1023/A:1020631532202

Veinger L, Diamant S, Buchner J, Goloubinoff P (1998). The small heat shock protein IbpB from Escherichia coli stabilizes stress-denatured proteins for subsequent refolding by a multichaperone network. J. Biol. Chem. 273:11032-11037. http://dx.doi.org/10.1074/jbc.273.18.11032

Walker DC, Girgis HS, Klaenhammer TR (1999). The groESL

chaperone operon of Lactobacillus johnsonii. Appl. Environ. Microbiol. 65: 3033-3041.

Whitaker RD, Batt CA (1991). Characterization of the heat shock response in Lactococcus lactis subsp. lactis. Appl. Environ. Microbiol. 57:1408-1412.

22

Use of real time polymerase chain reaction (PCR) and histopathological changes for detection of the *Toxoplasma gondii* parasite in male rats (experimental study)

Ghiadaa Abass Jassem

College of Veterinary Medicine, AL-Qadisiyia University, Iraq.

Toxoplasma gondii infection is wide spread in Iraq that is why this study was carried out to detect its presence in semen samples of infected male rats experimentally. Impression smears, real time polymerase chain reaction (PCR) and histopathological changes were used for detection of *T. gondii*. The current study included the isolation of local strains of *T. gondii* from placenta of women that experienced abortion that had toxoplasmosis history, 0.3 ml of suspension contained 100 tissue cyst per rat injected interperetonially in 40 rats. Eight weeks post inoculation, rats were sacrificed then dissected and epididymis was immediately removed. Impression smear were made from semen stained with eosin-nigrosin as initial diagnosis of infection confirmed by the presence of cysts of parasite. This confirmation is dependent on the molecular diagnosis by real time PCR which successfully detected the parasite in 90.3% of rats inoculated with aborted placenta suspension. Some histological changes in testicular tissue were collapse and shrinkage in somniferous tubules with multinucleated giant cells with vacuolar degeneration of lydig cell. These finding suggest that *T. gondii* infection cause a temporary impairment and insufficiency in male reproductive activity with probability of transmission of the parasite in semen to females or to other animals such as cats.

Key words: Testes, placenta, semen, real time polymerase chain reaction (PCR), impression, interperitonially, reproductive, male.

INTRODUCTION

Toxoplasma gondii is a protozoan parasite that is widespread globally and infect man and animals (Shi-guo et al., 2006). Most infections in humans are asymptomatic but at times the parasite can produce devastating disease, toxoplasmosis ranks high on the list of diseases which lead to death in patients with acquired immunodeficiency syndrome (AIDS) (Ahmad et al., 2012). In AIDS patients, any organ may be involved, including the testis, dermis and the spinal cord, while the brain is most organ frequently reported (Hill and Dubey, 2002).

T. gondii is horizontally transmitted to humans by the accidental ingestion of oocysts in cat feces or by eating raw or undercooked meat containing cysts. *T. gondii* can be transmitted by solid organ transplants to the recipient and transfusion (Dubey and Jones, 2008).

T. gondii can be transmitted by semen, milk, saliva and eggs (Dubey and Lindsay, 2006). The disease primarily occurs in children and adults (including pregnant women) which is asymptomatic in most patients (Afshari et al., 2013). In immunocompromised patients, reactivation of latent disease can cause life-threatening encephalitis. After a short phase of acute toxoplasmosis, the infection proceeds into its latent phase when cysts are formed and this cyst survive for long period of host's life, mainly in neural and muscular tissues of infected subjects. In immunocompetent subjects, the latent phase of infection is considered symptomatic and harmless, increased accidents toxoplasmosis induced changes in the behavior of humans and rodents may be dramatically different even if they are induced by the same process, e.g. by production or induction of production of the same neurotransmitter in the host brain tissue (Doubey and Jones, 2008).

Rats are the best model for human toxoplasmosis investigation, whereas toxoplasmosis in human is similar to rats, both mammals (Afshari et al., 2013), Therefore, the result of the study may be applied for human toxoplasmosis (Abdoi et al., 2012)

The detection of *T. gondii* organisms in clinical specimens in mouse inoculation and then the detection of *T. gondii*-specific antibodies is sensitive and specific but time-consuming, taking up to 6 weeks to obtain a diagnosis. Currently, cell culture is the most practical method for the detection of *T. gondii* parasitemia, but this is also relatively slow and may lack sensitivity, PCR has been found to be a sensitive, specific and rapid method for the detection of *T. gondii* DNA in amniotic fluid, blood tissue samples and cerebrospinal fluid. Thus, a more efficient method is needed to provide rapid and quantitative results for the diagnosis of *T. gondi* infection. Several groups have reported the use of PCR to screen human samples easily. PCR is more appropriate, as there is no need for cultivation of the parasite to diagnoses *T. gondii* by PCR amplification of *T. gondii* DNA which is present in peripheral blood. PCR assay was designed for the simultaneous detection of *T. gondii* DNA from the cerebrospinal fluid, and PCR may be the investigation of choice for brain biopsy (Lin et al., 2000)

The objective of this study was to detect *T. gondii* in semen samples of rats which were experimentally infected by using suspension prepared from placenta of women that experienced abortion due to toxoplasmosis.

MATERIALS AND METHODS

Laboratory animals

Mature male rats (*Rattus norvegicus*) were reared in plastic cages supplied with drinking water and food, and kept in animal house of the College of Veterinary Medicine of Al-qadisiya University. The animal house was supplied with ventilator fan and air conditioner in order to control the room temperature. Sterilized food and water were used for rats along the period of experiment.

Isolation of *T. gondii*

Parasite was isolated from placenta of women that experienced abortion with toxoplasmosis history. Placenta were cut into small pieces and mixed with an equal amount of normal saline and grinded by using mortar and pestle. This preparation (solution) was passed through a piece of gauze to avoid large particles and then centrifugation was done with 3000 rpm for 10 min. The supernatant was discarded and the sediment was suspended by normal saline and the process was repeated three times by washing the samples, 0.1 ml of compound 1,000 units of penicillin and streptomycin 100 mg was added to prevent contamination (Alkennay and Hassan, 2010). 0.3 ml suspension contains 100 tissues cyst of parasites used for experimental infection in 40 rats through intraperitoneal route, 10 rats without anything (as control). This method was used (AL-taie and Abdulla, 2008) for inoculation of the parasite from uncontaminated samples and it is a sensitive method .

Detection of parasite

After 8 weeks from the beginning of injecting the laboratory rats, the rats were anesthetized using 0.2 ml ketamine and 0.1 ml xylazine injected into intraperitoneal cavity by using a sterile syringe of 1 ml (Alves et al., 2011). Samples were collected from infected group and control group of rats to detect the parasite by using impression smear and to detect B1 gene by using qRT-PCR. After that, one testes was stored in formalin 10% for histopathological study.

Epididymis caudate for semen collection

The two epididymis were removed from each rat and immersed in 1 ml of warm physiological saline, then each epididymis caudate was minced using microsurgical scissor and kept at 37°C for later tests (Oyedeji et al., 2013).

Direct smears

Detection of *T. gondii* in experimentally inoculated rats by impression of smears which were obtained from semen was doned. Smears were dried, fixed in methanol and stained with eosin nigrosin stain (Hill and Dubey, 2002).

Testicular histology

Samples collected were fixed in 10% formalin for 48-72h, and the material was processed by histopathological technique method as published by Salibay and Claveria (2006). 5 μm semi-serial sections were placed on slides and stained with hematoxylin and eosin (H&E) for microscopic examination.

Samples collection

50 samples of semen from rats inoculated with parasite were collected with control. The samples stored in a refrigerator until use for genomic DNA extraction.

Table 1. Real-time PCR TaqMan probe and primers.

Primer	Sequence
B1 gene	F TCCCCTCTGCTGGCGAAAAGT
Primers	R AGCGTTCGTGGTCAACTATCGATTG
B1 gene Probe	FAM-TCTGTGCAACTTTGGTGTATTCGCAG-TAMRA

Table 2. The qPCR master mix.

qPCR master mix	Volume
Genomic DNA template	5 µL
B1 Forward primer (20pmol)	1 µL
B1 Reverse primer (20pmol)	1 µL
B1 TaqMan probe (25pmol)	2 µL
DEPC water	11 µL
Total volume	20 µL

Table 3. Thermocycler conditions.

qPCR step	Temperature	Time	Repeat cycle
Initial denaturation	95°C	5 min	1
Denaturation	95°C	15 s	
Annealing\extension Detection (scan)	60°C	1 min	45

Genomic DNA extraction

Genomic DNA was extracted from frozen semen samples of experimental rats by using (Genomic DNA mini Extraction kit. Geneaid. USA). 200 µL semen sample placed in 1.5 ml microcentrifuge and 20 µL proteinase K (10 mg/mL) was added for cell lysis. Then genomic DNA extracted according to kit instructions. The purified DNA was eluted in elution buffer provided with kit and store at -20°C, and the extracted DNA was checked by Nanodrop spectrophotometer.

Real-Time PCR

Real-Time PCR based TaqMan probe was performed for rapid detection of *T. gondii* according to method described by Mei-hui-lin *et al.* (2000). Real-Time PCR TaqMan probe and primers were used for amplification of conserved region B1 gene in *T. gondii*. These primers were provided by Bioneer Company, Korea as shown in Table 1.
 The Real-Time PCR amplification reaction was done by using (AccuPower® DualStar™ qPCR PreMix Bioneer. Korea) and the qPCR master mix were prepared for each sample according to company instruction as shown in Table 2.
 These qPCR master mix reaction components mentioned in Table 2 were added into AccuPower® DualStar™ qPCR PreMix tubes which contain Taq DNA polymerases, dNTPs, 10X buffer for TaqMan probe amplification. Then tubes were placed in Exispin vortex centrifuge at 3000 rpm for 3 min, after that transferred into MiniOpticon Real-Time PCR system and the thermocycler conditions shown in Table 3 was applied.

RESULTS AND DISCUSSION

After eight weeks of inoculating rats, there was impression smear of stained semen with eosin-nigrosin in the tissue cyst of *T. gondii* as initial detection of infection, the result revealed that the percentage of tissue cysts were 80% in different size and numerous bradyzoites (Figure 2A and B). This study is similar to that of AL-Kennany and Hassan (2010) who isolated the *T. gondii* in same way from placenta of aborted ewes and AL-Khaffaf and Abdullah (2005) who isolated the *T. gondii* bradyzoites from placenta of women that experienced abortion and confirmed by the appearance in different size and involved numerous number. Abdoli et al. (2012) referred that *T. gondii* parasite can infect the male genital tract. Dalimi and Abdoli (2013) reported presence of parasites in rats seminal fluid by using impression smear method. *T. gondii* can be found in different fluids and tissues of infected animals (Miller et al., 2000). Among the reproductive complication caused by toxoplasmosis in aborted sheep, it is one of the most important causes of economic losses (Moreno et al., 2012). The occurrence of infection forms in the semen of human and bulls (Scarpelli et al., 2009), semen and milk of goat (Figueiredo et al., 2001), semen of sheep (Lopes et al., 2011) was through possible routes of transmission of disease by ventral rout with sours infection. The data

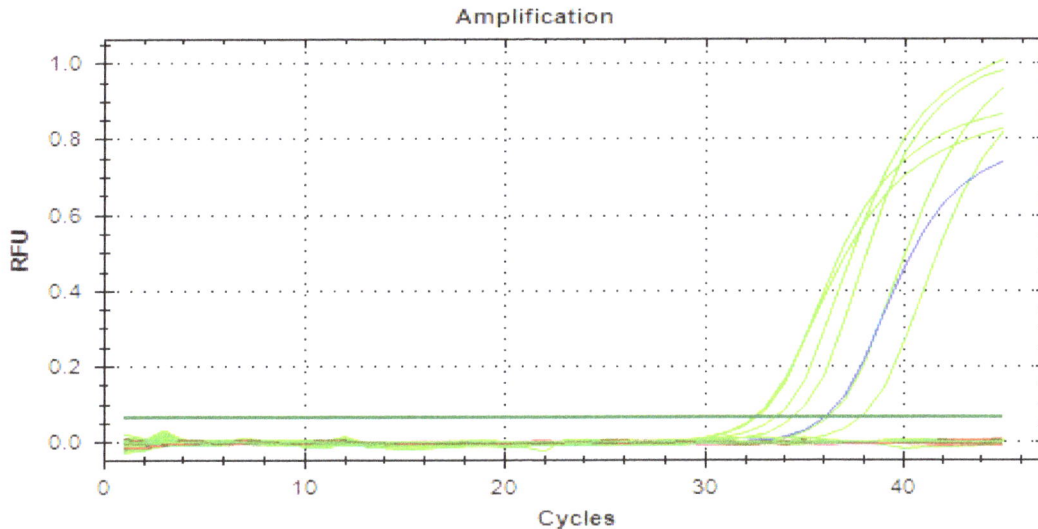

Figure 1. Real-time PCR amplification plot for B1 gene for the detection of *T. gondii* DNA in rat semen samples that show (green plot) positive result (Blue), positive control DNA *T. gondii* (Genekam .Germeny), (red) negative result.

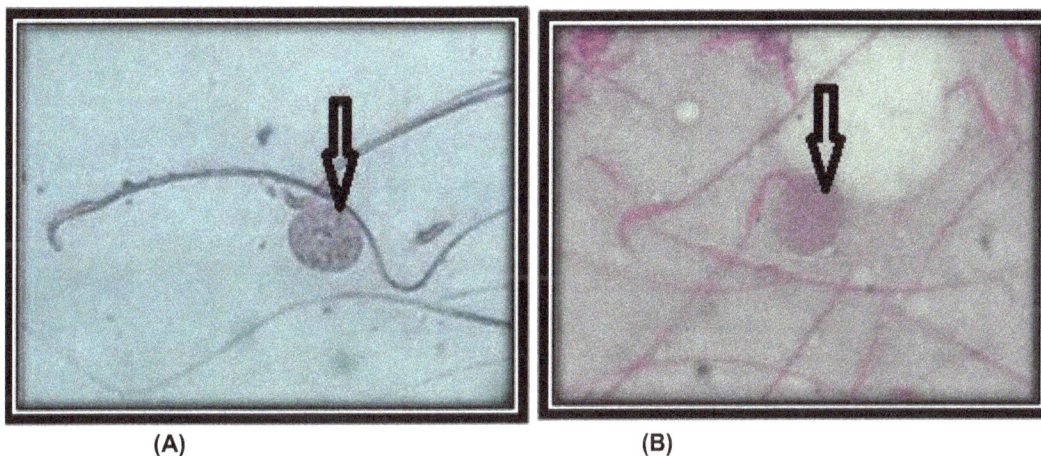

(A)   (B)

Figure 2. A and B: Toxoplasma cyst in the semen of infected rats stained with Eosin-Nigrosin.

obtained from limited studies performed in animal models reported presence of parasite in rat's brain (Terpsidis et al., 2009). While AL-Taie and Abdulla (2008) found *T. gondii* in brain, liver, heart, spleen, lymph nodes, kidney of the rats. The study was through possible routes of transmission of disease by ventral rout with sours infection.

The real-time PCR technique is very specific and sensitive method in detection of *T. gondii* when compared with direct impression smear test. The results of real-time PCR technique for semen showed the presence of parasite in high rate (90.3%) in rats inoculated with suspension of placenta (Figure 1). This is similar to the result of Shi-guo et al. (2006) who reported that the rate of infection was 100% in male rabbits. Also, this tech-

nique was used for detection of parasite in many samples of tissue and body fluid and in brain of sheep (Donovan et al., 2012). Mesquita et al. (2010) found *T. gondii* in blood and cerebrospinal fluid in AIDS patients while Amerizadeh et al. (2013) found *T. gondii* in serum samples and in amniotic fluid of pregnant women and in blood samples of newborns with congenital toxoplasmosis (48%) (Gunel et al., 2012).

Histologically *T. gondii* was detected as a cyst of parasite in some testicular tissue and parasites which was found in other section, there is sever congestion and hemorrhage in the somniferous tubules, presence of multinucleated giant cells, suppression of spermato-genesis vaculation of spematogonia with absence of spermatozoa, thickening in the interstitial tissue among

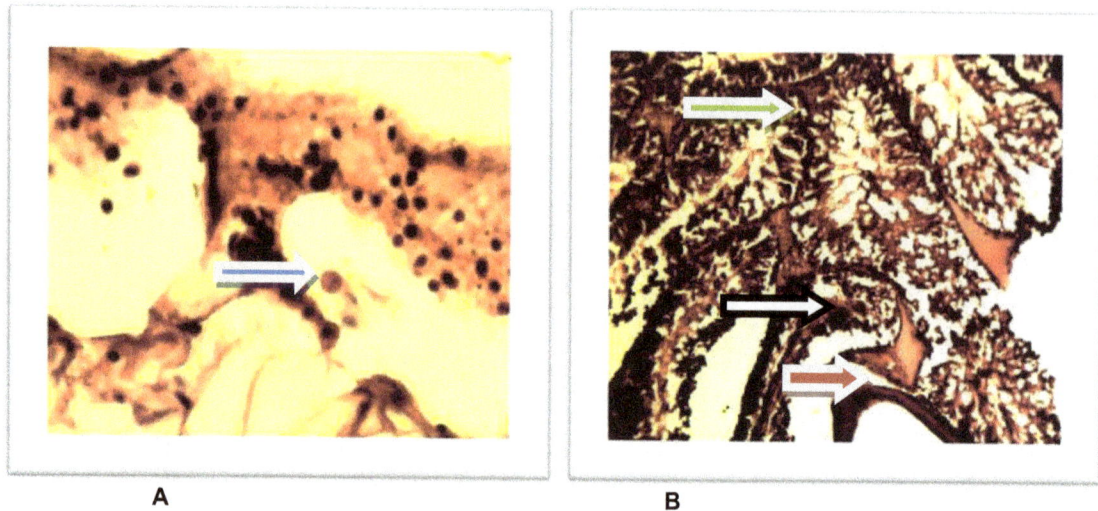

Figure 3. (A): *Toxoplasma gondii* cyst in rat testis (blue arrow). (B) Many histological changes of testicular (200X H&E) tissue in rat vaculation of spematogonia with absence of spermatozoa (green arrow), thickening in the interstitial tissue (white arrow), hemorrhage in the seminefrous tubules (red arrow) (H&E 200X).

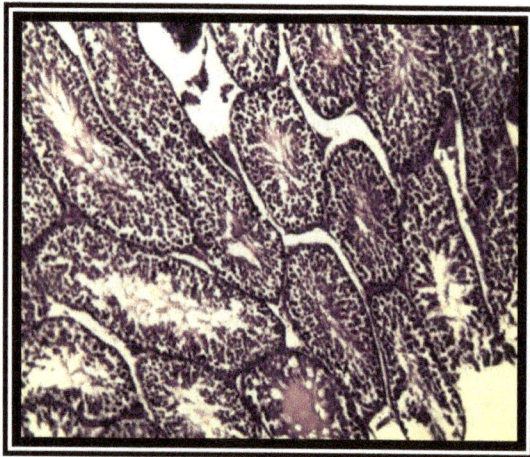

Figure 4. Histological section of rat testes. Control group. There are normal semniferous tubules with normal spermatogenesis, normal spermatogonia and spermatid. 10X H&E.

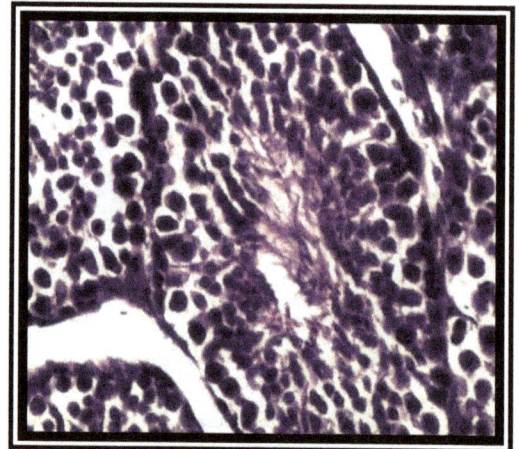

Figure 5. Histological section of rat testes. Control group. Higher magnification, there are normal seminiferous tubules with normal spermatogonia, normal primary and secondary spermatocytes with spermatid. 40X H&E.

somniferous tubules, many histological changes in testicular tissue collapse and shrinkage in somniferous tubules with multinucleated giant cell with vacuolar degeneration of lydig cells. These changes are not observed in the control rats, suggesting that these changes occur due to *T. gondii* infection (Figure 3A and B). Figures 4 and 5 show normal tissue and structure of spermatogenesis and somniferous tubules. Lopes et al. (2011) detected the parasite in parenchymal tissue of the male reproductive system of sheep by immunohisto-chemistry and the changes were multifollical mono-nuclear interstitial inflammatory infiltrate and diffuse testicular degeneration with calcified foci, while Gheoca et al. (2009) reported rare case of *T. gondii* cyst in testicle

sample of small mammals species, *Apodemus flavicollis* (mammal). Terpsidis et al. (2009) and Abdoli et al. (2012) reported that no histological changes were detected in examined testes in their findings suggesting that *T. gondii* infection can cause temporary impairment of the reproductive system by insufficient male production with the parasite been transferred through semen of animals as different source of infection, this may occur with human toxoplasmosis.

Conflict of Interests

The author(s) have not declared any conflict of interests.

Use of real time polymerase chain reaction (PCR) and histopathological changes for detection of the Toxoplasma...

155

REFERENCES

Abdoli A, Dalimi A, Movahedin M (2012). Impaired reproductive function of male rats infected with *Toxoplasma gondii*. Inst. Andrologia . 44:679-687

Afshari F, Imani A, Asl S, Farhang H, Ghasempour K, Ezzatzadeh A, Ainechi N (2013). Evaluation of Testosterone and Alkaline Phosphatase Activity Changes in Epidydimis of *Toxoplasma gondii* Infected Rats. Int. J. Women's Health Reprod. Sci. 1(2):1-8.

Ahmad M, Maqbool A, Mahmood-ul-Hassan M, Anjum A (2012). Prevalence of *Toxoplasma gondii* antibodies in human beings and commensally rodents trapped from Lahore, Pakistan, J. Anim. Plant Sci. 22(1):51-53.

Alkennay E, Hassan S (2010). Pathological study on some tissues of rat experimentally infected with *Toxoplasma gondii* . J. Edu. Sci. 23(3):8-17.

Al-Khaffaf F, Abdullah B (2005). Isolation of *Toxoplasma gondii* Parasite from Placental Tissue of Aborted Women in Nenavah Governerate. Alrafedian J. Sci. 6(16):93-99.

Al-taie A, Abdulla S (2008). *Toxoplasma gondii*: Experimental infection of Isolated local strain in Sulaimani Province Iraqi. J. Med. Sci. 6(3):60-71.

Alves M, Silva A, Bianchi L, Araujo E, Sant'Ana D (2011). *Toxoplasma gondii* induces death of gastric myenteric neurons in Rats. Int. J. Morphol. 29(1):293-298.

Amerizadeh A, Khoo B, The A, Golkar M, Karim IZ, Osman S, Yunus M, Noordin R (2013). Identification and real-time expression analysis of selected *Toxoplasma gondii* in-vivo induced antigens recognized by IgG and IgM in sera of acute toxoplasmosis patients, BMC Infect. Dis.13(287):1-12.

Dalimi A, Abdoli A (2013). *Toxoplasma gondii* and Male Reproduction Impairment: A new aspect of Toxoplasmosis Research, Jundishapur J. Microbiol. 6(8):e7184.

Donovan J, Proctor A, Gutierrez J, Worrell S, Nally J, Marques P, Brady C, McElroy M, Sammin D, Buxton D, Maley S, Bassett H and Markey B (2012). Distribution of lesions in fetal brains following experimental infection of pregnant sheep with *Toxoplasma gondii*, Vet. Pathol. 49(3):462-469.

Dubey J, Jones J (2008).*Toxoplasma gondii* infection in humans and animals in the United States, Int. J. Parasitol. 38:1257-1278.

Dubey J, Lindsay D (2006). Neosporosis, *Toxoplasmosis*, and *Sarcocystosis* in ruminants. Vet. Clin. Food Anim. 22:645-671.

Figueiredo J, Silva D, Cabral D, Mineo J (2001). Seroprevalence of *Toxoplasma gondii* Infection in goats by the indirect haemagglutination, Immunofluorescence and immunoenzymatic tests in the region of Uberlândia, Brazil .MemInst Oswaldo Cruz, Rio de.Janeiro .96(5):687-692.

Gheoca D, Haranglavean A, Gheoca V (2009). Aspect of *Toxoplasma* *Toxoplasma*-like parasitism in small mammal species in Transylvania and there role in *Toxoplasma* dissemination,VasileGoldis, Seria Ştiinele Vieţii . 19(1):69-73.

Gunel T, Kalelioglu I, Ermis H, Has R, Aydinli K (2012). Large scale pre-diagnosis of *Toxoplasma gondii* DNA genotyping by real-time PCR on amniotic fluid, Biotechnol. Biotechnol. 26(2):2913-2915.

Hill D, Dubey J (2002). *Toxoplasma gondii*: transmission, diagnosis and prevention,Clin . Microbiol. Infect. 8:634-640.

Lin M, Chen T, Kuo T, Tseng C, Tseng C (2000). Real-Time PCR for Quantitative Detection of *Toxoplasma gondii* . J. Clin. Microbiol . 38(11):4121-4125.

Lopes W, Santos T, Luvizotto M, Sakamoto C, Oliveira G, Costa A (2011). Histopathology of the reproductive system of male sheep experimentally infected with *Toxoplasma gondii*,.Parasitol. Res. 109:405-409.

Mei-Hui L, Tse-Ching C, Tseng Tong K, Chung T, Ching – Ping T (2000). Real- Time PCR for Quantitative Detection of *T. gondii* . J. Clin. Microbiol. 38(11):4121-4125.

Mesquita R, Ziegler A, Hiramoto R, Vidal J, Chioccola V (2010). Real-time quantitative PCR in cerebral toxoplasmosis diagnosis of Brazilian human immunodeficiency virus-infected patients, J .Med .Microbial. 59:641-647.

Miller D, Davis J, Rosa R, Diaz M, Perez E (2000). Utility of Tissue Culture for Detection of *Toxoplasma gondii* in Vitreous Humor of Patients Diagnosed with Toxoplasmic Retinochoroiditis. J. Clin. Microbiol. 38(10):3840-3842.

Moreno B, Fernández E, Villa A ,Navarro A, Cerrillo J, Mora L (2012). Occurrence of *Neospora cranium* and Toxoplasma gondii infections in ovine and caprine abortions. Vet. Parasitol. pp. 1-7.

Oyedeji K, Bolarinwa A, Oladosu I (2013). Effect Of Isolated Lupone Constitunet of Portulacaoleracea on Reproductive parameters in male rats .Int. J. Pharm. Sci. 5(2):684-690.

Salibay C, Claveria F (2006). *Toxoplasma gondii* infection in Philippines *Rattu* sspp. Confirmed through bioassay in *Musmusculus* .Vet. Arhiv. 76(4):351-361.

Scarpelli L, Lopes D, Migani M, Bresciani K, Costa A (2009). *Toxoplasma gondii* in experimentally infected *Bosaurus* and *Bosindicus* semen and tissues ,Pesq. Vet. Bras. 29(1):59-64.

Shi-guo L, Hai-zhu Z, Xin L, Zhe Z, Bin H (2006). Dynamic observation of polypide in semen and blood of rabbits infected with *Toxoplasma tachyzoites*. Chin. Med. J. 119(8):701-704.

Terpsidis K, Papazahariadou M, Taitzoglou I, Papaioannou N, Georgiadis M, Theodoridis I (2009).*Toxoplasma gondii* Reproductive parameters in experimentally infected male rats. Exp. Parasitol. 121:238-241.

23

Incidence and antibiotic susceptibility pattern of *Vibrio* species isolated from sea foods sold in Port-Harcourt, Nigeria

Chijioke A. Nsofor[1] , Samuel T. Kemajou[2] and Chibuzor M. Nsofor[2]

[1]Department of Biotechnology, Federal University of Technology, Owerri, Imo State, Nigeria.
[2]Department of Medical Laboratory Science, Madonna University, Elele, Rivers State, Nigeria.

Most *Vibrio* infections are associated with the consumption of raw or undercooked sea foods or exposure of wounds to warm seawater. In this study, 63 samples of a variety of sea foods viz: shrimps, periwinkle, prawn, crayfish collected from two major household markets in Port-Harcourt were cultivated using standard bacteriological method on Thiosulphate Citrate Bile Sucrose (TCBS) agar. A total of 63 *Vibrios* belonging to seven different species were isolated with *Vibrio fluvialis* recorded as the highest percentage frequency occurrence and the most predominant species, 30 (47.6%), followed by *Vibrio paraheamolyticus* with 19 (30.2%), *Vibrio vulnificus* with 4 (6.34%), *Vibrio metschnikovi* with 4 (6.34%), *Photobacterium* spp. with 3 (4.76%), *Vibrio cholerae* with 2 (3.17%) and the least was *Vibrio mimicus* with 1 (1.60%). Statistically, there was a significant different ($P < 0.05$) in prawn sold in Mile 1 market as compared to prawn sold in Mile 3 markets. These isolates were subjected to their susceptibility patterns using agar diffusion method, which recorded susceptible to most antibiotics used. The presence of these pathogenic strains of *Vibrios* in commonly consumed sea foods is of public concern.

Key word: Vibrios, sea foods, Nigeria.

INTRODUCTION

Vibrio species are Gram negative, facultative anaerobic motile asporogenous rod or curved rod-shaped bacteria with a single polar flagellum. The genus contains at least twelve species pathogenic to human, eight of which can cause or are associated with food-borne illness (Dickinson et al., 2013). The majority of the food-borne illness is caused by *Vibrio paraheamolyticus, Vibrio vulnificus, Vibrio fluvialis* and *Vibrio cholerae*. *V. vulnificus* is responsible for 95% of sea food related deaths while immune suppressed individuals are most susceptible to other *Vibrio* infection (Miyoshi , 2013). Marine *Vibrios* are ubiquitous in the marine environment; therefore, it is not surprising that many are pathogenic for various seafood hosts which are harvested for human consumption (Kaysner and DePaola, 2004).

Most countries have guidelines for detecting *V. parahaemolyticus, V. cholera* 0_1 and 0_{139} in seafood, whereas few have guidelines for *V. vulnificus*. Thus, there are routine microbiological analysis of seafood includes testing for *V. paraheamolyticus, V. cholera* 0_1 and 0_{139},

Table 1. Mean *Vibrio* species from sea foods bought in Mile 1 and Mile 3 markets in Port Harcourt represented as $\log_{10}$ cfu/ml.

Sample	Mile 1	Mile 3
Periwinkle	67.0±44.2	20.4±14.5
Shrimp	52.0±21.6	21.0±11.1
Prawn	24.4±13.9	49.4±41.6
Crayfish	15.0±9.8	17.7±11.5

but seldom for *V. vulnificus* (Park et al., 2013).

Some species are primarily associated with gastrointestinal illness (*V. cholerae* and *V. parahaemolyticus*) while others can cause non-intestinal illness, such as septicemia (*V. vulnificus*). In tropical and temperate regions, disease-causing species of *Vibrio* occur naturally in marine, coastal and estuarine environments and are most abundant in estuarine. Pathogenic *Vibrios,* in particular *V. cholerae,* can also be recovered from freshwater reaches of estuaries (Lutz et al., 2013). The occurrence of these bacteria does not generally correlate with numbers of fecal coli forms and depurations of shellfish may not reduce their numbers. However, a positive correlation between fecal contamination and levels of *V. cholerae* may be found in areas experiencing cholera outbreaks (Amita et al., 2003).

Cholera epidemics outbreaks have killed millions of people and continue to be the major public health concern worldwide (Islam et al., 2013). Fluid and electrolytes replacements are the major treatment of cholera patients, however, patients with severe disease require antibiotic treatment to reduce the duration of illness and reduce replacement fluid intake. Like most bacteria of clinical and public health significance, *V. cholerae* is continuously becoming resistance to a variety of antimicrobial agents, necessitating use of newer drugs which are more expensive and have more adverse effect. Spread of cholera epidemics worldwide has been associated with the emergence of multiple drug resistance among a large number of *V. cholera* strains (Dunstan et al., 2013).

Sea foods especially shellfish is a substrate for some zoonotic *Vibrios* of which these microorganisms, cause food poisoning and diarrhea in human. Sea foods are prone to bacterial contamination and could cause health risk to human consumers (Lutz et al., 2013). *Vibrios* are associated with live sea foods as they form part of the indigenous micro flora of the marine environment. *V. parahaemolyticus* is the top causative agent among all the reported seafood poisoning outbreaks in Nigeria (Eyisi et al., 2013). There is large number of public frozen seafood processing services distributed in the country, where a considerable number of people buy their fresh sea food products daily. Serious consequences relating

to national productivity and development can arise from lack of hygiene (Eyisi et al., 2013). In this study, we evaluated the incidence and antibiotic susceptibility pattern of *Vibrio* spp. isolated from commonly consumed sea foods in Port-Harcourt, Nigeria.

MATERIALS AND METHODS

Sample collection, cultivation and identification of *Vibrio* spp.

Fresh samples of variety of sea foods including prawn, periwinkles, crayfish and shrimp were collected randomly from different sellers at the two major household markets (Mile 1 and Mile 3) in Port-Harcourt city, Rivers State, Nigeria. The samples were transported to the laboratory immediately in a screw cap bottles containing enrichment medium (alkaline peptone water) with alkalinity ranging from 8.6- 9.0 and incubated at 37°C for 8-10 h. The samples was divided into 2 sets, one set of the samples was directly cultured onto the Thiosulphate Citrate Bile Salt Agar and the other set was carried out with a pour plate technique using a 10 fold serial dilution method with thiosulphate citrate bile salt agar. The inoculated plates were incubated overnight at 37°C, no antibiotics were added to the TCBS plates. *Vibrio* spp. were identified using conventional microbiological tests: Oxidase positive, rapid and darting motility, sugar fermentation (which differentiates the various *Vibrio* spp.) and salt tolerance (Cheesbrough, 2000).

Antibiotics susceptibility testing

The antibiotics susceptibility pattern of the isolates was determined using the disk diffusion method (Cheesbrough, 2000), on Mueller-Hinton agar (Oxoid, England). Inhibition zone diameter values were interpreted using standard recommendations of the Clinical Laboratory Standard Institute (CLSI, 2006). Susceptibility was tested against Ofloxacin (10 µg), Gentamycin (10 µg), Ciprofloxacin (10 µg), Ampicillin (30 µg), Streptomycin (30 µg), Pefloxacin (10 µg), Augmentin (30 µg), Septrin (30 µg), Ceporex (10 µg) (Oxoid, England). *Escherichia coli* ATCC 25922 was included as a reference strain.

Statistical analysis

Analysis of variance (ANOVA) and students T-test were used to compare means, and values were considered significant at $p< 0.05$. Post hoe multiple comparisons for the ANOVA were done using least significant difference (LSD).

RESULTS AND DISCUSSION

In this study, 63 samples of sea foods: 15 crayfishes, 18 prawns, 15 shrimps and 15 periwinkles were analyzed following standard microbiological method on TCBS Agar (Oxoid, England). The mean bacterial counts of the sea foods obtained from the two markets are shown in Table 1. Table 2 shows that there was no significant difference (P> 0.05) in the frequency of occurrence of *Vibrio* species in the sea food collected from Mile 1 and Mile 3 markets. Table 3 shows the antibiotic susceptibility pattern of the *Vibrio* isolates from sea foods bought in Mile 1 and Mile 3 market were significantly (P< 0.05) more susceptible to Ciprofloxacin, Ofloxacin, Pefloxacin, Gentamycin and

Table 2. Frequency of occurrence of *Vibrio* species isolated from prawns, periwinkle, crayfish and shrimp purchased in Mile 1 and Mile 3 markets.

Isolates	Mile 1					Mile 3			
	TNI	PR	PE	CR	SH	PR	PE	CR	SH
V. cholerae	2	0(0.00)	0(0.00)	0(0.00)	1(50)	1(50)	0(0.00)	0(0.00)	0(0.00)
V. fluvialis	30	5(16.7)	4(13.3)	3(10)	2(6.67)	4(13.3)	5(16.7)	4(13.3)	3(10)
V. paraheamolyticus	19	3(15.9)	1(5.3)	3(10)	2(10.5)	3(15.9)	3(15.9)	1(5.3)	3(15.9)
V. vulnificus	4	1(25)	0(0.00)	0(0.00)	0(0.00)	0(0.00)	1(50)	1(50)	0(0.00)
V. mimicus	1	0(0.00)	1(50)	0(0.00)	0(0.00)	0(0.00)	0(0.00)	0(0.00)	0(0.00)
V. metschnikovi	4	0(0.00)	0(0.00)	0(0.00)	0(0.00)	0(0.00)	0(0.00)	0(0.00)	0(0.00)
Photobacterium spp.	3	1(33.3)	0(0.00)	0(0.00)	0(0.00)	0(0.00)	0(0.00)	1(33.3)	0(0.00)

TNI = Total number of isolates; PR = Prawn;   PE = periwinkle;   CR = crayfish; SH = shrimp.

Table 3. The antibiotic susceptibility pattern of *Vibrio* species isolated from sea foods bought from Mile 1 and Mile 3 markets.

Isolates	TNI	OFX	PEF	CN	AU	CPX	SXT	S	PN	CEP
V. cholerae	2	2(100)	2(100)	2(100)	2(100)	2(100)	2(100)	2(100)	2(100)	1(50)
V. fluvialis	30	27(91.3)	28(93.3)	27(90)	29(96.4)	30(100)	23(76.7)	30(100)	26(86.7)	26(86.7)
parahemolyticus	19	19(100)	19(100)	17(89.5)	16(84.4)	19(100)	15(78.9)	19(100)	12(63.2)	17(89.5)
V. vulnificus	4	4(100)	4(100)	4(100)	2(50)	4(100)	4(100)	3(75)	2(50)	4(100)
V. mimicus	1	1(100)	1(100)	1(100)	1(100)	1(100)	0(0.00)	1(100)	1(100)	0(0.00)
V. metschnikovi	4	2(50)	2(50)	4(100)	2(50)	4(100)	0(0.00)	4(100)	2(50)	2(50)

TNI: Total number of isolates; OFX: ofloxacin (10 µg); CN: Gentamycin (10 µg); CPX: Ciprofloxacin (10 µg); PN: Ampicillin (30 µg)   S: Streptomycin (30 µg); PEF: Pefloxacin (10 µg); AU: Augmentin (30 µg); SXT: Septrin (30 µg); CEP: Ceporex (10 µg).

Streptomycin as compared to Augmentin, Septrin, Ampicillin and Ceporex.

This study indicates high occurrence of *V. fluvialis* and *V. paraheamolyticus* in the sea foods studied. Similar observations were made by Nwachukwu (2006) in studies on the pathogenic characteristics of non O_1 Vibrio cholera isolated from sea foods in Nigeria and in Malaysia by Elhadi et al. (2004). High occurrence of V. cholerae has also been reported in Bangladesh by Faraque et al. (2003). This high prevalence may be attributed to their high salt tolerance ability. In contrast, Caldini et al. (1997) reported that 150 *Vibrio* species isolated from the Aron River basin in Italy, recorded *V. cholerae* non O_1 as the most prevalent specie with 82% occurrence. This was attributed to the fact that River Basin is fresh water.

In Tanzania, data collected during cholera epidemics in 1990 and 1991 showed that all *V. cholerae* O_1 strains were sensitive to Streptomycin, Gentamycin, Ciprofloxacin (Webber et al., 1998). Rajkarnitar (2000) reported that 25 strains of *V. fluvialis* were susceptible to Tetracycline, Streptomycin while 84% were resistance to Ampicillin, the results are in line with this research.

However, it is important to note that *V. fluvialis, V. paraheamolyticus, V. vulnificus, V. metschnikovi, V. cholerae, V. mimicus* and *Photobacterium* spp. in periwinkles, prawns, crayfish and shrimp and other sea foods that are rampant in Port-Harcourt might be significantly responsible of gastroenteritis and severe diarrhea, wound infection cases among the population of the city considering the sources and high occurrence among the sea foods that were studied in this research.

Based on this research, the *Vibrio* species isolated showed different susceptibility to the antibiotics used. It was recorded that the *Vibrio* species isolated from the sea food sample showed 100% susceptibility to Streptomycin and Ciprofloxacin, followed by Ofloxacin, Pefloxacin, Gentamycin, Ceporex, Septrin and Augmentin and the least susceptible was Ampicillin. The antimicrobial susceptibility pattern of *Vibrio* species can be predicted easily, due to control in prescription of antimicrobial agents. Therefore it is clearly necessary to put more emphasis on food hygiene; thereby surveillance of potential contaminant bacteria in harvested seafood is crucial for sustenance of public health.

To prevent *Vibrio* infections, proper hygiene cooking, treatment of water supply and avoidance of eating raw food should be encouraged seriously; there should be establishment of a well separated sewage treatment infrastructure. Warning on possible cholera or other *Vibrio* spp. contamination should be reported around contaminated water sources with direction on how to decontaminate the water and sea foods for possible use. Effective sanitations practices should be practiced, if instituted and adhered to in time they are usually sufficient to stop an epidemic. Patients who present diarrhea symptoms should be referred to the health centre or hospital for immediate diagnosis and treatment.

In conclusion, the *Vibrio* species isolated from sea foods (periwinkles, prawns, crayfish and shrimp) showed high incidence in *V. fluvialis*, followed by *V. paraheamolyticus* while the least incidence was observed in *V. mimicus* among sea foods sold in some market in Port-Harcourt, Nigeria. It is important to note that this research is laying serious emphasis on the need for high level of hygiene and proper cooking of sea foods before eating. It is also important to note the need to establish regular nationwide antibiotic susceptibility surveillance of *Vibrio* species in different parts of the country so as to provide guidance on the best option in different situation.

REFERENCES

Amita F, Choudhury SR, Thungapathr M, Ramamurthy T, Nair GB, Gosh A (2003). Etiological studies in hospital on patients with secretory diarrhea in Calcutta J. Indian Med Assoc. 96:14-15.

Caldini G, Neri A, Cresti S, Boddi V, Rossolini GM, Lanciotti E (1997). High prevalence of *Vibrio cholerae* non-O1 carrying heat-stable enterotoxin-encoding genes among *Vibrio* isolates from a temperate-climate river basin of Central Italy. J. Appl. Environ.Microbiol. 63:2934-2939.

Cheesbrough M (2000). District Laboratory Practice in Tropical Countries, Part 2. Cambridge   University Press, Cambridge, UK; pp. 434.

Clinical Laboratory Standards Institute (2006). Performance standards for Antimicrobial susceptibility testing. National committee for clinical laboratory standards, Wayne pa.

Dickinson G, Lim KY, Jiang SC (2013). Quantitative Microbial Risk Assessment of Pathogenic Vibrios in Marine Recreational Waters of Southern California. Appl. Environ. Microbiol. 79(1):294-302.

Dunstan RA, Heinz E, Wijeyewickrema LC, Pike RN, Purcell AW, Evans TJ, Praszkier J, Robins-Browne RM, Strugnell RA, Korotkov KV, Lithgow T (2013). Assembly of the   Type II Secretion System such as Found in *Vibrio cholerae* Depends on the Novel Pilotin AspS. PLoS Pathog. 9(1):303-309.

Elhadi N, Radu S, Chen CH, Nishibuchi M (2004). Prevalence of potentially pathogenic *Vibrio* species in the seafood marketed in Malaysia. J. Food Protocol 67:1469-1475.

Eyisi OA, Nwodo UU, Iroegbu CU (2013). Distribution of Vibrio species in shellfish and water   samples collected from the Atlantic coastline of south-east Nigeria. J. Health Popul. Nutr. 31(3):314-320.

Faraque SM, Chowdlury N, Kamaruzzamam M, Ahmed QS, Faraque AS, Salam MA (2003). Re-emergence of epidemic *Vibrio cholerae* 0139, Bangladesh. Emerg. Infect. Dis. 9:1116-1122.

Islam A, Labbate M, Djordjevic SP, Alam M, Darling A, Melvold J, Holmes AJ, Johura FT, Cravioto A, Charles IG, Stokes HW (2013) Indigenous *Vibrio cholerae* strains   from a non-endemic region are pathogenic Open Biol. 3(2):120-181.

Kaysner CA, DePaola JA (2004). Vibrios In Food and Drug Administration Bacteriological AnalyticalManual. Retrieved from on 3rd October, 2006.

Lutz C, Erken M, Noorian P, Sun S, McDougald D (2013). Environmental reservoirs and   mechanisms of persistence of *Vibrio cholerae*. Front Microbiol. 4:375-380.

Miyoshi SI (2013). Extracellular proteolytic enzymes produced by human pathogenic *Vibrio* species, Front Microbiol. 4:339-345.

Nwachukwu E (2006). Studies on the pathogenic characteristics of non O_1 Vibrio cholera isolated from seafoods in Nigeria. J. Eng. Appl. Sci. 1:284-287.

Park JY, Jeon S, Kim JY, Park M, Osong SK (2013). Multiplex Real-time Polymerase Chain Reaction Assays for Simultaneous Detection of *Vibrio cholerae*, *Vibrio parahaemolyticus*, and *Vibrio vulnificus*. Public Health Res. Perspect. 4(3):133-139.

Rajkarnitar A (2000). Antibiotics resistant of *Vibrio cholerae* isolated from Kathmandu valley and characterization of the isolates by biotyping and serotyping. A dissertation presented to control Dept of Microbiology.

Webber JT, Minta ED, Canizaries R, Semigua A, Gomez I, Sempertugi R, Davila A, Greene KD, Cameron DN (1998). Epidemic cholera in Ecuador: Multi drug resistance and transmission by water and sea foods. Epidemiol. Infect. 112:1-11.

24

Detection of biofilm formation of a collection of fifty strains of *Staphylococcus aureus* isolated in Algeria at the University Hospital of Tlemcen

GHELLAI Lotfi[1] , HASSAINE Hafida[1], KLOUCHE Nihel[1], KHADIR Abdelmonaim[1], AISSAOUI Nadia[1], NAS Fatima[1] and ZINGG Walter[2]

[1]Laboratory of Applied Microbiology in Food, Biomedical and Environment (LAMAABE), Department of Biology, University of Tlemcen, 13000 Tlemcen, Algeria.
[2]Service de Prévention et de Contrôle de l'Infection. Hôpitaux Universitaires de Genève (HUG) Suisse.

The burden of disease caused by *Staphylococcus aureus* continues to grow; this organism has the ability to form biofilm and it is also a frequent cause of medical device and implant-related infections. The objective of this study was to evaluate the biofilm-forming ability of a collection of clinical isolates of *S. aureus*. In a total of 240 *Staphylococcus* spp. isolated from catheters, retrieved at five services (neonatology, internal medicine, pneumology, pediatric and neurology), only 50 (20.83%) strains were identified by conventional microbiological methods as *S. aureus* species; these strains were screened by microtiter plate assay for detection of biofilm formation. Of the 50 clinical isolates, 16 (32%) were non adherent, 20(40%) weakly, 10 (20%) moderately and 4(8%) strongly adherent. The quantitative method of microtiter plate can be involved as a simple, rapid, inexpensive and reproducible assay to assess biofilm formation which is further an important feature of pathogenecity of *S. aureus* in the clinical setting.

Key words: Microbial biofilm, *Staphylococcus aureus*, catheter, microtiter plate assay.

INTRODUCTION

Staphylococci are most often associated with chronic infections of implanted medical devices (Dunne, 2002; Raad, 2000). Such infections are predominately caused by *Staphylococcus aureus* and *Staphylococcus epidermidis*. The first one is known as an ubiquitous bacteria. It also has an inherent ability to form biofilms on biotic and abiotic surfaces (McCann et al., 2008; Begun et al., 2007). The biofilms protect the cells not only from host immune response but also from antimicrobial agents (Donlan et al., 2002). Indeed, biofilm formation is a major concern in nosocomial infections because it protects microorganisms from opsonophagocytosis and anti-biotics, leading to chronic infection and sepsis (Martí et al., 2010). These qualities have converged to make *S. aureus* a significant burden on our current health care system (Hobby et al., 2012). One of the patient populations most vulnerable to *Staphylococcus aureus* infection are those with implanted medical devices such as central venous catheters, cardiac valves and pace-makers, artificial joints and various orthopedic devices (Hobby et al., 2012). Therefore, once biofilm-associated *S. aureus* infections occur, they are difficult to be treated by conventional procedures (Trampuz and Widmer, 2006).

In fact, the biofilm formation involves the production of a polysaccharide intracellular adhesion (PIA) (Ziebuhr et al., 2001; Mack et al., 1996) which is the formal name of slime. This polysaccharide depends on the expression of the intercellular adhesion (icaADBC) operon, which encodes three membrane proteins (IcaA, IcaD and IcaC) with enzymatic activity and one extracellular protein (IcaB) (Djordjevic et al., 2002; Christensen et al., 1985). The icaADBC gene locus has also been detected in S. aureus and a range of other coagulase-negative staphylococci (Allignet et al., 2001; Cramton et al., 1999; Knobloch et al., 2002; McKenney et al., 1999). In addition, several surface proteins have been involved in the biofilm formation process, including biofilm asso-ciated protein (BAP) (Cucarella et al., 2001), S. aureus surface protein G (SasG) (Montanaro et al., 2011; Corrigan et al., 2007), Fibronectin-binding proteins (FnBPs) (Vergara-Irigaray et al., 2009; O'Neill et al., 2008) or Staphylococcal protein A (Spa). It is now suggested that protein-mediated biofilm formation under in vivo conditions is also an important virulence factor (Merino et al., 2009).

It is estimated that approximately 65% of all bacterial infections in humans are caused by biofilms (Costerton and Stewart, 2000) and Christensen et al. (1982) showed that 63% of the pathogenic strains produced slime, and only 37% of the nonpathogenic strains produced slime (Costerton et al., 1995). In the laboratory, Christensen et al. (1982) demonstrated that only one slime-producing cell per 16 000 non-slime-producing cells results in a culture that produces a gross amount of slime. Furthermore, there is increasing recognition that biofilm growth gives rise to a significant population of bacteria with a diverse set of phenotypes, often termed "variants" (Yarwood et al., 2007). This phenomenon has been explained by the "insurance hypothesis," which posits that the presence of diverse subpopulations increases the range of conditions in which the community as a whole can thrive (McCann, 2000; Yachi and Loreau, 1999).

A biomaterial can be defined as any substance, natural or synthetic, used in the treatment of a patient that at some stage, interfaces with tissue (Wollin et al., 1998). Although, any medical device easily inserted and removed (catheters, contact lenses, endotracheal and nasogastric tubes) or long-term implants (cardiac valves, hip joints and intraocular lenses) represents potentially a favorable support to microbial biofilms formation. Whereas, it is now well documented that biofilms are notoriously difficult to eradicate (Diani et al., 2014) and are often resistant to systemic antibiotic therapy and removal of infected device becomes necessary (Lewis, 2001; Souli and Giamarellou, 1998). Anyway, the skin surrounding the catheter insertion site has been implicated as the most common source of central venous catheters (CVC) colonization (Raad et al., 1993).

In order to study bacterial biofilms, a large variety of experimental direct (including microscopy techniques) and indirect observation methods have been developed. The microtiter plate procedure is an indirect method for estimation of bacteria in situ and can be modified for various biofilm formation assays (An and Friedman, 2000). This method has been investigated using many different organisms and stains (Hobby et al., 2012; Ramage et al., 2001; Stepanovic et al., 2000; Christensen et al., 1985; Deighton and Balkau, 1990; Miyake et al., 1992) in which the optical density (OD) of the stained bacterial film is measured with an automatic spectrophotometer.

In this study, we screened our original collection of 50 clinical isolates of S. aureus from intravenous catheter-associated infections by the polypropylene microtiter plate method for determining their ability to form biofilm. Parallelly, it is known that the genes that are crucial for biofilm formation are a subset of the genes involved in pathogenesis. This work was realized for the first time at the university hospital of Tlemcen. Our aim was to assess biofilm-forming ability of our collection, knowing that this organism is difficult to control and causes several constraints in different services of the hospital.

MATERIALS AND METHODS

Staphylococcus aureus isolates

In a total of 240 clinical isolates of Staphylococcus spp. isolated from catheters from four different services (neonatology, internal medicine, pneumology, pediatric and neurology service) at the university hospital of tlemcen (North-West Algeria) during a period of two years (from 2009 to 2011), 50 strains were identified as S. aureus on the basis of standard and conventional microbiological techniques including Gram stain, catalase and coagulase tests. The identification was completed with API Staph gallery (bioMérieux, Marcy l'Etoile, France).

Microtiter plate assays

In the present study, we screened the fifty clinical isolates of S. aureus for their ability to form biofilm by microtiter plate method according to the works of Christensen et al. (1985) with some modifications.

Strains from fresh agar plates were inoculated in 3 ml of brain heart infusion (BHI) with 1% glucose (Mathur et al., 2006) and incubated for 24 hours at 37°C in stationary conditions and diluted 1 in 20 with fresh medium. Individual wells of sterile, propylene, 96 well Microplate were filled with 200 µl of the diluted cultures and 200 µl aliquots of only BHI + 1% glucose were dispensed into each of eight wells of the column 12 of microtiter plate to serve as a control (to check non-specific binding and sterility of media). After incubation (24 h at 37°C), the microtiter plates content of each well was removed by tapping the bottom plates. The wells were washed four times with 200 µL of phosphate buffer saline (1 ×PBS pH 7.2) to remove planktonic bacteria. The plates were then inverted and blotted on paper towels and allowed to air dry for 15 min (Broschat et al., 2005). Adherent organisms forming-biofilms in plate were fixed with sodium acetate (2%) and stained with crystal violet (0.1% w/v) (Borucki et al., 2003; Mathur et al., 2006) and allowed to incubate at room temperature for 15 min. After removing the crystal

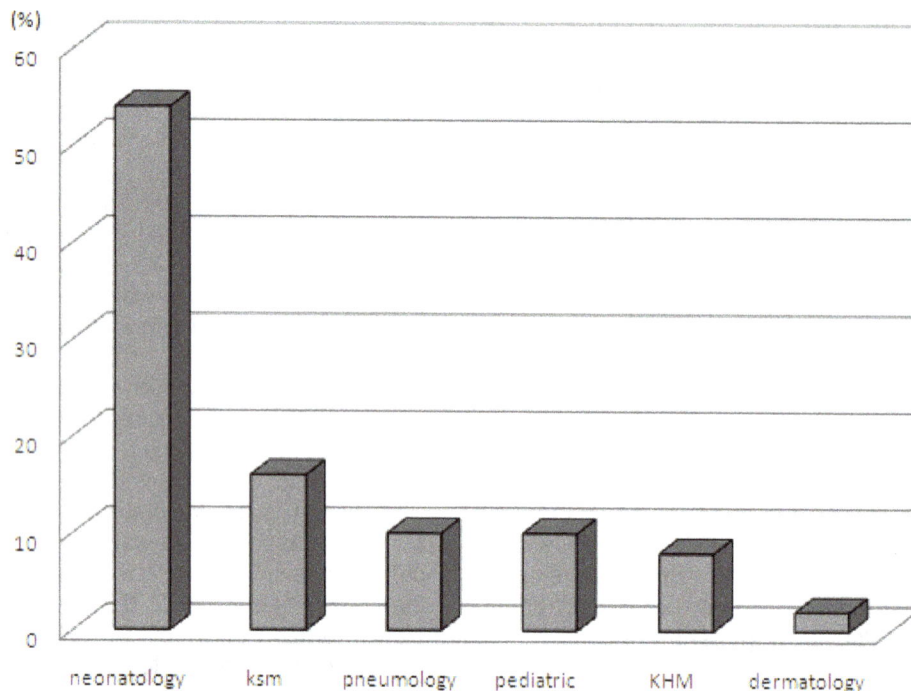

Figure 1. Distribution of the fifty studied clinical isolates of *S. aureus* according to different services of the university hospital of Tlemcen during a period of two years.

violet solution, wells were washed three times with 1 × PBS to remove unbound dye. Finally, all wells were filled by 200 µl ethanol (95%) to release the dye from the cells. Optical density (OD) of stained adherent bacteria was determined with an Absorbance Microplate Reader (model EL×800) at wavelength of 630 nm. To correct background staining, the OD values of the eight control wells were averaged and subtracted from the mean OD value obtained for each strain. The experiment was repeated three times separately for each strain and the average values were calculated with standard deviation (SD).

Classification of adherence

The mean values of OD obtained for blank tests were subtracted from the mean values of OD obtained for each test strain to correct the background staining of microtiter plate. The Absorbance Microplate Reader (model EL×800) used in this study has a dynamic range from 0 to 3.0 OD. According to the classification of Christensen et al. (1985) using the microtiter-plate, strains are divided into three categories: non-adherent, weakly adherent and strongly adherent. However, our clinical isolates were classified into four categories (Stepanovic et al., 2000): non-adherent (OD < ODc); weakly-adherent (ODc < OD < 2xODc); moderately-adherent (2xODc < OD < 4xODc); strongly-adherent (4xODc < OD); with ODc: the cut-off OD (three standard deviations above the mean OD of the blank test). The averaged OD values and standard deviations were made by Excel computer software.

RESULTS

As can be shown in Figure 1, of the fifty (20.83%) clinical strains of *S. aureus*: 27 (54%), 9 (18%), 5 (10%), 5 (10%)

and 4 (8%), were respectively isolated from the following services: Neonatology, pneumology, pediatric, neurology, and internal medicine.

The results of microtiter plate assay used for assessment of biofilm-forming ability of the fifty clinical isolates of *S. aureus* are presented in Figure 2. The method applied in this study allowed us to measure biofilm formation after growth in BHI 1% glucose for 24 h at 37 °C. Spectrophotometric measurement of optical densities (OD) of adherent cells enabled us to classify our clinical isolates collection into four categories (Figure 2); non adherent (OD ≤0.2), weakly (0.2<OD≤0.4), moderately (0.4<OD≤0.8) and strongly (0.8<OD) adherent strains (Figure 3). Of the 50 clinical isolates studied, 16 (32%) were designated as non adherent, 20 (40%) as weakly 10 (20%) as moderately and 4 (8%) as strongly adherent.

DISCUSSION

The *Staphylococcus* genus acquires a huge importance in implant-related infections (Campoccia et al., 2006). Elsewhere, the number of diseases caused by *S. aureus* continues to grow. One of the reasons why *S. aureus* is such a ubiquitous pathogen is that it colonizes the anterior nasopharynx in 10 to 40% of humans and can be easily transferred to the skin (Williams, 1963). Biofilm-forming ability is one of the crucial ways that enable this microorganism to express it pathogenecity. It was found

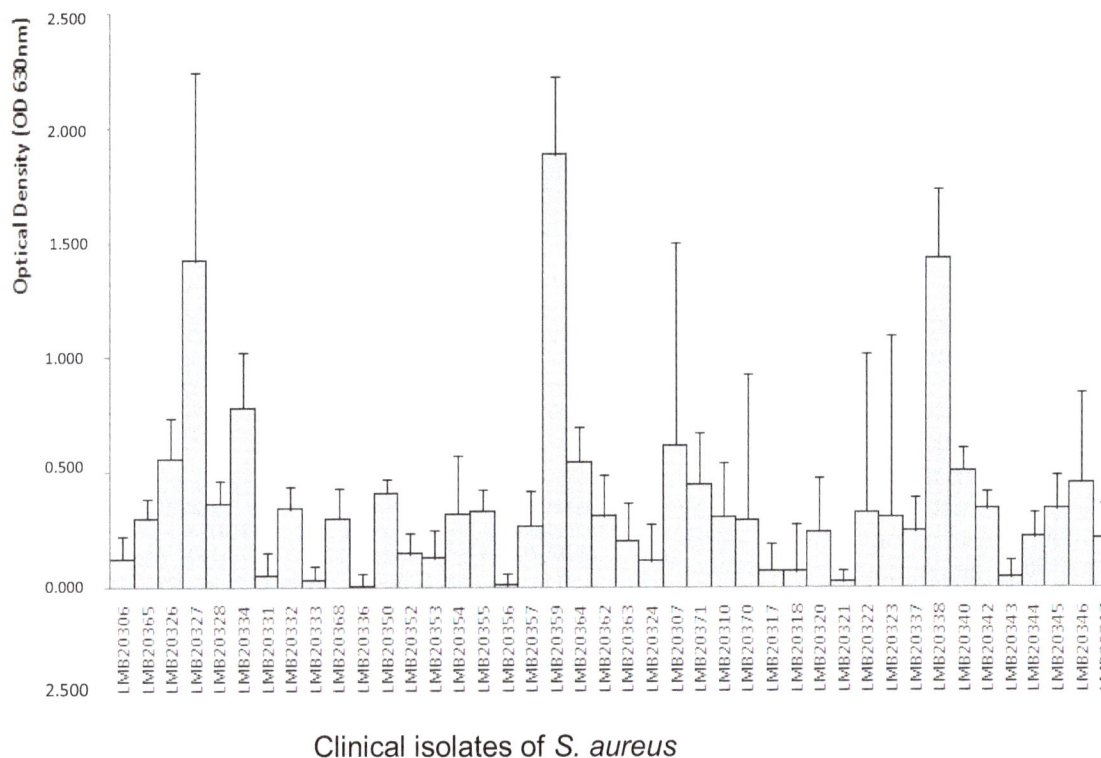

Clinical isolates of *S. aureus*

Figure 2. Biofilm-forming ability on polypropylene microtiter plate of the fifty clinical isolates of *S. aureus* following growth for 24 h at 37°C in brain heart infusion 1% glucose. Bars represent mean values of OD (measured at wavelength of 630 nm) and their standard deviations.

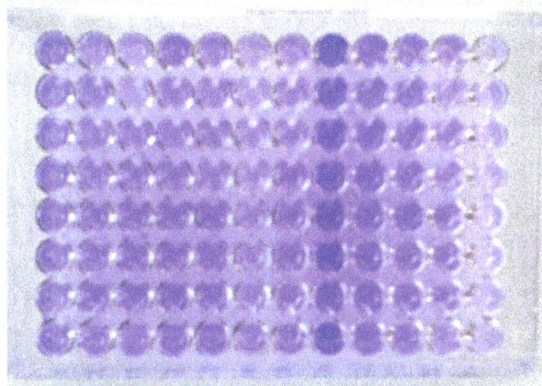

Figure 3. Screening of biofilm formation with crystal violet staining by the 96 well microtiter plate: (I) high, (II) moderate (III) weak and (IV) non adherent.

that the virulence of the organism does indeed vary with its ability to adhere to plastic tissue culture plates (Baddour et al., 1984). Furthermore, as the process of adherence is the initial event in the microbial patho-genesis of infection, failure to adhere will result in removal of the microorganism from the surface of an implanted medical device and avoidance of device-related infection (Ofek and Beachey, 1980). Moreover,

biofilm formation by *S. aureus* is influenced by environ-mental factors like sugars (glucose and/or lactose) or proteases present in the growth medium and depends also on the genetic make-up of a particular *S. aureus* isolate (Melchior et al., 2009). Therefore, according to several researches it was supposed that assessing for biofilm formation could be a useful marker for the pathogenicity of staphylococci. Their active adhesion mechanisms are currently regarded as crucial virulence factors and frequently considered for the characterization of the clinical isolates in studies of molecular pathogenesis and epidemiology (Campoccia et al., 2006). However, some authors considered that there is a little or no correlation between biofilm formation *in vitro* and the clinical outcome of the infection (Kotilainen, 1990; Perdreau-Remington et al., 1998).

In this study, the largest number of clinical isolates of *S. aureus* was collected from neonatology services (n=27), followed by internal medicine (n=9), pneumology and pediatric services (n=5) and finally the neurology services (n=4) (Figure 1). Furthermore, investigation of the correlation between the isolation sites and biofilm-forming ability was not highlighted in this work but it would be efficient to note that among the four strains of *S. aureus* recognized as strongly adherents, two are from the neonatology services.

Various methods have been used to quantify adhesion

of microorganisms to different surfaces. Direct methods allow the *in situ* observation of microbial colonization, including microscopy techniques (laser-scanning confocal, transmission electron and scanning electron microscopy) and indirect methods such as microtiter plate assay, Tube method (TM) and Congo red agar (CRA). Among these various methods, we have used in this study a simple *in vitro* microtiter pate method to quantify the biofilm formation of 50 clinical isolates of *S .aureus*. This method has the advantage of enabling researchers to rapidly analyze adhesion of multiple bacterial strains or growth conditions within each experiment (Djordjevic et al., 2002).

It is known that the direct observation by microscopic techniques is the most important method to study adhesive cells and biofilms, but we think that the microtiter plate assay can be used alternatively as an accurate, rapid, reproducible and inexpensive primer screening method. Thus, this simple quantitative method enables us to assess simultaneously a big number of strains for their biofilm-forming ability. However, in order to complete and enhance the final results obtained in this study, it would be efficient to carry out other experiments, such as PCR for detection of *ica*ADBC genes in the isolates and comparison with the microtiter plate assay results; and animal infection test especially among the four strongly adherent stains to assess the relationship between the biofilm formation and the pathogenicity.

ACKNOWLEDGEMENT

We are very thankful for co-operation received from Dr. Zingg Walter.

REFERENCES

Allignet J, Aubert S, Dyke KG, El Solh N (2001). *Staphylococcus caprae* strains carry determinants known to be involved in pathogenicity: a gene encoding an autolysin-binding fibronectin and the *ica* operon involved in biofilm formation. Infect. Immun. 69:712-718.

An YH, Friedman RJ (2000). Handbook of bacterial adhesion: principles, methods, and applications. Humana Press, Totowa, N.J.

Baddour LM, Christensen GD, Hester MG, Bisno AL (1984). Production of experimental endocarditis by coagulase-negative staphylococci: variability in species virulence. J. Infect. Dis. 150:721-727.

Begun J, Gaiani JM, Rohde H, Mack D, Calderwood SB, Ausubel FM, Sifri CD (2007). Staphylococcal biofilm exopolysaccharide protects against Caenorhabditis elegans immune defenses. PLoS Pathog. 3:e57.

Borucki MK, Peppin JD, White D, Loge F, Call DR (2003). Variation in Biofilm Formation among Strains of Listeria monocytogenes. Appl. Environ. Microbiol. 69:7336-7342.

Broschat SL, Call DR, Kuhn EA, Loge FJ (2005). Comparison of the reflectance and Crystal Violet assays for measurement of biofilm formation by Enterococcus. Biofilms. 2:177-181.

Campoccia D, Montanaro L, Renata Arciolaa C (2006). The significance of infection related to orthopedic devices and issues of antibiotic resistance. Biomaterials. 27:2331-2339.

Christensen GD, Simpson WA, Bisno AL, Beachey EH (1982). Adherence of slime-producing strains of *Staphylococcus epidermidis* to smooth surfaces. Infect. Immun. 37:318-326.

Christensen GD, Simpson WA, Younger JJ, Baddour LM, Barrett FF, Melton DM, Beachey EH (1985). Adherence of coagulase-negative staphylococci to plastic tissue culture plates: a quantitative model for the adherence of staphylococci to medical devices. J. Clin. Microbiol. 22:996-1006.

Corrigan RM, Rigby D, Handley P, Foster TJ (2007). The role of *Staphylococcus aureus* surface protein SasG in adherence and biofilm formation. Microbiol. 153:2435-2446.

Costerton JW, Stewart PS (2000). Biofilm and device-related infections. In: Persistent Bacterial Infections (Eds. Nataro, JP, Blaser, MJ and Cunningham- Rundles, S), ASM Press, Washington, pp. 423-439.

Costerton JW, Lewandowski Z, Caldwell D, Korber DR, Lappin-Scott HM (1995). Microbial biofilms. Ann. Rev. Microbiol. 49:711-745.

Cramton SE, Gerke C, Schnell NF, Nichols WW, Goötz F (1999). The intercellular adhesion (*ica*) locus is present in *Staphylococcus aureus* and is required for biofilm formation. Infect. Immun. 67:5427-5433.

Cucarella C, Solano C, Valle J, Amorena B, Lasa I, Penades JR, Bap (2001). a *Staphylococcus aureus* surface protein involved in biofilm formation. J. Bacteriol. 183:2888-2896.

Deighton MA, Balkau B (1990). Adherence measured by microtiter assay as a virulence marker for *Staphylococcus epidermidis* infections. J. Clin. Microbiol. 28:2442-2447.

Diani M, Esiyok OG, Nima Ariafar M, Yuksel FN, Altuntas EG, Akcelik N (2014). The interactions between *esp*, *fsr*, *gelE* genes and biofilm formation and pfge analysis of clinical *Enterococcus faecium* strains. 8(2):129-137.

Djordjevic D, Wiedmann M, McLandsborough LA (2002). Microtiter plate assay for assessment of Listeria monocytogenes biofilm formation. Appl. Environ. Microbiol. 68:2950-2958.

Donlan RM, Costerton JW (2002). Biofilms. Survival mechanisms of clinically relevant microorganisms. Clin. Microbiol. Rev. 15:167-193.

Dunne WM (2002). Bacterial adhesion:Seen any good biofilms lately. Clin Microbiol Rev.15:155-166.

Hobby GH, Quave CL, Nelson K, Compadre CM, Beenken KE, Smeltzer MS (2012). Quercus cerris extracts limit *Staphylococcus aureus* biofilm formation. J. Ethnopharmacol. 144:812-815.

Knobloch JKM, Horstkotte MA, Rohde H, Mack D (2002). Evaluation of different detection methods for biofilm formation in *Staphylococcus aureus*. Med. Microbiol. Immunol. 191:101-106.

Kotilainen P (1990). Association of coagulase-negative staphylococcal slime production and adherence with the development and outcome of adult septicemias. J. Clin. Microbiol. 28: 2779-2785.

Lewis K (2001). Riddle of biofilm resistance. Antimicrob. Agents Chemother. 45:999-1007.

Mack D, Fischer W, Korbotsch A, Leopold K, Hartmann R, Egge H, Laufs R (1996). The intercellular adhesin involved in biofilm accumulation of *Staphylococcus epidermidis* is a linear β-1,6-linked glucosaminoglycan: Purification and structural analysis. J. Bacteriol.178(1):175-183.

Martí M, Trotonda MP, Tormo-Más MA, Vergara-Irigaray M, Cheung AL, Lasa I, Penadés JR (2010). Extracellular proteases inhibit protein-dependent biofilm formation in *Staphylococcus aureus*. Microbes Infect. 12:55-64.

McCann KS (2000). The diversity-stability debate. Nature. 405:228-233.

McCann MT, Gilmore BF, Gorman SP (2008). *Staphylococcus epidermidis* devicerelated infections: pathogenesis and clinical management. J. Pharm. Pharmacol. 60:1551-1571.

Mathur T, Singhal S, Khan S, Upadhyay DJ, Fatma T, Rattan A (2006) Detection Of Biofilm Formation Among The Clinical Isolates Of Staphylococci: An Evaluation Of Three Different Screening Methods. Indian J. Med. Microbiol. 24(1):25-29.

McKenney D, Pouliot KL, Wang Y, Murthy V, Ulrich M, Doring G, Lee JC, Goldmann DA, Pier GB (1999). Broadly protective vaccine for *Staphylococcus aureus* based on an in vivo-expressed antigen. Sci. 284:1523-1527.

Melchior MB, van Osch MHJ, Graat RM, van Duijkeren E, Mevius DJ, Nielen M, Gaastra W, Fink-Gremmels J (2009). Biofilm formation and genotyping of *Staphylococcus aureus* bovine mastitis isolates: Evidence for lack of penicillin resistance in Agr-type II strains. Vet. Microbiol. 137:83-89.

Merino N, Toledo-Arana A, Vergara-Irigaray M, Valle J, Solano C, Calvo E, Lopez JA, Foster TJ, Penades JR, Lasa I (2009). Protein A-mediated multicellular behavior in Staphylococcus aureus. J. Bacteriol. 191:832-843.

Miyake Y, Fujiwara S, Usui T, Suginaka H (1992) Simple method for measuring the antibiotic concentration required to kill adherent bacteria. Chemother. 38:286-290.

Montanaro L, Speziale P, Campoccia D, Ravaioli S, Cangini I, Pietrocola G, et al (2011). Scenery of Staphylococcus implant infections in orthopedics. Future Microbiol. 6:1329-1349.

Ofek I, Beachey EH (1980). General concepts and principals of bacterial adherence in animals and man. In Bacterial Adherence, Ed. E.H. Beachey. Chapman and Hall, London.pp.1-29.

O'Neill E, Pozzi C, Houston P, Humphreys H, Robinson DA, Loughman A, Foster TJ, O'Gara JP (2008). A novel Staphylococcus aureus biofilm phenotype mediated by the fibronectin-binding proteins, FnBPA and FnBPB. J. Bacteriol. 190:3835-3850.

Perdreau-Remington F, Sande MA, Peters G, Chambers HF (1998). The abilities of a Staphylococcus epidermidis wild-type strain and its slime-negative mutant to induce endocarditis in rabbits are comparable. Infect. Immunol. 66:2778–2781.

Raad I (2000). Management of intravascular catheter–related infection. J Antimicrob. Chemother. 45:267-270.

Raad II, Costerton W, Sabharwal U, Sacilowski M, Anaissie E, Bodey GP (1993). Ultrastructural analysis of indwelling vascular catheters: a quantitative relationship between luminal colonization and duration of placement. J. Infect. Dis. 168:400-407.

Ramage G, Vande Walle K, Wickes BL, Lôpez-Ribot JL (2001). Standardized method for in vitro antifungal susceptibility testing of Candida albicans biofilms. Antimicrob. Agents Chemother. 45:2475-2479.

Souli M, Giamarellou H (1998). Effects of Slime produced by clinical isolates of coagulase negative staphylococci on activities of various antimicrobial agents. Antimicrob. Agents Chemother. 42:939-941.

Stepanovic S, Vukovic D, Dakic I, Savic B, Svabic-Vlahovic M (2000). A modified microtiter-plate test for quantification of staphylococcal biofilm formation. J. Microbiol. Methods .40:175-179.

Trampuz A, Widmer AF (2006). Infections associated with orthopedic implants. Curr Opin Infect Dis. 19:349-356.

Vergara-Irigaray M, Valle J, Merino N, Latasa C, Garcia B, Ruiz de Los Mozos I, Solano C, Toledo-Arana A, Penades JR, Lasa I (2009). Relevant role of FnBPs in Staphylococcus aureus biofilm associated foreign-body infections. Infect. Immun. 77: 3978-3991.

Williams R (1963). Healthy carriage of Staphylococcus aureus: its prevalence and importance. Bacteriol. Rev. 27:56-71.

Wollin TA, Tieszer C, Riddell JV, Denstedt JD, Reid G (1998). Bacterial biofilm formation, encrustation, and antibiotic adsorption to ureteral stents indwelling in humans. J. Endourol. 12:101-111.

Yachi S, Loreau M (1999). Biodiversity and ecosystem productivity in a fluctuating environment: The insurance hypothesis. Proc. Natl. Acad. Sci. USA, 96:1463-1468.

Ziebuhr W, Lößner I, Krimmer V, Hacker J (2001). Methods to detect and analyse phenotypic variation in biofilm-forming staphylococci. Methods Enzymol. 336:195-203.

Yarwood JM, Paquette KM, Tikh IB, Volper EM, Greenberg EP (2007). Generation of Virulence Factor Variants in Staphylococcus aureus Biofilms. J. Bacteriol. 189:7961-7967.

25

Global review of meningococcal disease: A shifting etiology

J. Leimkugel[1], V. Racloz[1*], L. Jacintho da Silva[2] and G. Pluschke[1]

[1]Swiss Tropical Institute, Socinstrasse 57, Basel, Switzerland.
[2]Novartis Vaccines and Diagnostics, Sidney Street 45, Cambridge, U.S.A.

Despite expansive studies over the past century, the epidemiology of invasive meningococcal disease (IMD) has remained elusive in some of its aspects. The following review attempts to summarize the past and current trends in the etiology of IMD. Data was collected through the analysis of peer-reviewed studies and surveillance data on national, sub-national and regional levels performed using various search engines such as pubmed (www.ncbi.nlm.nih.gov/pubmed/), regional WHO homepages (www.who. int) and department of health websites. Despite the establishment of improved surveillance, the reasons for the differences in IMD epidemiology between endemic and epidemic settings are not fully understood. Factors influence the timing and distribution of epidemics including climatic, socio-economic and cultural factors involving changes in human lifestyle, natural growth of the human population, crowding and increased mobility. These have also strongly affected the global population structure of *Neisseria meningitides* and are still currently responsible for changing patterns in IMD epidemiology. In recent years, much interest has arisen on the subject due to both the development of conjugate vaccines and to the continuing occurrence of outbreaks, many of them in industrialized countries. With antimicrobial resistance on the rise, effective and affordable vaccines along with continued surveillance are needed to help combat this complex disease.

Key words: *Meningococcal meningitis*, epidemiology, vaccines, antimicrobial resistance, surveillance.

INTRODUCTION

Despite expansive studies over the past century, the epidemiology of invasive meningococcal disease (IMD) has remained elusive in some of its aspects. In recent years much interest has arisen on the subject due to both the development of conjugate vaccines and to the continuing occurrence of outbreaks and epidemics, many of them in industrialized countries, in spite of the availability of these more efficacious vaccines. The following review attempts to summarize the past and current trends in the etiology of IMD.

Data were collected through the analysis of peer-reviewed studies and surveillance data on national, sub-national and regional levels performed using various scientific and non-scientific search engines such as pubmed (www.ncbi.nlm.nih.gov/pubmed/), scielo (www.scielo.org) and scholar google, regional WHO homepages (www.who.int) and department of health websites. Search vocabulary included terms such as "meningitis", "meningococcal disease", "neisseria meningitidis", "invasive meningococcal disease", "outbreak", "incidence", "serogroups-A, B, C, W135, Y, E29, X", "clonal complexes", "meningococcal vaccines" as well as individual country names. Allowing for differences in surveillance methods and limitations in ascertainment, we contrast major trends of meningococcal disease epidemiology in different regions with a historic perspective.

N. meningitidis, a gram-negative diplococcal bacterium, is a commensal of the human *nasopharynx* and causative agent of *meningococcal meningitis* and *septicemia*. It is one of the most significant human pathogens and is together with *Streptococcus pneumoniae* and *Haemophilus influenzae,* one of the major causes of bacterial meningitis. The disease burden associated with bacterial meningitis is comparable to tropical diseases such as *trypanosomiasis, Chagas* disease, *schistosomiasis, leishmaniasis, lymphatic filariasis, onchocerciasis,* leprosy and

---

*Corresponding author. E-mail: vracloz@gmail.com.

dengue all combined (Mathers et al., 2002).

Transmitted through droplets of respiratory secretions, *N. meningitidis* carriage rates reach up to 10% of the general population at any point in time (Yazdankhah and Caugant, 2004) whilst invasive disease, developing in a small percentage of carriers is regarded as a medical emergency. Case-fatality rates critically depend on access to health services and despite the availability of antibiotic therapy, 10% of patients die within two days of disease onset whilst *sequelae* such as hearing loss, brain damage and learning disabilities can affect up to 20% of the survivors (Welch and Nadel, 2003). *Meningococcal septicemia*, a complication of the disease, often causes hemorrhagic rashes and rapid circulatory collapse and is associated with higher case-fatality and sequelae rates. The annual incidence of meningococcal disease ranges in many industrialized countries from 1 - 5 cases per 100'000 (Greenwood, 2007). Incidence rates (IR) are usually highest during early childhood, in teenagers and young adults, although this also depends on a variety of epidemiological factors. Particularly during epidemics, a broad age spectrum may be observed (Greenwood et al., 1979; Moore, 1992).

Whilst limited outbreaks of disease are often associated with overcrowding, resulting in high transmission of *meningococci* in schools, student dormitories, refugee camps or military facilities, epidemic meningococcal disease claims many lives especially in the so called African *meningitis* belt (Greenwood, 1999). Outbreaks of meningococcal disease in the belt are generally confined to the dry season whilst in Europe, Northern America and other temperate regions, a seasonal increase of meningococcal cases is usually recorded during winter months (Cartwright, 1995; Tikhomirov et al., 1997). Apart from direct case contact and overcrowding, smoking and exposure to smoke (Hodgson et al., 1997; Stanwell-Smith et al., 1994) have also been identified as possible risk factors for infection and invasive disease.

MENINGOCOCCAL CLASSIFICATION AND TYPING

Serological differentiation of meningococcal isolates began in the early 20th century. From a first subclassification of *meningococci* into main types associated with epidemics, versus rare types associated with endemic disease, four decades were needed to acquire a more comprehensive understanding of antigenic diversity and to develop a standardized serological typing system which was introduced in 1950 (Vedros, 1987). Nowadays, 13 meningococcal serogroups reflecting the chemical structure of the polysaccharide capsule are known (Yazdankhah and Caugant, 2004), of which six (A, B, C, W135, X and Y) are responsible for the vast majority of IMD cases. Historically, epidemics of serogroup, a *meningococci* have caused the highest disease burden worldwide. Currently in many industrialized regions

serogroup B, responsible for endemic disease but also local and intercontinental outbreaks is the most common cause of IMD. Serogroup C has been responsible for causing severe outbreaks in Europe and North America in the past 30 years with a high disease burden in adolescents. Although the remaining meningococcal serogroups are regarded as less virulent and usually only cause symptomless colonization of healthy individuals. serogroups W135 [13-15], Y [16] and X [17-20] have recently also been associated with outbreaks of IMD.

Serotyping and serosubtyping rely on the detection of distinct epitopes present on different classes of outer membrane proteins of *N. meningitidis*. Serotyping epitopes are found on class 2 or class 3 outer membrane proteins (OMP) (P or B) of *N meningitidis*. Serosubtyping epitopes are present on class 1 OMP (P or A). Although serological typing provides essential information on antigenic diversity in particular for potential OMP-based vaccination strategies, typing reagents are not available for all serotypes or subtypes. Additionally, due to phase variation or loss of the corresponding genes many strains do not react serologically (Claus et al., 2005).

More recently, several molecular methods have been developed which complement serological typing techniques (Caugant, 2008). Multi locus enzyme electrophoresis (MLEE) using the natural variation in electrophoretic mobility of various enzymes has been the standard typing method used until the late 1990s. In the last decade, it was replaced by multilocus sequence typing (MLST) based on the nucleotide sequences of 400 - 500 bp long fragments of seven housekeeping genes (Maiden et al., 1998). For each gene, the sequence variants observed are assigned as distinct alleles which define the sequence type (ST) based on the seven loci of every isolate. Evolutionary developments can be modeled using MLST data in combination with the e-Burst (Based upon related sequence types) application (Feil et al., 2004).

Clonal complexes are defined as closely related groups of isolates in which all STs are linked to a single locus variant of at least one other ST. The clonal complexes differ substantially in their pathogenic potential and only 11 of the 37 so far identified complexes are strongly associated with invasive disease (Caugant, 2008). In particular, serogroup A *meningococci* have a clonal population structure (Olyhoek et al., 1915 and 1983; Morelli et al., 1997; Achtman, 2004) with the majority of isolates in the MLST database assigned to one of three major lineages, the ST-1, ST-4 and ST-5 clonal complexes. The other meningococcal serogroups show a greater genotypic and serologic diversity and clonal complexes commonly include isolates of different capsular serogroups (Yazdankhah et al., 2004) (http://mlst.net).

The genetic factors that determine virulence of *N. meningitidis* strains have yet to be clearly identified. Avirulence of unencapsulated strains shows that expression of a polysaccharide capsule is an important factor. Epidemiological data indicate that certain capsule polysaccharides such as A, B and C may in particular contribute to resistance against host defense mechanisms. However, recent outbreaks caused by strains belonging to serogroups W135, X or Y demonstrate that strains expressing other capsules also present epidemic potential. It is mostly assumed that complex constellations of genetic factors determine virulence (Caugant, 2008).

FIRST REPORTS OF MENINGOCOCCAL DISEASE IN THE 19TH CENTURY

Whilst accounts of "epidemic convulsions" in China date back a thousand years (Zhen, 1987), the first reliable records of epidemic meningococcal disease originate from the early 19th century, specifically from Geneva (Switzerland) in 1805 (Vieusseux, 1805) and Massachusetts (U.S.A) in 1806 (Harrison and Broome, 1987). The causative agent was not identified until 1884 (Marchiafava and Celli, 1884) and *N. meningitidis* was cultivated for the first time by Weichselbaum in Vienna in 1887 who named it *Diplococcus meningitidis intracellularis* (Weichselbaum, 1887). Early accounts and reports of the disease were compiled by Hirsch in his "Handbook of geographical and historical pathology" which appeared in 1886. Hirsch distinguished three periods of IMD spread in Europe and Northern America, prior to the emergence of *N. meningitidis* as a globally distributed pathogen in the 1880s (Hirsch, 1886). First outbreaks were seen in various regions of the USA and Canada in the years 1805 - 1816 and in Europe, particular in France and southern Italy from 1805 - 1830.

During the second period between 1837 - 1850 outbreaks were reported from France, Italy, Denmark, Spain, Ireland and Germany. In the USA, meningitis outbreaks mostly affected several eastern states during that period (1842 - 1850) (Hirsch, 1886). During the third period between 1854 - 1875, IMD became widely distributed in the USA and in Europe with severe outbreaks in Sweden and Germany. Hirsch described many characteristic features of IMD such as the strong seasonality, the role of overcrowding as risk factor, the age distribution of cases and age shifts in certain epidemic situations (Hirsch, 1886). In response to the emergence of epidemic meningitis in Europe, the first notification system for meningococcal disease was implemented in 1875 in Sweden, even before the causative agent was described (Jones and Abbott, 1987; Peltola, 1987; Cartwright, 1995).

On the African continent, first outbreaks were seen, most likely imported by French soldiers among soldiers in Algiers around 1840 (Hirsch, 1886) and in South Africa by 1888 (Greenwood, 1999). First accounts from the near east comes from Jerusalem in 1872 and Persia 1874/75 (Hirsch, 1886). In South America, the disease was first reported amongst Portuguese immigrants in Sao Paolo in 1906 (de Moraes and Barata, 2005). Other early reports dated back to 1879 in the Indian subcontinents, 1897-98 in China, Hong Kong and the Yangtze Valley and 1900 in

Australia and New Zealand (Hansman, 1987).

1900 - 1950: Large serogroup A epidemics associated with the world wars

In the early 20th century, several larger outbreaks were recorded in Europe and Northern America including Germany where the industrial and mining regions were severely affected with over 10'000 cases being reported between 1904 and 1907. Similarly, Norway, Sweden and Scotland experienced serious outbreaks in the period preceding the first world war (Jones and Abbott, 1987; Peltola, 1987; Peltola, 1983; Kuzemenska and Kriz, 1987). During the first world war, England was heavily affected with altogether more than 10,000 cases in civilian and military populations (1915 - 1919) but also Italy reported an epidemic wave of IMD from 1915 - 1918 (Ballada et al., 1990). In the USA, major outbreaks of IMD were seen in the early 20th century in particular during the recruitment and mobilization of military forces. This included outbreaks associated with the occupation of Cuba which began in 1899 as well as in Mexico in 1914. The first large epidemic with over 5,000 cases occurred during the winter of 1916 - 1917. Additionally, US Army camps located in France and Britain during the first world war reported over 5,000 cases with a case fatality rate of 39%. Post-war, further outbreaks were reported in 1928 - 1929 and 1936 - 1937, affecting US Army personnel as well as spilling over into the civilian population (Harrison and Broome, 1987; Brundage and Zollinger, 1987). In Canada, two outbreaks were reported in 1929 and 1941 with overall IRs of 3.1 and 12.8 /100,000 respectively (Varughese and Acres, 1987).

The most extended outbreaks in Europe occurred during the second world war throughout the continent, primarily affecting the countries directly involved in the war.

Between 1940 and 1942, England reported 30'000, Germany 15,000 and Italy nearly 9,000 cases. Outbreaks were also reported from Austria, France, Portugal, Spain, Bulgaria, Romania and Greece (WHO, 1945). In Scandinavia, meningococcal outbreaks were introduced by occupying troops and IRs peaked in 1941 (Norway, Iceland) and 1944 (Denmark). Whilst Finland, occupied by the Russians, experienced a small outbreak, meningococcal meningitis epidemics were absent from Sweden in the 1940s (Peltola, 1987). Yugoslavia and Hungary each reported epidemics of greater than 3'000 cases in 1940 before entering the war in 1941 (WHO, 1945). In the USA, an epidemic occurred during the mobilization for the second world war in the winter of 1942 - 1943 with more than 13'000 cases being reported in army training camps mostly affecting fresh recruits (Brundage and Zollinger, 1987).

Comparable to Europe and Northern America, serogroup A epidemics occurred in Australia on a continent-wide basis during the two world wars. Overall IRs peaked at 30 - 35/100'000. Relatively high notification rates of

IMD attributed to extensive population movements and immigrations (Patel, 2007) were also recorded in several states in the 1950's with IRs of up to 7/100'000 in western Australia and Victoria.

In Asia, following outbreak reports as early as 1879 in British-occupied India, larger incidents were recorded in prisons located in Shikarpur and Alipore as well as in China and Hong Kong where local outbreaks were described in 1917/1918 (Hong Kong), 1919 - 1920 (Anhui), 1923 (Taiwan), 1932 (Guangzhou), Macao (1932) and 1934/1935 (Hunan). First nationwide epidemics occurred in 1938 and 1949 (Zhen, 1987; Cadbury, 1934). In Russia two serogroup A epidemics occurred in 1931 and 1940 (Achtman et al., 1969 – 1997).

In South America, cases of IMD were reported in Sao Paolo since 1906 and a first epidemic started in 1920 with the majority of cases being attributed to serogroup A and about 25% to serogroup C (de Moraes and Barata, 2005). After the peak in 1923 with IRs of circa 12/100'000, the rate returned to endemic levels of 1.3 - 4.1/100'000 by 1926. During the second world war, the region was again seriously affected by epidemics followed by a large outbreak in Sao Paulo in 1947. During this period, outbreaks were also noted in Chile (1941 - 1942) (Cruz et al., 1990) and Mexico (1945-1949) (Almeida-Gonzalez et al., 2004).

Already in 1963, Lapeyssonnie defined the so called 'African Meningitis Belt' a region south of the Sahara and north of the tropical rain forest within the 300 – 1100 mm mean annual rainfall isohyets and stretching from the Gambia to Ethiopia and Sudan. He reported that endemic and epidemic IMD occurs in this area on a strictly seasonal basis during the dry and hot season of the year (December - May) and rapidly declines with the onset of the rains (Moore, 1992). These large epidemics were characterized as occurring within 8 - 14 year cycles and having IRs of up to 1% of the general population (Lapeyssonnie, 1963). The first epidemics in West Africa occurred in Nigeria in 1905 (McGahey, 1905) and in Ghana between 1906 - 1908. The *meningococci* causing the 1906 - 1908 outbreak were possibly introduced by pilgrims returning from Mecca and were also implicated in an epidemic in Sudan a few years earlier (Greenwood, 1999). Subsequently, three epidemic waves set off in 1934, 1942 and 1949. The latter causing over 50'000 cases in 1950 and 1951 respectively in Sudan and Nigeria and 16'000 cases in Burkina Faso in 1957. Virtually all bacterial isolates obtained in the first decades of the 20th century showed serogroup A capsule agglutination except for a few isolates from Sudan where serogroup B was seen or the capsule could not be determined with the available antisera (Lapeyssonnie, 1963).

1950 - 1990: Changes in serogroup distribution

After the end of the second world war, stabilizing social

and economic conditions were associated with a decreease in large serogroup A outbreaks previously reported in the industrialized countries (Schwartz et al., 1989). The introduction of both capsule carbohydrate vaccines and chemoprophylactic measures contributed further to a decrease in IRs (Harrison and Broome, 1987; Brundage and Zollinger, 1987). Although epidemics be-came rarer in these regions. Certain foci, such as nor-thern parts of Europe and the USA remained hotspots for IMD. During 1973 - 1976, a serogroup A epidemic occurred in Finland accounting for over 1'500 cases (Peltola, 1978), whilst a serogroup B outbreak in Iceland and Norway (1976 - 1986) claimed 3'000 cases in Norway alone with high case-fatality rates associated with ele-vated rates of septicemia (Peltola, 1983; Caugant et al., 1986; Bovre and Gedde-Dahl, 1980). In the USA, a serogroup A outbreak in skid row populations in Alaska, Seattle and Portland was recorded in 1975 (Filice et al., 1984). In Canada, case clusters of serogroup A followed by serogroup C caused elevated IRs in the early 1970s (Varughese and Acres, 1987).

In Australia, epidemics of serogroup A also subsided after the second world war with the exceptions of two localized outbreaks among the Aboriginal population in Alice springs in 1971 - 1973 (Creasey, 1991) and 1987 - 1991 (Patel et al., 1993). In Canberra a sharp increase in the proportion of serogroup C disease appeared in the 1980s, preceded by clusters of serogroup C disease in 1968 (Hansman, 1987). Rising IRs in the 1980s was attributed to both serogroups B and C.

Serogroup A outbreaks were also reported from New Zealand in the Maori population between 1985 - 1987 representing one of many examples that distinct ethnic groups may be affected differently. In New Zealand IRs of 68/100'000 and 17/100'000 were recorded in 1968 in Maori tribes and non-Maori populations respectively (Knights, 1972). Apart from these serogroup A outbreaks, IRs generally dropped continuously during the 1960s and 1970s to <0.5/100'000 in 1981 with a dominance of serogroup B.

Although the quality of surveillance data is poor for many regions in Asia, certain trends in incidence and serogroup distribution can be deduced. In China the IR of IMD increased in the early 1950s continuously until the end of the 1980s to hyperendemic levels between 8.7 in low and >25/100'000 in high incidence regions. In addition, nationwide epidemic waves peaked in 1957/1958, 1967, 1977 and 1985 (Wang et al., 1992). In Russia, an epi-demic introduced by Vietnamese factory workers entering through China, started in 1968 and peaked in Moscow in 1970 (Wang et al., 1992). In India, various outbreaks of meningococcal meningitis occurred in particular in Delhi and its surrounding in 1966 and in 1985 as well as in other regions such as Surat and Gujarat in 1985 - 1987 (Manchanda et al., 2006). Mongolia suffered from IMD outbreaks in 1974 - 1975 (Ebright et al., 2003). In Viet-nam, an epidemic occurred in Ho Chi Minh city in

1972 - 1973 and a serogroup C outbreak associated with a high case fatality rate was reported in 1977 - 1978 from the southern provinces (Oberti et al., 1981). In Bhutan 250 cases were notified in 1985 - 1986 (Manchanda et al., 2006), and in Nepal a severe serogroup A epidemic in the Kathmandu valley in 1982 - 1984 (Manchanda et al., 2006; Cochi et al., 1987). In contrast, in Japan (Takahashi et al., 2004) and Taiwan (Hsueh et al., 2004), IMD IRs have been comparably low during the past 50 years, although more than 4'000 annual cases were reported in pre-world war II Japan whereby serogroups B (57%) and Y (21%) have dominated strain collections in the past 30 years (Takahashi et al., 2004).

In the near and middle east, three major serogroup A epidemics were reported between 1967 - 1989 with a seasonal pattern similar to the African meningitis belt. However, IRs have never reached comparable dimensions as in the Belt (Girgis et al., 1993; Sippel and Girgis, 1987) whilst reports from Israel indicate a major contribution of serogroup B in the general population. Serogroup A has been isolated at a higher frequency in the Arab community. Compared to the national average, IRs were higher in communities with lower socio-economic status and in Jerusalem (2.45 vs. 1.13 /100'000) (Block et al., 1993; Stein-Zamir et al., 2007).

In South America, low endemic levels (IRs of 0.3 - 0.6/100'000) of IMD were reported from Chile and Brazil (Cruz et al., 1990) and many other countries (WHO, 1977) throughout the 1950s and 1960s.

In Sao Paolo, a serogroup C epidemic starting in 1971 was replaced by a second wave caused by serogroup A. IRs exceeded 100/100'000 (Souza et al., 1974) with 20'000 cases being reported in that year alone (WHO, 1977). In Argentina, an outbreak with a national IR of 8.3/100'000 and 2,144 notified cases occurred in 1974 with a dominance of serogroup C (82%) and some contribution of serogroup A (18%) (WHO, 1975). Similarly, an outbreak reported in southern Chile in 1978 - 1979 saw a serogroup shift from C to A associated with increasing case fatality rates. Since 1988, hyperendemic IRs of IMD have been notified in Sao Paolo (Santos and Ruffino-Netto, 2005), along with an increase in serogroup B disease. The most prominent serogroup B outbreaks occurred in Cuba in the early 1980s (Caugant et al., 1986; Rico et al., 1996) with a peak IR of 14/100'000 and in Iquique, Chile with an IR of >20/100'000 between 1983 - 1987, rated 20 times higher than in the rest of the country (Cruz et al., 1990). In Mexico, IMD has been rare since an outbreak in the 1940s (Almeida-Gonzalez et al., 2004).

Epidemic meningococcal disease on the African continent continued to threaten countries in the meningitis belt in particular, occurring in frequent intervals (Figure 1).

Virtually all epidemics were caused by serogroup A until the 1990s with the exception of some reported serogroup C outbreaks in the 1970s (Broome et al., 1983; Whittle et

al., 1975). Data from outside the meningitis belt are rare. Some hospital based studies indicate that in many regions *S. pneumoniae* and *H. influenzae* meningitis (particular in children) are of higher significance than IMD (Cadoz et al., 1981; Muhe and Klugman, 1999; Gordon et al., 2000; Peltola, 2001). Longitudinal data from South Africa show a decrease of IRs from 5 - 10/100'000 to < 2/100'000 since 1945 with hyperendemic waves (Coulson et al., 2007).

Since 1990: Improved surveillance and the introduction of conjugate vaccines

Since 1990, several regional surveillance networks have been established including the EU-IBIS network in Europe (http://euibis.org), the SIREVA network in South America, the WHO-EMRO network in the countries of North Africa and the near and Middle East, including national surveillance centers distributed throughout the world such as the active bacterial core surveillance and the center for disease control (CDC) in the United States (http://www.cdc.gov/ncidod/DBMD/abcs/). Even though this has substantially increased the ascertainment of meningococcal disease in these areas, many factors for example different surveillance methods or case definitions still limit the comparability of epidemiological datasets for different regions (Trotter et al., 2005).

In Europe about 90% of IMD is currently caused by serogroups B and C, yet the exact distribution of serogroups and the relative importance of different clonal complexes vary significantly between countries. Interestingly, Sweden has a relatively high proportion of serogroup W135 and Y isolates (EU-IBIS Network, 2003/2004). Although it has largely disappeared from Europe since the 1950s, serogroup A IMD is still sporadically reported. With an exception of the finish outbreak in the 1970s, most of these cases are imported cases from endemic areas and not associated with major spread into the general population (Zhu et al., 2001). However, in Eastern Europe and Russia, serogroup A seems to persist and has been responsible for more than 30% of all analyzed IMD cases in two Romanian districts between 2000 and 2002 (Luca et al., 2004).

Overall IRs in the USA has stabilized at about 1/100'000 per year, although there is some spatial and temporal variation between states and regions (Harrison and Broome, 1987). The contribution of serogroup Y increased from 2% during 1989 - 1991 to 37% during 1997 - 2002 (Bilukha and Rosenstein, 2005). Hyperendemic rates of serogroup B disease are found in the state of Oregon since 1987 (Bilukha and Rosenstein, 2005).

In Canada, the overall IR has remained between 0.5 and 2 cases /100'000 since the early 1940s (Varughese and Acres, 1987; Pollard and Scheifele, 2001). Recent trends include two waves of increased serogroup C meningitis in several provinces in the early 1990s and 2001 (De Wals, 2004) and a cluster of serogroup B

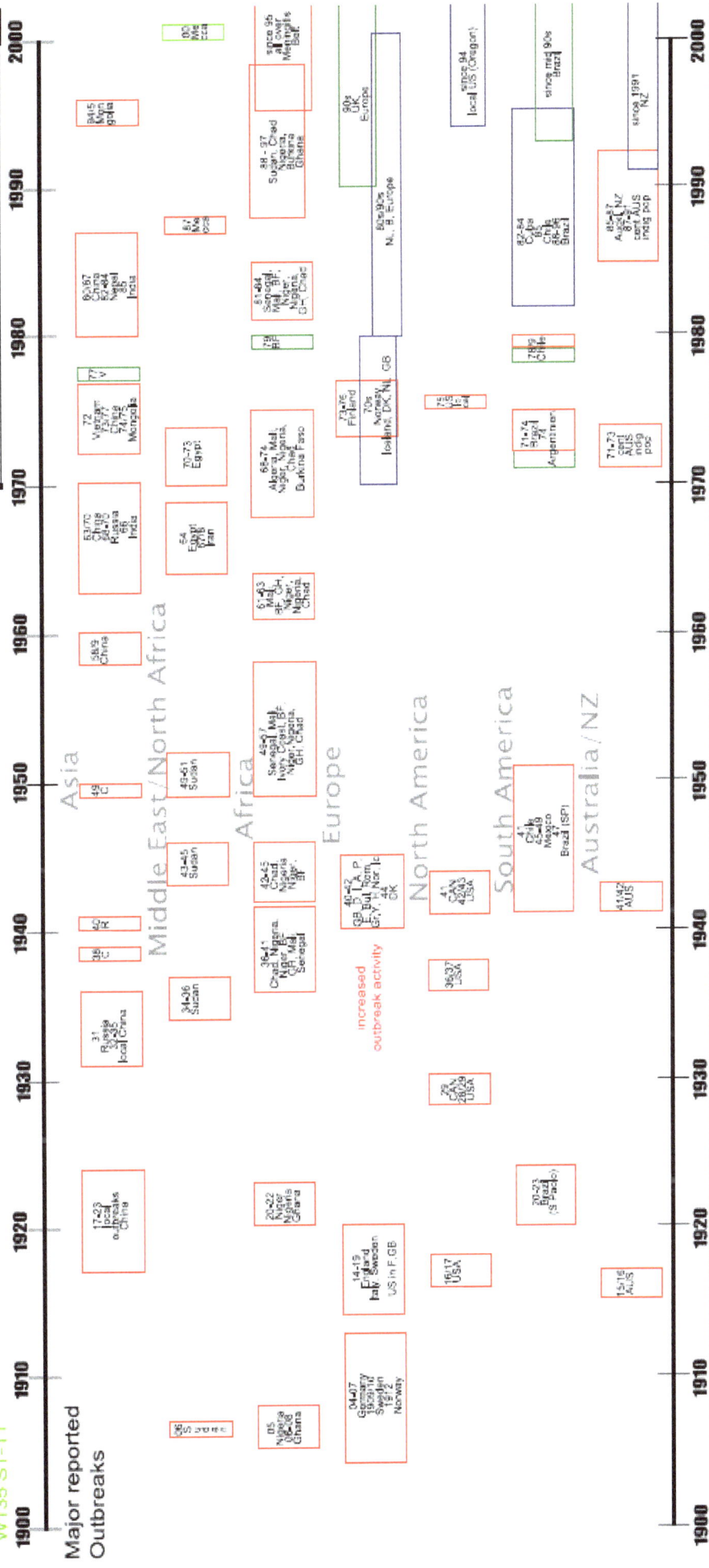

Figure 1. Large meningococcal meningitis outbreaks of the 20th century. Pandemic meningococcal clones with their period of isolation and most important reported outbreaks of the 20th century. (Abbreviations; BF= Burkina Faso/Obervolta, C=China, F=France, GB= Great Britain, GH= Ghana/Goldcoast, R=Russia, V=Vietnam).

meningococcal cases in 2003 (Law et al., 2006). Invasive meningococcal disease in Australia and New Zealand has persisted at a hyperendemic level since the 1990s. In Australia, serogroup B meningococci dominate, although serogroup C meningococci have also caused outbreaks in several regions motivating the introduction of the serogroup C conjugate vaccine (Patel, 2007). In New Zealand, an extended serogroup B epidemic started in the early 1990s and reached its peak between 1996 - 2000 with IRs of 13.9/100'000. A strain-specific OMP- based vaccine was used to control the epidemic (Kelly et al., 2007).

In the larger countries of Asia such as India, China and Russia serogroup A IMD is still very important with outbreaks reported in Mongolia in 1994/1995 (WHO, 1995), Moscow in 2003 (Communicable Disease and Health Protection Quarterly Review, 2003) and Delhi in 2005 (Manchanda et al., 2006). In China, the strong decrease of the annual national IR to ~0.2/100'000 was attributed to regular vaccination campaigns introduced in the 1980s (Zhang et al., 2007). Recently however, reports of serogroup C outbreaks together with rising IR in certain regions have been published (Zhang et al., 2007; Shao et al., 2006) suggesting a gradual replacement of serogroup A by serogroup C (Ni et al., 2008). In Japan and Taiwan serogroups B and Y have dominated for a long period of time (Takahashi et al., 2004). Since 2001 an increasing number of cases caused by imported serogroup A, C and W135 strains have contributed to an increase in national IRs (Hsueh et al., 2004; Yang et al., 2006; Chiou et al., 2006).

In Morocco and Tunisia, serogroup A has lost its significance in particular since the introduction of vaccination for the Hajj in Mecca and serogroup B has dominated since the 1990s (Zerouali et al., 2002; Maalej et al., 2006). Reports from Saudi Arabia have indicated a predominance of serogroup A between 1995 and 1999 while serogroup W135 caused 13% of IMD cases (Lingappa et al., 2000). The overall national IR is low during endemic periods (Mahmoud et al., 2002; Almuneef et al., 1998) but during a W135 outbreak in 2000, it rose to 2.5/100'000. For Pakistan, IRs of >4 cases/100'000 have been reported in the 1990s (http://gis.emro.who.int/HealthSystemObservatory/Database/Forms.aspx).

In South America, after a long period of low IR in the 1970s and 80s, increasing IR and outbreak activities were reported since the late 1980s. In Brazil, a serogroup B outbreak, starting in 1988 and peaking in 1996 was mainly observed in the large cities, for example, Sao Paolo and Rio de Janeiro. Furthermore, serogroup C caused severe outbreaks in different parts of the country as seen in Rio de Janeiro between 1993 and 95 (Barroso and Rebelo, 2007) and in Sao Paolo since 2002 (de Lemos et al., 2007) as well as other parts of the country (Baethgen et al., 2008). The recent emergence of serogroup W135 IMD has also been observed (Barroso and Rebelo, 2007; Baethgen et al., 2008). In Argentina in contrast, IR levels decreased from 2.6 to 0.6/100'000

between 1993 and 2005. The dominance of serogroup B meningococcal disease was interrupted by an intermediate peak of serogroup C between 1996 and 2000. Furthermore, an increase of serogroups Y and W135 has been observed (Chiavetta et al., 1993-2005). In Uruguay a local serogroup B outbreak threat the city of Santa Lucia in 2001, but was answered with immediate vaccination campaign (Pírez et al., 2004). Even from Mexico, increasing numbers of serogroup C have been reported (Almeida-Gonzalez et al., 2004).

The meningitis belt of Africa still carries the heaviest meningococcal disease burden, although changes in the epidemiology have recently been observed. Serogroup A continues to play a major role but substantial outbreaks have also been caused by W135 meningococci. There seems to be an extension of the meningitis belt to the east and the south which may be associated with climate change (Savory et al., 2006; Molesworth et al., 2002). Outbreaks of serogroup W135 (Taha et al., 2000; Fonkoua et al., 2002; Traore et al., 2006) and serogroup X (Djibo et al., 1995-2000; Gagneux et al., 2002; Boisier et al., 2007) are of concern for control strategies based on monovalent serogroup A conjugate vaccines. In South Africa, IR of 0.64/100'000 were reported in 2002 whereby 41% of the cases were attributed to serogroup B, 23% to serogroup A, 21% to serogroup Y, 8% serogroup C and 5% serogroup W135 (Coulson et al., 2007). Similarly to above mentioned areas, a highly lethal serogroup W135 clone caused increasing IR particularly in the Gauteng Province (von Gottberg et al., 2008). Together with serogroup C (Faye-Kette et al., 2003; Newman, 2004), serogroup B seems to play a major role in African countries outside of the meningitis belt such as Cameroon, Uganda, Madagascar (http://www.mlst.nt) and Angola (Gaspar et al., 2001).

Global spread of hypervirulent clones

With improved surveillance systems in place, growing strain collections from laboratories worldwide and advances made in molecular epidemiology, global trends in IMD can more readily be attributed to the spread of hypervirulent clones (Figure 1). Three pandemic waves caused by serogroup A meningococci are well documented. The first involved North Africa and the Mediterranean in 1967 and has mainly been associated with the ST1 clonal complex. In addition, clonally related non-epidemic isolates are available from Pakistan, the Philippines, the USA, South Africa, several European countries, Russia and Australia (Olyhoek et al., 1915 and 1983; WHO, 1977). The ST5 clonal complex caused the highest serogroup A disease burden in the second half of the 20th century and was responsible for the second and third pandemic waves. The second pandemic started in the early 1980s in China and Nepal and spread to many countries throughout the world facilitated by pilgrims returning from Mecca where an outbreak occurred in 1987 and extended to epidemics in the African meningitis belt between 1988 - 1995. A third wave began in Asia in the early 1990s and continued in Africa since 1995. While the second pandemic was mainly associated with ST5 bacteria, the third was caused by ST7 (a single locus variant of ST5) meningococci (Zhu et al., 2001; Achtman, 1997; Caugant and Nicolas, 2007). Recently, outbreak strains belonging to a new serogroup A clone associated with ST 2859 (a single locus variant of ST7) have been isolated in Burkina Faso, Ghana and Sudan (Sie et al., 2008).

The few serogroup A meningococcal strains available from world war I and II associated epidemics in Europe and the USA belong to the ST4 clonal complex. Since bacterial isolates from earlier serogroup A outbreaks in South America, Australia and China are not available, the extent of the spread of the ST4 meningococci is not known. However, ST4 complex bacteria have been isolated in the African meningitis belt for over 20 years from epidemic waves in 1960 -1963, 1968 -1974 and 1981 - 1983 as well as from the inter-epidemic periods (Crowe et al., 1987). ST4 complex isolates are very homogenous by MLST and belong almost exclusively to one single sequence type, ST4. The latest ST4 complex isolates in the MLST database dated from 1992, suggesting a loss of significance of this subgroup since then. In contrast, ST1 complex bacteria have continuously been isolated as seen for example during recent outbreaks in Moscow (www.mlst.net).

Apart from the dramatic serogroup A pandemics, serogroup B, C and W135 have also contributed substantially to the global burden of IMD. While one has to take into account that the MLST database is highly biased in terms of geographical coverage of the isolates, the entered datasets nevertheless shows that the most important non-serogroup A clonal complexes of the 2nd half of the 20th century are the ST32, ST11 and ST44/41 clonal complexes. The ST32 (ET 5) clonal complex emerged in Norway in 1975 causing a serogroup B epidemic with an IR of 24/100'000. Subsequent outbreaks were recorded in Iceland, Denmark, the Netherlands and Great Britain (Poolman et al., 1986).

Intercontinental spread of this clone was responsible for outbreaks in Cuba, Chile and Brazil (Cruz et al., 1990; Schwartz et al., 1989; Caugant et al., 1986). The first extended serogroup B outbreak in the US in the 1990s has also been attributed to this clonal complex (Diermayer et al., 1999).

The ST11 (ET-37) clonal complex is associated with serogroups B, C, W135 and Y. Most serogroup C outbreaks described have been caused by strains belonging to this complex. The first isolates date back to 1917 and the first outbreak were reported in the 1960s among US army personnel. It has also been attributed to the 1971 outbreak in Brazil (Caugant, 1998). Since the 1980s, the ET15 clone of the ST11 complex emerged with increasing contributions to IMD in Europe and North America in particular in adolescents and young adults encouraging

the implementation of serogroup C polysaccharide conjugate vaccines (Hubert and Caugant, 1997; Jelfs et al., 2000). Interestingly, the W135 meningococci responsible for the outbreaks in Mecca and the African meningitis belt since the year 2000 also belong to the ST11 clonal complex. Causing sporadic IMD all over the world including the African meningitis belt since the 1970s, a hypervirulent clone of this serogroup has caused serious outbreaks in different geographic regions after the outbreak in Mecca in 2000 (Aguilera et al., 2000; Traore et al., 2006; Kwara et al., 1998; Mayer et al., 2000).

With over 1,000 associated STs, the ST44/41 clonal complex (Lineage III) dominated by serogroup B is according to the MLST database the most diverse meningococcal clonal complex. In contrast to the other complexes, it is also frequently isolated from healthy carriers. First isolates of a particular lineage III clone were found in the 1960s followed by a strong increase in the frequency of observed cases in the Netherlands around 1980s and developed into one of the most important meningococcal groups in the 1990s (Scholten et al., 1994). Later, closely related meningococci caused increasing IMD case numbers in Belgium (Van Looveren et al., 1998) and have been responsible for an extended outbreak in New Zealand since 1991 (Dyet and Martin, 2006). Currently in Europe, the ST41/44 is the most dominant clonal complex causing IMD in Ireland, Netherlands, Belgium, and Italy (EU-IBIS Network, 2003/2004).

Furthermore, the ST8 (cluster A4) clonal complex seen worldwide since 1960 and the ST18 clonal complex mostly isolated in Eastern Europe and Russia since the 1970 have been described (Caugant, 1998; Caugant et al., 1987). Since the introduction of MLST, four additional hypervirulent clonal complexes have been identified, the ST174 complex mostly associated with serogroup W135 disease, the ST269 complex, an important cause of IMD in the UK and recently associated with serogroup B outbreaks in Canada (Law et al., 2006), the ST334 clonal complex and the ST461 clonal complex (Caugant, 2008).

Antimicrobial resistance

As with many other pathogens, *N. meningitidis* has developed resistance mechanisms against a wide range of antimicrobials (Manchanda et al., 2006; Jorgensen et al., 2005). Although most reported resistance has been linked to serogroups B and C, antibiotic resistant serogroup A isolates have also been described. Tetracycline, sulfisoxazole and trimethoprim-sulfamethoxazole resistance has been observed in Africa, Asia, and the USA with tetracycline resistance being attributed to the drug efflux mechanism encoded by *tet*(B) (Crawford et al., 2005). Resistance of serogroup B strains to chloramphenicol, streptomycin and sulphonamides in isolates from Vietnam and France (Galimand et al., 1998) were caused by internal deletions affecting the catP gene and directly influencing the production of the enzyme chloramphenicol acetyltransferase. In strains from Portugal and several other European countries, penicillin and sulphadiazine resistance caused by a reduced affinity to penicillin binding protein type 2 has been described (Ferreira et al., 2001-2002).

Impact of meningococcal meningitis vaccination

Attempts to prevent IMD through vaccination started as early as 1912, yet the first polysaccharide-based vaccines against serogroups A and C were developed in the 1960s (Gotschlich et al., 1969) and proved effective in military recruits (Artenstein et al., 1970) as well as against epidemic serogroup A meningitis in Africa (Erwa et al., 1973; Ettori et al., 1977) and Europe (Makela et al., 1977). Until the 1990s, only pure polysaccharide-based vaccines were available against serogroups A, C, Y and W135. In spite of a vaccine efficacy of 95% against serogroup A disease in older children and adults in the first year after immunization (Patel and Lee, 2005), the polysaccharide antigens do not induce long lasting antibody responses nor immunological memory. Furthermore, immune responses in infants are poor (Jodar et al., 2002) and only serogroup A polysaccharide vaccines elicit a limited and short lived immunity in infants below 2 years of age (MacLennan et al., 1999; Reingold et al., 1985). Due to these limitations, polysaccharide vaccines have mainly been applied in response to outbreaks and epidemics (Cochi et al., 1987; Jamba et al., 1979; Peltola et al., 1976; Lennon et al., 1992). It was often criticized that particularly in the African meningitis belt, vaccination campaigns were often carried out too late to prevent the majority of cases (Robbins et al., 2003). Additionally, the effect of preventive immunization in Africa has also raised some controversy (Hassan et al., 1998; Lengeler et al., 1995; WHO, 2003). Only few countries have adopted polysaccharide vaccine routine immunization schemes for the control of epidemic IMD. In Egypt, a measurable reduction in the serogroup A meningococcal disease after the introduction of a school-based immunization program has been reported (Nakhla et al., 2005). In China, large scale preventive immunization programs were carried out in order to prevent serogroup A epidemics (Zhen, 1987). In Brazil a mass immunization was performed in response to the severe A and C epidemics in the early 1970s.

For a better immunogenicity and immunological memory as well as satisfactory effectiveness in infants, polysaccharide-protein conjugate vaccines against serogroups A, C, Y and W135 have been developed. The first meningococcal conjugate vaccine against serogroup C was introduced into routine immunization in the UK in 1999 (Jodar et al., 2002), resulting in an overall decrease in serogroup C invasive disease by over 85% between 1999 and 2001 in the targeted age groups (Balmer et al., 2002). It was shown that in contrast to pure polysaccharide vaccines, capsule conjugate vaccines also

affected carriage and transmission (Maiden and Stuart, 2002) and might therefore contribute to the development of herd immunity in the non-vaccinated population. Yet the limited data available indicates that this effect is restricted to the vaccine serogroup and has only little impact on the overall meningococcal colonization (Maiden and Stuart, 2002; Trotter et al., 2005). Thereafter, the meningococcal C conjugate vaccine was also introduced into the national immunization scheme of other European countries. Additionally, a quadrivalent conjugate vaccine (A,C,Y,W135) for targeted use in populations was licensed in the USA and Canada in 2005 and 2006 respectively (Vu et al., 2006). A serogroup A polysaccharide conjugate vaccine for use in mass vaccination campaigns in Africa is currently being developed (www.meningvax.org). Since the poly-sialic acid capsule of serogroup B meningococci is not suitable for the design of a polysaccharide–based vaccine, vaccines targeting OMPs are currently being investigated but are strongly restricted by the high diversity of OMPs. However, strain-specific outer membrane vesicle (OMV)-based vaccines already developed to combat epidemics in Norway (Bjune et al., 1991) and Cuba (Sierra et al., 1991), showed limited efficacy. The Cuban vaccine was used further to control outbreaks in Chile, Brazil (de Moraes et al., 1992) and Uruguay (Pírez et al., 2004). The most recent example of a "tailor made" strain-specific OMV vaccine was used in the New Zealand epidemic and has been shown to be effective in reducing the incidence of IMD during a serogroup B epidemic (Kelly et al., 2007).

DISCUSSION

Despite the establishment of improved surveillance in many parts of the world, the reasons for the differences in IMD epidemiology between endemic and epidemic settings are not fully understood rendering it difficult to predict the emergence of epidemics. Several factors influence the timing and distribution of epidemics including climatic, socio-economic and cultural factors involving changes in human lifestyle, natural growth of the human population, crowding and increased mobility. These have also strongly affected the global population structure of N. meningitidis and are still currently responsible for changing patterns in IMD epidemiology. It will be crucial to analyse meningococcal colonization patterns in greater detail, in particular after the introduction of new capsule conjugate and OMV vaccines in order to understand the consequences of these interventions. Traditionally in Northern America, IMD during endemic periods was primarily caused by serogroup B and C, whereas serogroup A dominated during epidemics (Branham, 1956). This suggests that low level endemic IMD may have existed even before the emergence of serogroup A associated epidemic disease in the beginning of the 19th century. During the last sixty years serogroup A epidemics have largely disappeared from many developed regions and only endemic IMD primarily caused by sero-group B and C meningococci is found. However, the recent emergence or detection due to altered surveillance performances of new or previously unrecognized threats such as epidemics caused by serogroup W135 and X in Africa, the increasing significance of serogroup Y in Northern America and of serogroup C in China indicate that prevention and control of IMD will remain a challenge. With antimicrobial resistance on the rise, effective and affordable vaccines along with continued surveillance are needed to help combat this complex disease.

ACKNOWLEDGEMENTS

This publication made use of the Neisseria multi locus sequence typing website (http://pubmlst.org/ neisseria/) developed by Keith Jolley and Man-Suen Chan and sited at the University of Oxford (Branham, 1956). The development of this site has been funded by the Wellcome trust and the European Union.

REFERENCES

Achtman M (1997). Microevolution and epidemic spread of serogroup A Neisseria meningitidis--a review. Gene 192: 135-140.

Achtman M (2004) Population structure of pathogenic bacteria revisited. Int. J. Med. Microbiol. 294: 67-73.

Achtman M, van der EA, Zhu P, Koroleva IS, Kusecek B, Morelli G (2001). Molecular epidemiology of serogroup a meningitis in Moscow, 1969 to 1997. Emerg. Infect. Dis. 7: 420-427.

Aguilera JF, Perrocheau A, Meffre C, Hahne S (2002). Outbreak of serogroup W135 meningococcal disease after the Hajj pilgrimage, Europe, 2000. Emerg. Infect. Dis. 8: 761-767.

Almeida-Gonzalez L, Franco-Paredes C, Perez LF, Santos-Preciado JI (2004). [Meningococcal disease caused by Neisseria meningitidis: epidemiological, clinical, and preventive perspectives]. Salud Publica Mex 46: 438-450.

Almuneef M, Memish Z, Khan Y, Kagallwala A, Alshaalan M (1998). Childhood bacterial meningitis in Saudi Arabia. J. Infect. 36: 157-160.

Artenstein MS, Gold R, Zimmerly JG, Wyle FA, Schneider H, Harkins C (1970). Prevention of meningococcal disease by group C polysaccharide vaccine. N. Engl. J. Med. 282: 417-420.

Baethgen LF, Weidlich L, Moraes C, Klein C, Nunes LS, Cafrune PI Lemos AP, Rios SS, Abreu MF, Kmetzsch C, Sperb AF, Riley LW, Rossetti ML, Zaha A (2008). Epidemiology of meningococcal disease in southern Brazil from 1995 to 2003, and molecular characterization of Neisseria meningitidis using multilocus sequence typing. Trop. Med. Int. Health 13: 31-40.

Ballada D, Ventura G, Tumbarello M, Cauda R, Toti L, Ortona L (1990). [Meningococcal meningitis in Italy: 1887-1986]. Medicina (Firenze), 10: 129-133.

Balmer P, Borrow R, Miller E (2002). Impact of meningococcal C conjugate vaccine in the UK. J. Med. Microbiol. 51: 717-722.

Barroso DE, Rebelo MC (2007). Recognition of the epidemiological significance of Neisseria meningitidis capsular serogroup W135 in the Rio de Janeiro region, Brazil. Mem. Inst. Oswaldo Cruz 102: 773-775.

Bilukha OO, Rosenstein N (2005). Prevention and control of meningococcal disease. Recommendations of the Advisory Committee on Immunization Practices (ACIP). MMWR Recomm. Rep. 54: 1-21.

Bjune G, Hoiby EA, Gronnesby JK, Arnesen O, Fredriksen JH, Halstensen A, Holten E, Lindbak AK, Nokleby H, Rosenqvist E (1991). Effect of outer membrane vesicle vaccine against group B meningococcal disease in Norway. Lancet 338: 1093-1096.

Block C, Roitman M, Bogokowsky B, Meizlin S, Slater PE (1993). Forty years of meningococcal disease in Israel: 1951-1990. Clin Infect Dis, 17: 126-132.

Boisier P, Djibo S, Sidikou F, Mindadou H, Kairo KK, Djibo A, Goumbi K, Chanteau S (2005). Epidemiological patterns of meningococcal meningitis in Niger in 2003 and 2004: under the threat of N. meningitidis serogroup W135. Trop. Med. Int. Health 10: 435-443.

Boisier P, Nicolas P, Djibo S, Taha MK, Jeanne I, Mainassara HB Tenebray B, Kairo KK, Giorgini D, Chanteau S (2007). Meningococcal meningitis: unprecedented incidence of serogroup X-related cases in 2006 in Niger. Clin. Infect. Dis. 44: 657-663.

Bovre K, Gedde-Dahl TW (1980). Epidemiological patterns of meningococcal disease in Norway 1975-1979. NIPH Ann 3: 9-22.

Branham SE (1956). Milestones in the history of the meningococcus. Can. J. Microbiol. 2: 175-188.

Broome CV, Rugh MA, Yada AA, Giat L, Giat H, Zeltner JM Sanborn WR, Fraser DW (1983). Epidemic group C meningococcal meningitis in Upper Volta, 1979. Bull World Health Organ 61: 325-330.

Brundage JF, Zollinger WD(1987). Evolution of meningococcal disease epidemiology in the U.S. army. In Evolution of Meningococcal Disease I. Edited by Vedros N.A. Boca Raton: CRC Press pp. 5-26.

Cadbury WW(1934). Epidemic Cerebrospinal Meningitis in China. American Journal of Public Health and the Nation's Health 24: 925-935.

Cadoz M, Denis F, Mar ID (1981). [An epidemiological study of purulent meningitis cases admitted to hospital in Dakar, 1970-1979]. Bull World Health Organ 59: 575-584.

Cartwright K (1995). Introduction and Historical Aspects. In Meningococcal Disease. Edited by Cartwright Ked. West Sussex: John Wiley & Sons. pp. 1-20.

Cartwright K (1995). Meningococcal Carriage and Disease. In Meningococcal Disease. Edited by Cartwright Ked. West Sussex: John Wiley & Sons. pp. 115-147.

Caugant DA (1998). Population genetics and molecular epidemiology of Neisseria meningitidis. APMIS 106: 505-525.

Caugant DA (2008).Genetics and evolution of Neisseria meningitidis: Importance for the epidemiology of meningococcal disease. Infect. Genet. Evol.

Caugant DA, Froholm LO, Bovre K, Holten E, Frasch CE, Mocca LF Zollinger WD, Selander RK (1986). Intercontinental spread of a genetically distinctive complex of clones of Neisseria meningitidis causing epidemic disease. Proc Natl Acad Sci U S A 83: 4927-4931.

Caugant DA, Mocca LF, Frasch CE, Froholm LO, Zollinger WD, Selander RK (1987). Genetic structure of Neisseria meningitidis populations in relation to serogroup, serotype, and outer membrane protein pattern. J. Bacteriol. 169: 2781-2792.

Caugant DA, Nicolas P (2007). Molecular surveillance of meningococcal meningitis in Africa. Vaccine, 25 Suppl. 1: A8-11.

Chiavetta L, Chavez E, Ruzic A, Mollerach M, Regueira M (2007). [Surveillance of Neisseria meningitidis in Argentina, 1993-2005: distribution of serogroups, serotypes and serosubtypes isolated from invasive disease]. Rev Argent Microbiol, 39: 21-27.

Chiou CS, Liao JC, Liao TL, Li CC, Chou CY, Chang HL Yao SM, Lee YS (2006). Molecular epidemiology and emergence of worldwide epidemic clones of Neisseria meningitidis in Taiwan. BMC Infect. Dis. 6: 25.

Claus H, Maiden MC, Wilson DJ, McCarthy ND, Jolley KA, Urwin R (2005) Genetic analysis of meningococci carried by children and young adults. J. Infect. Dis. 191: 1263-1271.

Hessler F, Frosch M, Vogel U (2005). Genetic analysis of meningococci carried by children and young adults. J. Infect. Dis. 191: 1263-1271.

Cochi SL, Markowitz LE, Joshi DD, Owens RC, Jr., Stenhouse DH, Regmi DN (1987). Control of epidemic group A meningococcal meningitis in Nepal. Int. J. Epidemiol. 16: 91-97.

Communicable Disease and Health Protection Quarterly Review: October to December 2003. From the Health Protection Agency, Communicable Disease Surveillance Centre. J. Public Health (Oxf) 2004, 26: 205-213.

Coulson GB, von Gottberg A, du PM, Smith AM, de Gouveia L, Klugman KP (2007). Meningococcal disease in South Africa, 1999-2002. Emerg. Infect. Dis. 13: 273-281.

Crawford SA, Fiebelkorn KR, Patterson JE, Jorgensen JH (2005). International clone of Neisseria meningitidis serogroup A with tetracycline resistance due to tet(B). Antimicrob Agents Chemother 49:

1198-1200.

Creasey SA (1991). Epidemic meningococcal meningitis in central Australia in the 1970s. Med. J. Aust. 155: 725-726.

Crowe BA, Olyhoek T, Neumann B, Wall B, Hassan-King M, Greenwood B Achtman M (1987). A clonal analysis of Neisseria meningitidis serogroup A. Antonie Van Leeuwenhoek 53: 381-388.

Cruz C, Pavez G, Aguilar E, Grawe L, Cam J, Mendez F, Garcia J, Ruiz S, Vicent P, Canepa I (1990) Serotype-specific outbreak of group B meningococcal disease in Iquique, Chile. Epidemiol. Infect. 105: 119-126.

de Lemos AP, Yara TY, Gorla MC, de Paiva MV, de Souza AL, Goncalves MI, de Almeida SC, do Valle GR, Sacchi CT (2007). Clonal distribution of invasive Neisseria meningitidis serogroup C strains circulating from 1976 to 2005 in greater Sao Paulo, Brazil. J. Clin. Microbiol. 45: 1266-1273.

de Moraes JC, Barata RB (2005) [Meningococcal disease in Sao Paulo, Brazil, in the 20th century: epidemiological characteristics]. Cad Saude Publica 21: 1458-1471.

de Moraes JC, Perkins BA, Camargo MC, Hidalgo NT, Barbosa HA, Sacchi CT Landgraf IM, Gattas VL, Vasconcelos HG (1992). Protective efficacy of a serogroup B meningococcal vaccine in Sao Paulo, Brazil. Lancet 340: 1074-1078.

De Wals P (2004). Meningococcal C vaccines: the Canadian experience. Pediatr. Infect. Dis. J. 23: S280-S284.

Diermayer M, Hedberg K, Hoesly F, Fischer M, Perkins B, Reeves M Fleming D (1999). Epidemic serogroup B meningococcal disease in Oregon: the evolving epidemiology of the ET-5 strain. JAMA 281: 1493-1497.

Djibo S, Nicolas P, Alonso JM, Djibo A, Couret D, Riou JY Chippaux JP (1995-2000). Outbreaks of serogroup X meningococcal meningitis in Niger. Trop Med Int Health 2003 8: 1118-1123.

Dyet KH, Martin DR (2006). Clonal analysis of the serogroup B meningococci causing New Zealand's epidemic. Epidemiol. Infect. 134: 377-383.

Ebright JR, Altantsetseg T, Oyungerel R (2003). Emerging infectious diseases in Mongolia. Emerg. Infect. Dis. 9: 1509-1515.

Erwa HH, Haseeb MA, Idris AA, LAPEYSSONNIE L, Sanborn WR, Sippel JE (1973). A serogroup A meningococcal polysaccharide vaccine: studies in the Sudan to combat cerebrospinal meningitis caused by Neisseria meningitidis group A. Bull World Health Organ 49: 301-305.

Etienne J, Sperber G, Adamou A, Picq JJ (1990). [Epidemiological notes: meningococcal meningitis of serogroup X in Niamey (Niger)]. Med Trop (Mars) 50: 227-229.

Ettori D, Salidou P, Renaudet J, Stoeckel PH (1977). Le vaccin antimeningococcique polysaccharidique du type A: Premiers essais controles en afrique de l'ouest. Medecine Tropicale 37: 225-230.

EU-IBIS Network (2006). Invasive Neisseria meningitidis in Europe 2003/2004. 1-86.. London, Health Protection Agency.

Faye-Kette H, Doukou ES, Boni C, Akoua-Koffi C, Diallo-Toure K, Kacou-N'Douba A, Bouzid S, Dossso M, Timité-Konan M (2003). [Agents of community acquired purulent meningitis in the child: epidemiologic trends in Abidjan, Cote d'Ivoire, from the year 1995 to 2000]. Bull Soc. Pathol. Exot. 96: 313-316.

Feil EJ, Li BC, Aanensen DM, Hanage WP, Spratt BG (2004). eBURST: inferring patterns of evolutionary descent among clusters of related bacterial genotypes from multilocus sequence typing data. J. Bacteriol. 186: 1518-1530.

Ferreira E, Dias R, Canica M (2006). Antimicrobial susceptibility, serotype and genotype distribution of meningococci in Portugal, 2001-2002. Epidemiol. Infect. 134: 1203-1207.

Filice GA, Englender SJ, Jacobson JA, Jourden JL, Burns DA, Gregory D Counts GW, Griffiss JM, Fraser DW (1984). Group A meningococcal disease in skid rows: epidemiology and implications for control. Am J Public Health. 74: 253-254.

Fonkoua MC, Taha MK, Nicolas P, Cunin P, Alonso JM, Bercion R Musi J, Martin PM (2002). Recent increase in meningitis Caused by Neisseria meningitidis serogroups A and W135, Yaounde, Cameroon. Emerg. Infect. Dis. 8: 327-329.

Gagneux SP, Hodgson A, Smith TA, Wirth T, Ehrhard I, Morelli G Genton B, Binka FN, Achtman M, Pluschke G (2002). Prospective study of a serogroup X Neisseria meningitidis outbreak in northern

Ghana. J. Infect. Dis. 185: 618-626.

Galimand M, Gerbaud G, Guibourdenche M, Riou JY, Courvalin P(1998). High-level chloramphenicol resistance in Neisseria meningitidis. N. Engl. J. Med. 339: 868-874.

Gaspar M, Leite F, Brumana L, Felix B, Stella AA (2001). Epidemiology of meningococcal meningitis in Angola, 1994-2000. Epidemiol. Infect. 127: 421-424.

Girgis NI, Sippel JE, Kilpatrick ME, Sanborn WR, Mikhail IA, Cross E Erian MW, Sultan Y, Farid Z (1993). Meningitis and encephalitis at the Abbassia Fever Hospital, Cairo, Egypt, from 1966 to 1989. Am. J. Trop. Med. Hyg. 48: 97-107.

Gordon SB, Walsh AL, Chaponda M, Gordon MA, Soko D, Mbwvinji M, Molyneux ME, Read RC (2000). Bacterial meningitis in Malawian adults: pneumococcal disease is common, severe, and seasonal. Clin Infect Dis, 31: 53-57.

Gotschlich EC, Goldschneider I, Artenstein MS (1969). Human immunity to the meningococcus. IV. Immunogenicity of group A and group C meningococcal polysaccharides in human volunteers. J. Exp. Med. 129: 1367-1384.

Greenwood B (1999). Manson Lecture. Meningococcal meningitis in Africa. Trans. R. Soc. Trop. Med. Hyg. 93: 341-353.

Greenwood B (2007). The changing face of meningococcal disease in West Africa. Epidemiol. Infect. 135: 703-705.

Greenwood BM, Bradley AK, Cleland PG, Haggie MH, Hassan-King M, Lewis LS, Macfarlane JT, Taqi A, Whittle HC, Bradley-Moore AM, Ansari Q (1979). An epidemic of meningococcal infection at Zaria, Northern Nigeria. 1. General epidemiological features. Trans. R. Soc. Trop. Med. Hyg. 73: 557-562.

Hahne SJ, Gray SJ, Jean F, Aguilera, Crowcroft NS, Nichols T Kaczmarski EB, Ramsay ME (2000 and 2001) W135 meningococcal disease in England and Wales associated with Hajj. Lancet 2002, 359: 582-583.

Hansman D (1987). Epidemiology of Meningococcal Disease in Australia and New Zealand with a Note on Papua New Guinea. In Evolution of Meningococal Disease II. Edited by Vedros N.A. CRC Press. pp. 9-18.

Harrison LH, Broome CV (1987). The epidemiology of meningococcal meningitis in the U.S. civilian population. In Evolution of Meningococcal Disease I. Edited by Vedros N.A. CRC Press pp. 27-45.

Hassan J, Massougbodji A, Chippaux JP, Massit B, Josse R(1998). Meningococcal immunisation and protection from epidemics. Lancet 352: 407-408.

Hirsch A (1886). Epidemic cerebro-spinal meningitis. In Handbook of Geographical and Historical Pathology. Hirsch A. pp. 547-590.

Hodgson A, Smith T, Gagneux S, Enos KE, Adjuik M, Pluschke G Mensah NK, Binka F,Genton B (2002). Meningococcal meningitis in northern Ghana: epidemiological features of the 1997 outbreak in the Kassena-Nankana district. University of Basel.

Hsueh PR, Teng LJ, Lin TY, Chen KT, Hsu HM, Twu SJ, Ho SW, Luh KT (2004). Re-emergence of meningococcal disease in Taiwan: circulation of domestic clones of Neisseria meningitidis in the 2001 outbreak. Epidemiol. Infect. 132: 637-645.

Hubert B, Caugant DA (1997). Recent changes in meningococcal disease in Europe. Euro. Surveill. 2: 69-71.

Jamba G, Bytchenko B, Causse G, Cvjetanovic B, Ocirvan G, Tsend N Kupul J, Tseren S, Batsuri A, Hisigdorj A, Burian V (1979). Immunization during a cerebrospinal meningitis epidemic in the Mongolian People's Republic, 1974-75. Bull World Health Organ 57: 943-946.

Jelfs J, Munro R, Ashto FE, Caugant DA (2000). Genetic characterization of a new variant within the ET-37 complex of Neisseria meningitidis associated with outbreaks in various parts of the world. Epidemiol. Infect. 125: 285-298.

Jodar L, Feavers IM, Salisbury D, Granoff DM (2002). Development of vaccines against meningococcal disease. Lancet 359: 1499-1508.

Jolley KA, Chan MS, Maiden MCJ(2004). mlstdbNet – distributed multi-locus sequence typing (MLST) databases. BMC Bioinformatics, 5:86

Jones DM, Abbott JD (1987). Meningococcal Disease in England and Wales. In Evolution of Meningococcal Disease I. Edited by Vedros N.A. CRC Press. pp. 65-90.

Jorgensen JH, Crawford SA, Fiebelkorn KR (2005). Susceptibility of

Neisseria meningitidis to 16 antimicrobial agents and characterization of resistance mechanisms affecting some agents. J. Clin. Microbiol. 43: 3162-3171.

Kelly C, Arnold R, Galloway Y, O'Hallahan J (2007). A prospective study of the effectiveness of the New Zealand meningococcal B vaccine. Am J. Epidemiol. 166: 817-823.

Knights HT (1972). Meningococcal meningitis in New Zealand with special reference to carrier rates in military trainees. N. Z. Med. J. 76: 16-22.

Kuzemenska P, Kriz B (1987). Epidemiology of meningococcal disease in central and eastern Europe. In Evolution of Meningococcal Disease I. Edited by Vedros N.A. CRC Press. pp. 03-137.

Kwara A, Adegbola RA, Corrah PT, Weber M, Achtman M, Morelli G Caugant DA, Greenwood BM (1998). Meningitis caused by a serogroup W135 clone of the ET-37 complex of Neisseria meningitidis in West Africa. Trop Med Int Health, 3: 742-746.

Lapeyssonnie L (1963). La méningite cérébro-spinale en Afrique. Bull World Health Organ 28: SUPPL-114.

Law DK, Lorange M, Ringuette L, Dion R, Giguere M, Henderson AM Stoltz J, Zollinger WD, De Wals P, Tsang RS (2006). Invasive meningococcal disease in Quebec, Canada, due to an emerging clone of ST-269 serogroup B meningococci with serotype antigen 17 and serosubtype antigen P1.19 (B:17:P1.19). J. Clin. Microbiol. 44: 2743-2749.

Lengeler C, Kessler W, Daugla D (1995). The 1990 meningococcal meningitis epidemic of Sarh (Chad): how useful was an earlier mass vaccination? Acta. Trop. 59: 211-222.

Lennon D, Gellin B, Hood D, Voss L, Heffernan H, Thakur S (1992). Successful intervention in a group A meningococcal outbreak in Auckland, New Zealand. Pediatr. Infect. Dis. J. 11: 617-623.

Lingappa JR, Al Rabeah AM, Hajjeh R, Mustafa T, Fatani A, Al Bassam T Badukhan A, Turkistani A, Makki S, Al-Hamdan N, Al-Jeffri M, Al Mazrou Y, Perkins BA, Popovic T, Mayer LW, Rosenstein NE (2000). Serogroup W-135 Meningococcal Disease during the Hajj. Emerg. Infect. Dis. 2003 9: 665-671.

Luca V, Gessner BD, Luca C, Turcu T, Rugina S, Rugina C Ilie M, Novakova E, Vlasich C (2004). Incidence and etiological agents of bacterial meningitis among children <5 years of age in two districts of Romania. Eur. J. Clin. Microbiol. Infect. Dis. 23: 523-528.

Maalej SM, Kassis M, Rhimi FM, Damak J, Hammami A (2006). [Bacteriology of community acquired meningitis in Sfax, Tunisia (1993-2001)]. Med. Mal. Infect. 36: 105-110.

MacLennan J, Obaro S, Deeks J, Williams D, Pais L, Carlone G Moxon R, Greenwood B (1999). Immune response to revaccination with meningococcal A and C polysaccharides in Gambian children following repeated immunisation during early childhood. Vaccine, 17: 3086-3093.

Mahmoud R, Mahmoud M, Badrinath P, Sheek-Hussein M, Alwash R, Nicol AG (2002). Pattern of meningitis in Al-Ain medical district, United Arab Emirates--a decadal experience (1990-99). J. Infect. 44: 22-25.

Maiden MC, Bygraves JA, Feil E, Morelli G, Russell JE, Urwin R Zhang Q, Zhou J, Zurth K, Caugant DA, Feavers IM, Achtman M, Spratt BG (1998). Multilocus sequence typing: a portable approach to the identification of clones within populations of pathogenic microorganisms. Proc Natl Acad Sci USA 95: 3140-3145.

Maiden MC, Stuart JM: Carriage of serogroup C meningococci 1 year after meningococcal C conjugate polysaccharide vaccination. Lancet 2002 359: 1829-1831.

Makela PH, Peltola H, Kayhty H, Jousimies H, Pettay O, Ruoslahti E Sivonen A, Renkonen OV (1977). Polysaccharide vaccines of group A Neisseria meningtitidis and Haemophilus influenzae type b: a field trial in Finland. J. Infect. Dis. 136 Suppl. pp. S43-S50.

Manchanda V, Gupta S, Bhalla P (2006). Meningococcal disease: history, epidemiology, pathogenesis, clinical manifestations, diagnosis, antimicrobial susceptibility and prevention. Indian J. Med. Microbiol. 24: 7-19.

Marchiafava E, Celli A (1884). Spra i micrococchi della meningite cerebrospinale epidemica. Gazz degli Ospitali p. 5.

Mathers CD, Bernard C, Moesgaard IK, Inoue M, Fat DM, Shibuya K, Stein C,Tomijima N and Xu H (2003). Global Burden of Disease in 2002: data sources, methods and results. pp. 1-54.

Mayer LW, Reeves MW, Al Hamdan N, Sacchi CT, Taha MK, Ajello GW

Schmink SE, Noble CA, Tondella ML, Whitney AM, Al-Mazrou Y, Al-Jefri M, Mishkhis A, Sabban S, Caugant DA, Lingappa J, Rosenstein NE, Popovic T (2002). Outbreak of W135 meningococcal disease in 2000: not emergence of a new W135 strain but clonal expansion within the electophoretic type-37 complex. J. Infect. Dis. 185: 1596-1605.

McGahey K (1905). Report on an outbreak of epidemic cerebro-spinal meningitis in Zungeru during February and March. J. Trop. Med. 8: 210-216.

Molesworth AM, Thomson MC, Connor SJ, Cresswell MP, Morse AP, Shears P Hart CA, Cuevas LE (2002). Where is the meningitis belt? Defining an area at risk of epidemic meningitis in Africa. Trans. R. Soc. Trop. Med. Hyg. 96: 242-249.

Moore PS (1992). Meningococcal meningitis in sub-Saharan Africa: a model for the epidemic process. Clin. Infect. Dis. 14: 515-525.

Morelli G, Malorny B, Muller K, Seiler A, Wang JF, del Valle J provide Achtman M (1997). Clonal descent and microevolution of Neisseria meningitidis during 30 years of epidemic spread. Mol. Microbiol. 25: 1047-1064.

Muhe L, Klugman KP (1999). Pneumococcal and Haemophilus influenzae meningitis in a children's hospital in Ethiopia: serotypes and susceptibility patterns. Trop. Med. Int. Health 4: 421-427.

Nakhla I, Frenck RW, Jr., Teleb NA, Oun SE, Sultan Y, Mansour H Mahoney F (2005). The changing epidemiology of meningococcal meningitis after introduction of bivalent A/C polysaccharide vaccine into school-based vaccination programs in Egypt. Vaccine 23: 3288-3293.

Newman MJ (2004) Bacterial Menigitis in Southern Ghana: A 3-Year Review of Bacterial Isolates. Ghana Med. J. 25: 349-354.

Ni JD, Jin YH, Dai B, Wang XP, Liu DQ, Chen X, Zheng Y, Ye DQ (2008). Recent epidemiological changes in meningococcal disease may be due to the displacement of serogroup A by serogroup C in Hefei City, China. Postgrad Med. J. 84: 87-92.

Oberti J, Hoi NT, Caravano R, Tan CM, Roux J (1981). [An epidemic of meningococcal infection in Vietnam (southern provinces)]. Bull World Health Organ 59: 585-590.

Olyhoek T, Crowe BA, Achtman M (1987). Clonal population structure of Neisseria meningitidis serogroup A isolated from epidemics and pandemics between 1915 and 1983. Rev. Infect. Dis. 9: 665-692.

Patel M, Lee CK(2005). Polysaccharide vaccines for preventing serogroup A meningococcal meningitis. Cochrane Database Syst. Rev. CD001093.

Patel MS (2007). Australia's century of meningococcal disease: development and the changing ecology of an accidental pathogen. Med. J. Aust. 186: 136-141.

Patel MS, Merianos A, Hanna JN, Vartto K, Tait P, Morey F Jayathissa S (1993). Epidemic meningococcal meningitis in central Australia, 1987-1991. Med. J. Aust. 158: 336-340.

Peltola H (1978) Group A meningococcal polysaccharide vaccine and course of the group A meningococcal epidemic in Finland. Scand. J. Infect. Dis. 10: 41-44.

Peltola H (1983). Meningococcal disease: still with us. Rev. Infect. Dis. 5:71-91.

Peltola H (1987). Meningococcal Disease: An old enemy in Scandinavia. In Evolution of Meningococcal Disease I. Edited by Vedros N.A. CRC Press. pp. 91-102.

Peltola H (2001). Burden of meningitis and other severe bacterial infections of children in africa: implications for prevention. Clin. Infect. Dis. 32: 64-75.

Peltola H, Makela PH, ELO O, Pettay O, Renkonen OV, Sivonen A (1976). Vaccination against meningococcal group A disease in Finland 1974-75. Scand. J. Infect Dis. 8: 169-174.

Pírez MC, Picón T, Galazka J, Rubio I, Montano A, Ferrari AM (2004). Control de un brote epidémico de enfermedad meningogócica por N. meningitidis serogrupo B. Rev Med Uruguay 20: 92-101.

Pollard AJ, Scheifele D (2001). Meningococcal disease and vaccination in North America. J. Paediatr. Child Health 37: S20-S27.

Poolman JT, Lind I, Jonsdottir K, Froholm LO, Jones DM, Zanen HC (1986). Meningococcal serotypes and serogroup B disease in north-west Europe. Lancet 2: 555-558.

Reingold AL, Broome CV, Hightower AW, Ajello GW, Bolan GA, Adamsbaum C Jones EE, Phillips C, Tiendrebeogo H, Yada A (1985). Age-specific differences in duration of clinical protection after vaccination with meningococcal polysaccharide A vaccine. Lancet 2: 114-118.

Rico Cordeiro O, Jimenez Barreras R., Pereira Colls C (1996). Enfermedad meningocócica y VA-MENGOC BC en menores de un ano. Cuba, 1983-1991. Revista Cubana de Medicina Tropical 48.

Robbins JB, Schneerson R, Gotschlich EC, Mohammed I, Nasidi A, Chippaux JP Bernardino L, Maiga MA (2003). Meningococcal meningitis in sub-Saharan Africa: the case for mass and routine vaccination with available polysaccharide vaccines. Bull World Health Organ 81: 745-750.

Santos ML, Ruffino-Netto A (2005). [Meningococcal disease: epidemiological profile in the Municipality of Manaus, Amazonas, Brazil, 1998/2002]. Cad Saude Publica 21: 823-829.

Savory EC, Cuevas LE, Yassin MA, Hart CA, Molesworth AM, Thomson MC (2006). Evaluation of the meningitis epidemics risk model in Africa. Epidemiol Infect. 1-13.

Scholten RJ, Poolman JT, Valkenburg HA, Bijlmer HA, Dankert J, Caugant DA (1994). Phenotypic and genotypic changes in a new clone complex of Neisseria meningitidis causing disease in The Netherlands, 1958-1990. J. Infect. Dis. 169: 673-676.

Schwartz B, Moore PS, Broome CV (1989). Global epidemiology of meningococcal disease. Clin Microbiol Rev 2 Suppl: pp. S118-S124.

Shao Z, Li W, Ren J, Liang X, Xu L, Diao B Li M, Lu M, Ren H, Cui Z, Zhu B, Dai Z, Zhang L, Chen X, Kan B (2006). Identification of a new Neisseria meningitidis serogroup C clone from Anhui province, China. Lancet, 367: 419-423.

Sie A, Pfluger V, Coulibaly B, Dangy JP, Kapaun A, Junghanss T Pluschke G, Leimkugel J (2008) ST2859 serogroup A meningococcal meningitis outbreak in Nouna Health District, Burkina Faso: a prospective study. Trop. Med. Int. Health.

Sierra GV, Campa HC, Varcacel NM, Garcia IL, Izquierdo PL, Sotolongo PF Casanueva GV, Rico CO, Rodriguez CR, Terry MH (1991). Vaccine against group B Neisseria meningitidis: protection trial and mass vaccination results in Cuba. NIPH Ann 14: 195-207.

Sippel JE, Girgis NI: Epidemiology of Meningococcal Disease in Northeastern Africa (1987). In Evolution of Meningococcal Disease II. Edited by Vedros N.A. CRC Press pp. 1-8.

Souza dM, Munford RS, Risi JB, Antezana E, Feldman RA (1974). Epidemic disease due to serogroup C Neisseria meningitidis in Sao Paulo, Brazil. J. Infect. Dis. 129: 568-571.

Stanwell-Smith RE, Stuart JM, Hughes AO, Robinson P, Griffin MB, Cartwright K (1994). Smoking, the environment and meningococcal disease: a case control study. Epidemiol. Infect. 112: 315-328.

Stein-Zamir C, Abramson N, Zentner G, Shoob H, Valinsky L, Block C (2007). Invasive meningococcal disease in children in Jerusalem. Epidemiol. Infect. pp. 1-8.

Taha MK, Achtman M, Alonso JM, Greenwood B, Ramsay M, Fox A Gray S, Kaczmarski E (2000). Serogroup W135 meningococcal disease in Hajj pilgrims. Lancet 356: 2159.

Takahashi H, Kuroki T, Watanabe Y, Tanaka H, Inouye H, Yamai S, Watanabe H (2004). Characterization of Neisseria meningitidis isolates collected from 1974 to 2003 in Japan by multilocus sequence typing. J. Med. Microbiol. 53: 657-662.

Tikhomirov E, Santamaria M, Esteves K (1997). Meningococcal disease: public health burden and control. World Health Stat Q 50: 170-177.

Tondella ML, Popovic T, Rosenstein NE, Lake DB, Carlone GM, Mayer LW Perkins BA (2000). Distribution of Neisseria meningitidis serogroup B serosubtypes and serotypes circulating in the United States. The Active Bacterial Core Surveillance Team. J. Clin. Microbiol. 38: 3323-3328.

Traore Y, Njanpop-Lafourcade BM, Adjogble KL, Lourd M, Yaro S, Nacro B Drabo A, Parent du Châtelet I, Mueller JE, Taha MK, Borrow R, Nicolas P, Alonso JM, Gessner BD (2006). The rise and fall of epidemic Neisseria meningitidis serogroup W135 meningitis in Burkina Faso, 2002-2005. Clin. Infect. Dis. 43: 817-822.

Trotter C, Samuelsson S, Perrocheau A, de Greeff S, de Melker H, Heuberger S, Ramsay M (2005). Ascertainment of meningococcal disease in Europe. Euro. Surveill. 10: 247-250.

Van Looveren M, Vandamme P, Hauchecorne M, Wijdooghe M, Carion F, Caugant DA Goossens H (1998). Molecular epidemiology of recent belgian isolates of Neisseria meningitidis serogroup B. J. Clin.

Microbiol. 36: 2828-2834.

Varughese PV, Acres SE (1987) Meningococcal Disease in Canada. In Evolution of Meningococcal Disease I. Edited by Vedros N.A. CRC Press pp. 47-63.

Vedros NA (1987). Development of Meningococcal Serogroups. In Evolution of Meningococcal Disease II. Edited by Vedros N.A. CRC Press. pp. 33-37.

Vieusseux M (1805). Memoire sur le maladie que a regne à Geneve au printemps de 1805. J. Med. Chird Pharmacol. p. 11.

von Gottberg A, du Plessis M, Cohen C, Prentice E, Schrag S, de Gouveia L, Coulson G, de Jong G, Klugman K (2008). Emergence of endemic serogroup W135 meningococcal disease associated with a high mortality rate in South Africa. Clin. Infect. Dis. 46: 377-386.

Vu DM, Welsch JA, Zuno-Mitchell P, Dela Cruz JV, Granoff DM (2006). Antibody persistence 3 years after immunization of adolescents with quadrivalent meningococcal conjugate vaccine. J. Infect. Dis. 193: 821-828.

Wang JF, Caugant DA, Li X, Hu X, Poolman JT, Crowe BA, Achtman M (1992). Clonal and antigenic analysis of serogroup A Neisseria meningitidis with particular reference to epidemiological features of epidemic meningitis in the People's Republic of China. Infect. Immun. 60: 5267-5282.

Weichselbaum A (1887): Ueber die Aetiologie der akuten Meningitis cerebrospinalis. Fortschritte der Medizin 5: 573-583.

Welch SB, Nadel S (2003). Treatment of meningococcal infection. Arch. Dis. Child 88: 608-614.

Whittle HC, Evans-jones G, Onyewotu I, Adjukiewich A, Turunen U, Crockford J Greenwood BM (1975). Letter: Group-C Meningococcal Meningitis in the Northern Savanna of Africa. Lancet 1: 1377.

WHO (1945) Cases and Deaths of Cerebro-Spinal Meningits Reported in Certain Countries in Europe, Africa and America, From 1939 to 1945. Weekly Epidemiological Record 39: 276-282.

WHO (1975). Meningitis - Argentina. Wkly. Epidemiol. Rec. 50: 161-168.

WHO (1977). Meningococcal Meningitis. Wkly Epidemiol. Rec. 48: 381-388.

WHO (1995). Meningocal Disease - Mongolia. Wkly Epidemiol. Rec. 70: 281-282.

WHO (2003). Round table discussion. Bull World Health Organ 81: 751-755.

Yang CY, Lee YS, Huang LS, Kuo YL, Liu YL, Lu CH (2006). Antigenic diversity of Neisseria meningitidis isolated in Taiwan between 1995 and 2002. Scand. J. Infect. Dis. 38: 273-280.

Yazdankhah SP, Caugant DA (2004). Neisseria meningitidis: an overview of the carriage state. J. Med. Microbiol. 53: 821-832.

Yazdankhah SP, Kriz P, Tzanakaki G, Kremastinou J, Kalmusova J, Musilek M Alvestad T, Jolley KA, Wilson DJ, McCarthy ND, Caugant DA, Maiden MC (2004). Distribution of serogroups and genotypes among disease-associated and carried isolates of Neisseria meningitidis from the Czech Republic, Greece, and Norway. J. Clin. Microbiol. 42: 5146-5153.

Zerouali K, Elmdaghri N, Boudouma M, Benbachir M (2002). Serogroups, serotypes, serosubtypes and antimicrobial susceptibility of Neisseria meningitidis isolates in Casablanca, Morocco. Eur J Clin Microbiol. Infect. Dis. 21: 483-485.

Zhang X, Shao Z, Yang E, Xu L, Xu X, Li M, Ren J, Zhu Y, Yang F, Liang X, Mayer LW, Xu J, Jin Q (2007). Molecular characterization of serogroup C Neisseria meningitidis isolated in China. J. Med. Microbiol. 56: 1224-1229.

Zhen H (1987). Epidemiology of Meningococal Disease in China. In Evolution of Meningococcal Disease II. Edited by Vedros N.A. CRC Press. pp.19-32.

Zhu P, van der EA, Falush D, Brieske N, Morelli G, Linz B, Popovic T, Schuurman IGA, Adegbola RA, Zurth K, Gagneux S., Platonov AE, Riou JY, Caugant DA, Nicolas P, Achtman M (2001). Fit genotypes and escape variants of subgroup III Neisseria meningitidis during three pandemics of epidemic meningitis. Proc. Natl. Acad. Sci. USA 98: 5234-5239.

26

Biodiversity of molecular profile of *Staphylococcus aureus* isolated from bovine mastitis cases in West Algeria

Nadjia BENHAMED and Mebrouk Kihal

Laboratory of Applied Microbiology, Department of Biology, Faculty of Science, Oran University, BP16, Es-senia 31000 Oran, Algeria.

Bovine mastitis is an inflammation of the mammary gland with local and or symptoms that occasionally result in a systemic infection. This disease has a profound impact on animal welfare and milk quality, and the most costly disease affecting dairy cows. The bacteria *Staphylococcus aureus* is one of the most frequently isolated pathogens from both subclinical and clinical infections. This study was conducted to investigate the phenotypic and genotypic characterization of *S. aureus* involved in dairy cow mastitis in West Algeria. A total of 141 isolates of *S. aureus* isolated from quarter milk samples were collected from dairy cows. All retained *S. aureus* species contained *gyr* gene and were identified by molecular typing. The presence of resistance was evaluated in *S. aureus*. Staphylococci antimicrobial resistance was performed by detection of *mecA* gene. Several virulence factors including toxin of the Pantin Valentine leukocidin coding gene (*pvl*) were also investigated by polymerase chain reaction (PCR). Only one strain of *S. aureus* was *mec*A - and *pvl*+ gene.

Key words: Mastitis, *S. aureus, Gyr A*, PCR, *mecA, pvl*.

INTRODUCTION

Mastitis, the most expensive disease of dairy cows, continues to be a persistent problem in the dairy industry (Barkema et al., 2009; Le Marechal et al., 2011). Mastitis, inflammation of the mammary gland with local and or general symptoms that occasionally result in a systemic infection, can be caused by a wide range of microorganisms, including gram-negative and gram-positive bacteria (Le Marechal et al., 2011).

Staphylococcus aureus is one of the most frequently isolated pathogens from both sub-clinical and chronic infections (Watts 1988). *S. aureus* still remains one of the most significant organisms associated with clinical and subclinical bovine mastitis, not only in Algeria but worldwide. This disease is considered to be the most frequent and most costly production disease in dairy herds of de-veloped countries (Fourichon et al., 2001). Mastitis is one of the dominant pathological dairy farming actually decreased milk production per cow due to the prevalence of clinical and subclinical mastitis (Le Marechal et al., 2011). The main etiology is infectious. It results in the majority of cases by cell type inflammatory response involving an increase in the cell concentration in milk, cell counts in the diagnosis of mastitis is essential and the results should be according to the results of the CMT (California Mastitis test).

Bovine mastitis produces a wide variety of problems in the dairy farm. The treatment of this disease is based on the use of antibiotics which are not always effective.

These drugs are also responsible for the presence of residues the milk the increase antibiotic-resistant strains.

Probiotic products were proposed as a valid alternative to antibiotic therapies and are also useful for the prevention of infectious syndromes (Espeche et al., 2012). *S. aureus* is the most predominant contagious pathogen responseble for clinical and subclinical infections in lactating cows (Kerro-Dego et al., 2002).

This study was conducted to investigate the phenotypic and genotypic characterization of *S. aureus* involved in dairy cow mastitis in West Algeria. The presence of resistance was evaluated in *S. aureus*. Staphylococci antimicrobial resistance was performed by detection of *mecA* gene and several virulence factors including toxin of the Pantin Valentine leukocidin coding gene (*pvl*) by polymerrase chain reaction (PCR).

MATERIALS AND METHODS

Bacteriological analysis

Milk samples were analyzed in Applied Microbiology Laboratory, Faculty of Sciences, Oran University. Milk samples were cultured on several media, on blood (5% sheep blood) agar plates, incubited at 37°C for 24 h. Growth on the plates was confirmed by additional laboratory tests in accordance. *S. aureus* was identified by means of typical colony and cells morphology, catalase reaction, by coagulase reaction using rabbit plasma (Quinn et al., 1994) (coagulase positive), or Pastorex (agglutination test) (Bio-rad, France) and biochemical characterization using the Api-staph system (Biomerieu, France). Strains expressed phenotypic resistance to cefoxitin confirmed by polymerase chain reaction detection of the *mec*A gene, typing of the accessory gene regulator (agr) and detection of control specific gene Gyr of *S. aureus*. Coagulase-negative staphylococci were identified by typical colony and cell morphology and coagulase reaction.

Susceptibility testing

Susceptibility testing was performed by disk (Bio-rad, France) diffusion method on Muller-Hinton agar plates. Testing was performed according to recommendation of Ca-SFM-veterinarian 2012 (Committee on Antimicrobial Company Information French Microbiology). The antibiotics tested were penicillin G (PG-6 µ g), kanamyin (K-30 IU), gentamicin (Gm-15 µ g), tetracyclin (Te-30UI), erythromycin (E-15UI), lincomycin (L-15 µ g), pristinamycin (PT-15 µg), chloramphenicol (C-30 µ g), pefloxacin (Pef-5 µ g), fosfomycin (Fos-5 µ g), cefoxitin (Fox-30 µ g), fusidic acid (FA-10 µ g), vancomycin (VA-30 µ g), oxacillin (Ox 5 cmg), Amikacin (10 mcg), carbenicillin (cB 100 µ g) and ciprofloxacin (5 mcg). For testing susceptibility in staphylococci, 2% NaCl was added to the broth and to Muller-Hinton agar plates. Control strains, *S. aureus* ATCC 43300 and *S. aureus* ATCC 25923 were tested in parallel with each batch of isolates (Smyth et al., 2001). The susceptibility to the cefoxitin of *S. aureus* was confirmed by polymerase chain reaction of the *mec*A gene (Elazhari et al., 2010).

Bacterial DNA extraction

DNA of each strain of *S. aureus* was extracted according to the standard protocol (Sambrook et al., 1989). The collected DNA was precipitated, described by electrophoresis on agarose gel and then stored at -20°C.

Detection of *mecA* gene, *GyrA* gene and *pvl* gene by PCR

The confirmation of *S. aureus* species was performed on the basis of standard biochemical tests. The isolates were further characterized by molecular analysis amplifying the gene *gyr* typing of the accessory gene regulator (Brakstad et al., 1992). A duplex PCR for the simultaneous fragment 533 base pairs (bp) specific *mec*A gene and another 280 bp fragment of the gene *GyrA* were used to prove the *S. aureus* species. The pathogenicity and virulence of *S. aureus* is associated with the capacity of this organism to produce several virulence factors including Panton-Valentine leukocidin (PVL) (Shittu et al., 2011). In addition, the Pantin-Valentine Leukocidin coding *pvl* gene was detected by simple PCR using specific fragment 433 pb (Sung et al., 2008); the primer used for mecA.

Typing of *agr* gene (accessory gene regulator)

Using PCR multiplex for search simultaneous fragment *agr* type (agr1; 440, agr2; 550, agr3; 300, agr4; 650), the research has been carried out for strains of *S. aureus;* the fragment used to define the type of *agr S. aureus* isolates. All amplification products were separated by electrophoresis on agarose gel 1.5% stained with ethidium bromide (0.5 µg/ml) in Tris-borate-EDTA TBE (at a rate of one to two drops added). Photographs of gels were taken under ultraviolet (UV) device (Gel Doc) (Sambrook et al., 1989).

RESULTS

Precise identification of *S. aureus* infected cows is important for successful implementation of a mastitis control program. Therefore, according to the phenotypic, biochemical properties as well as by amplification of the *gyr* gene, all of the isolates obtained in this study were identified as *S. aureus*. Data descriptive of Staphylococci strains isolated from quarter milk samples from clinical or subclinical mastitis cases in the study is shown in Table 1.

Detection of *pvl-luk* toxin by amplification of the *pvl* gene from extracted DNA of the *S. aureus* strains reveled that positive amplification of the 533 pb fragment of *pvl* gene from the extracted DNA of only one strains from 11, this strain had the following molecular profile *agr*3, *mec*A-, *pvl*+, the rest of strains had a similar molecular profile (Figure 3).

Antimicrobial susceptibility

Antimicrobial susceptibility testing reported a high susceptibility of *S. aureus* strains to antimicrobial agents which was confirmed by PCR by the absence of *mecA* gene. Our results shows absence of *mecA* gene for all *S. aureus* strains which were phenotypically susceptible to cefoxitin and oxacillin.

Results presented in Figure 2 shows the absence of the *mecA* gene from extracted DNA of *S. aureus* strains tested; this result confirmed the antibiogram results for susceptibility to methicillin.

DISCUSSION

S. aureus is still considered one of the most common

Table 1. Data descriptive of Staphylococci strains isolated from quarter milk samples from clinical or subclinical mastitis cases in the study.

Staphylococci strain	Data of strain					
	Clinical case %	subclinical case %	Winter %	Spring %	Summer %	Autumn %
S. aureus	58.86	41.14	26	17.78	40	16.22
SCN	0	100	2.33	40	47.67	10

SCN, Staphylococcus coagulase negative.

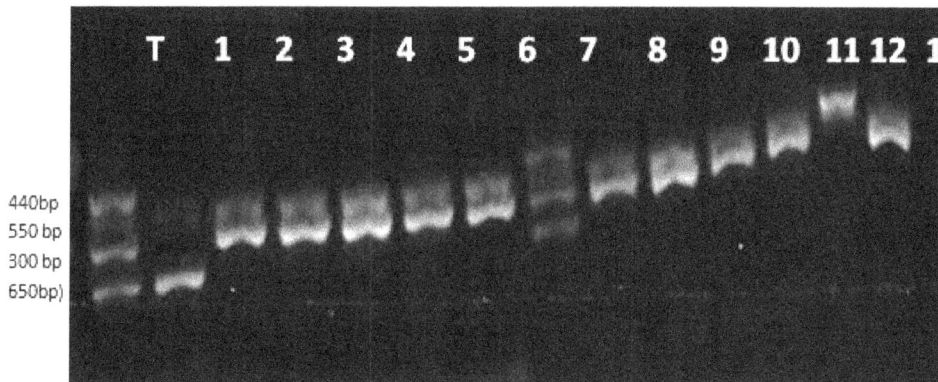

Figure 1. Agarose gel electrophoresis of polymerase chain reaction (PCR) amplification of agr1/2/3/4 gene.

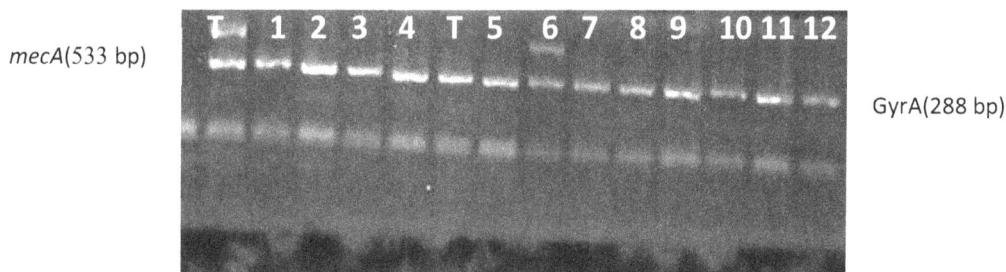

Figure 2. Agarose gel electrophoresis of polymerase chain reaction (PCR) double amplification of *mecA* and *gyr A* gene.

etiological agents associated with clinical and subclinical infections in lactating cows by Esmat and Bader (1996) and El-Seedy et al., 2010.

Precise identification of *S. aureus*-infected cows is important for successful implementation of a mastitis control program. Therefore, according to the phenotypic, biochemical properties as well as by amplification of the *gyr* gene (Figure 1), all of the isolates obtained in this study were identified as *S. aureus* (Dastmalchi, 2012).

Results presented in Figure 2 show the absence of the *mecA* gene from extracted DNA of *S. aureus* strains tested, this result confirmed the antimicrobial susceptibility testing results for susceptibility to methicillin. The strains of Staphylococcus aureus are (SASM) and showed high level of resistantance to eryhtromycine, ciprofloxacin, penicillin and a susceptibility to kanamycin. All strains

showed also a high resistantance to to tetracycline, gentamicin and bacitracin (Figure 4).

Detection of *pvl-luk* toxin by amplification of the *pvl* gene from extracted DNA of the *S. aureus* strains reveled that positive amplification of the 533 pb fragment of *pvl* gene from the extracted DNA of only one strain from 11, this strain had the following profiles *agr*3, *mecA*-, *pvl*+ (Figure 3). This finding was comparable to the study of Sung et al. (2008). One isolate was positive for gene encoding the components of the Pantin-Valentine-Leukocidin (*pvl-luk*); these results are similar to those obtained by Monecke et al. (2011) and Shittu et al.(2011).

Conclusions

Our results indicate that several bacteria species of S.

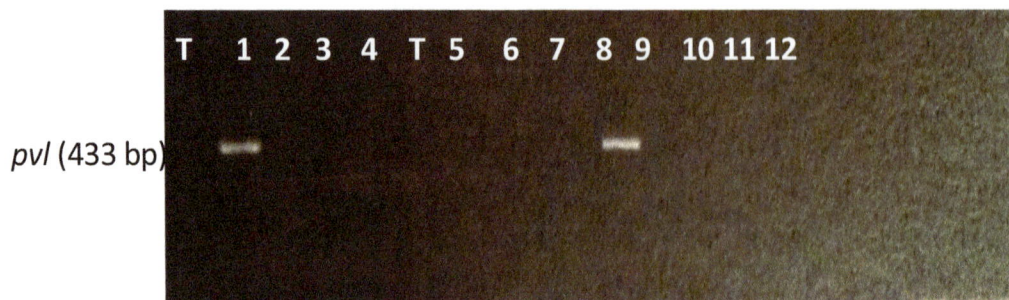

Figure 3. Agarose gel electrophoresis of polymerase chain reaction (PCR) double amplification of *PVL* gene.

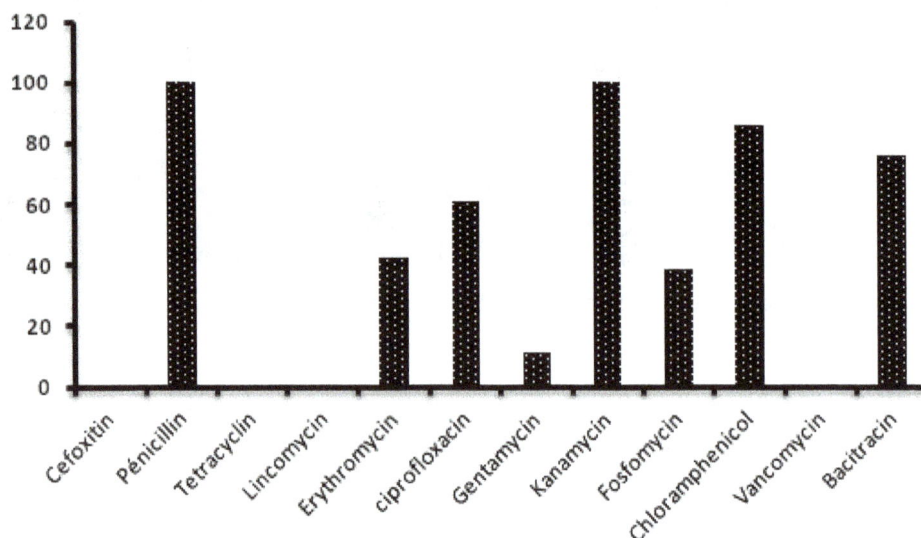

Figure 4. Antibiosusceptibility of strains of *Staphylococcus aureus* isolated.

aureus were found in mastitis cases. Genotypic and phenotypic identification of *Staphylococcus aureus* was confirmed by detection of *gene gyrA* and *agr* molecular typing. The antimicrobial susceptibility testing showed that *S. aureus* isolates from Oran Region West Algeria exhibited high susceptibility to all antimicrobial agents tested and due to the absence of *mecA* gene in all strains of *S. aureus* tested. The results of study show also that one strain of *Staphylococcus aureus* carried PVL-Luk toxin due to presence of *PVL* gene which showed a susceptibility to cefoxitin.

REFERENCES

Barkema HW, Green MJ, Bradley AJ, Zadoks RN (2009). Invited review:The Role of contagiousDisease In Udder. health. J. Dairy. Sci. 92:4717-4729.

Brakstad OG, Aasbakk K, Maeland JA (1992). Detection of *Staphylococcus aureus* by polymerase chain reaction amplification of the nuc gene. J. Clin. Microbiol. 30:1654-1660.

Dastmalchi HS (2012). *Coa* types and antimicrobial resistance profile of *Staphylococcus aureus* isolates from cases of bovine mastitis. Comp. Clin. Pathol. 21:301-307.

Elazhari M, Zerouali K, Elhzbchi D, Cohen N, Elmalki A, Dersi N, Hassar M, Timinouni M, Saile R (2010). Sensibilité aux antibiotiques des souches de *Staphylococcus aureus* communautaires à Casablanca (Maroc).Revue Tunisienne d'Infection. 4:134-140.

El-Seedy FR, El-Shabrawy M, Hakim AS , Dorgham SM, Nagwa S, Bakry MA, Osman NMN (2010).Recent Techniques used for isolation and characterization of *Staphylococcus aureus* from Mastitis Cows. J. Am. Sci. 6:(2).

Esmat M , Bader A (1996).Some studies on mastitis meteritis agalcti syndrome in cows. Vet. Med. J. Giza. 44(2):303-309.

Espeche MC, Pellegrino M , Frola I, Larriestra A, Bogni C, Macías F (2012). Lactic acid bacteria from raw milk as potentially beneficial strains to prevent bovine mastitis. Mol. Boil. Genet. Biotechnol. 18:103-109.

Fourichon C, Seegers H, Beaudeau F, Bareille N (2001). Economic losses consecutive to health disorders in dairy farms in pays de la loire (france),52nd Meeting of the European Association of Animal Production, Budapest (H). pp. 26-29.

Kerro-Dego O, van Dijk JE, Nederbragt H (2002). Factors involved in the early pathogenesis of bovine Staphylococcus aureus mastitis with emphasis on bacterial adhesion and invasion. A review. Vet. Quart. 24:181-198.

Le Marechal C, thiery R, Le loir Y(2011). Mastitis impact on technologie proprerties of milk and quality of milk products- review. Dairy Sci. Technol. 91:247-282.

Monecke S, Ehricht R, Slickers P, Wernery R, Johson B, Wernery U (2011).Microarray-based genotyping of S. aureus isplates from camels. Vet. Microbiol. 150:309-314.

Quinn PJ, Carter ME, Maeker BK, Carter GR (1994). Clinical Veterinary Microbiology. Mosby, London, p.648.

Sambrook J,Fritsch EF, Manitatis T(1989). Molecular Cloning: A Laboratory Manual. 2nd ed. Cold Spring Harbor Laboratory Press, Cold Spring.

Shittu AO, Okon K, Adesida S, Oyedara O, Witte W, Strommenger B, Layer F, Nübel U (2011). Antibiotic resistance and molecular epidemiology of Staphylococcus aureus in Nigeria. BMC Microbiol. 11:92

Smyth RW, Kahlmeter G, Olsson B, Liljequist B, Hoffman B (2001).Methods for identifying methicillin resistancein Staphylococcus aureus. J. Hosp. Infect. 48(2):103-107.

Sung JML, Lloyd DH, Lindsay JA (2008). Staphylococcus aureus host specificity: comparative genomics of human versus animal isolates by multi-strain microarray. Microbiology 154:1949-1959.

Watts JL (1988). Etiological agents of bovine mastitis. Vet. Microbiol. 16:41-66, PMID: 3354192.

27

Acinetobacter spp. in the patients and environment of University Hospital of Yopougon, Côte d'Ivoire, from 2007 to 2011

S. Méité[1,2], Boni-Cissé C.[1,2] , Mlan Tanoa A. P.[1], Zaba F. S.[1], Faye-Ketté H.[1,2] and Dosso M.[1,2]

[1]Laboratory of Bacteriology-Virology, University Hospital (CHU) of Yopougon, 22 BP 539, Abidjan, Côte d'Ivoire.
[2]Department of Microbiology, Faculty of Medical Sciences, University Hospital (CHU) of Yopougon, 22 BP 539, Abidjan, Côte d'Ivoire.

The objective of this work was to study the epidemiology and antibiotic resistance of strains of *Acinetobacter* spp. in the University Hospital of Yopougon Abidjan. This work studied the *Acinetobacter* strains isolated from humans and environment of the hospital; they were preserved in the culture collection of the Laboratory of Bacteriology from January 2007 to December 2011. Isolation and identification were made by conventional bacteriological methods, and antibiotic susceptibility was studied by the method of agar diffusion. Interpretation was made according to the standards of the CA-SFM. 110 strains of *Acinetobacter* spp. have been studied (61% of human strains and 39% of strains isolated from the hospital environment). *Acinetobacter baumannii* was the most isolated in 66% of cases. 52. 8% of strains were resistant to ceftazidime; 5.6% to imipenem; 21.2% to gentamicin and 35.2% to ciprofloxacin. 12.5% of human strains of *A. baumannii* were multi-resistant bacteria. *Acinetobacters* spp. are present in the hospital environment and patients with a predominance of *A. baumannii* species. The presence of imipenem-resistant strains is a major public health problem because their disclosure could lead to therapeutic impasse in hospital.

Key words: *Acinetobacter* spp., epidemiology, antibiotic resistance.

INTRODUCTION

Acinetobacter spp. are emerging opportunistic pathogenic bacteria that play an important role in hospitals worldwide (Munoz-Price and Weinstein, 2008; Schreckenberger et al., 2007). Indeed, in recent years, these bacteria have become a concern in hospital services in different countries.

Thus, they are responsible for nosocomial infections manifested by septicemia, urinary tract infections, secon-

dary meningitis and especially lung infections in patients under mechanical ventilation in intensive care units (Balkhy et al., 2014; Meite et al., 2011). Moreover, in recent years, *Acinetobacter* infections of the central nervous system, skin, soft tissue and bone have been observed (Bergogne-Berezin and Towner, 1996; French et al. 1980; Jaffar et al., 2007).

These infections are difficult to treat, due to the

increasing resistance of *Acinetobacter* strains growing on different families of antibiotics. Several studies have shown an appearance and an increase in the resistance of *Acinetobacter* spp. to carbapenem, choice antibiotics for the treatment of infections caused by these bacteria (Balkhy et al., 2014; Boni-Cisse et al., 2011; Chaisathaphol and Chayakulkeeree, 2014; Halstead et al., 2007; Unal and Garcia-Rodriguez, 2007).

In addition, antibiotic resistant strains of *Acinetobacter* spp. have a great ability to survive for several days in hospital environment, inert material or dust, thereby increasing the likelihood of transmission of the inter-human bacteria via a human or tank material (Boni-Cisse et al., 2011).

In Côte d'Ivoire, although *Acinetobacter* is responsible for nosocomial infections (Boni-Cisse et al., 2011; Meite et al., 2010, 2011), few data are available on circulating bacteria including species, the resistance profile of strains to antibiotics and the main hospital tank bacteria.

The objective of this work is to study *Acinetobacter baumanii* complex circulating in patients and hospital environment as well as their antibiotic resistance profile.

MATERIALS AND METHODS

The authors studied 110 strains of *Acinetobacter* spp. kept in the culture collection of the Laboratory of Bacteriology Virology of CHU Yopougon, from January 2007 to December 2011. These strains of *Acinetobacter* spp. were isolated from Biologic patients and hospital environment (sink, handle of door, sick bed and respirators).

Isolation of strains

Acinetobacter strains were revived in Brain Heart broth for 3 h at 37°C and then re-plated on nutrient agar, sheet blood agar and non selective lactose agar; they were incubated at 37°C for 18 to 48 h. The colonies were identified by standard bacteriology tests (culture on minimal agar) as the genus *Acinetobacter* ssp. which are Gram negative bacilli, non-motile; strictly aerobic, oxidase-negative and glucose non fermentative. Reference strains, *Pseudomonas aeruginosa* ATCC 27853 and *Escherichia coli* ATCC 25922 were used for the validation of rapid biochemical tests used.

Biotypique study

For identification of *Acinetobacter calcoaceticus-A. baumanii* complex, the following characters were investigated: the ability to grow at 41 and 44°C, using citrate as sole carbon source, the production of an ornithine decarboxylase (ODC) and the production of arginine dihydrolase (ADH). Reference strains ATCC 13883 *Klebsiella pneumoniae*, *Shigella sonnei* ATCC 25931 and ATCC 27853 *Pseudomonas aeruginosa* were used to validate the research of ODC and ADH.

Studies of antibiotic susceptibility

The search for antibiotic susceptibility was performed by the method of agar diffusion. The following antibiotics disk of biorad were tested: ticarcillin/clavulanate (75/10 µg); ceftazidime (30 µg);

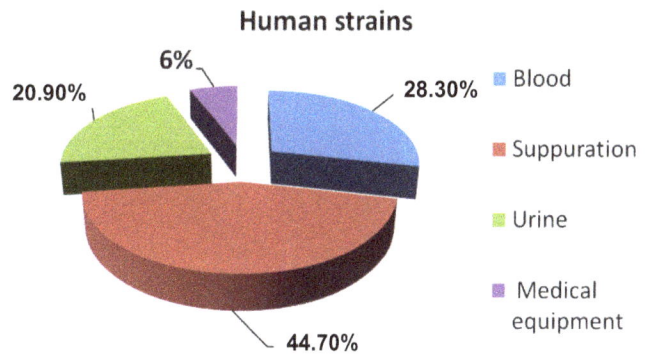

Human strains

28.30% Blood

44.70% Suppuration

20.90% Urine

6% Medical equipment

Figure 1. Distribution of human strains according to organic products.

imipenem (10 µg); aztreonam (30 µg); gentamicin (10 IU); amikacin (30 µg); netilmicin (30 µg); tobramycin (10 µg) and ciprofloxacin (5 µg). Interpretation of results was done according to the standards committee of the susceptibility of the French society of Microbiology (CA-SFM) recommendations 2012.

RESULTS

A total of 110 strains of *Acinetobacter* were analyzed, of which 61% were strains of human origin and 49% of environmental strains. 44.8% of the human strains were from suppurations, 28.3% from blood, 20.9% from urine and 6% from medical equipment (Figure 1). In the environment surface, 38.1% were isolated from sink, 23.8% from the patient bed, 21.4% from the handle of door and 12% from respirators. 100% of the strains gave positive cultures on isolation media used. About biotypique study, none of the strains yielded positive test for glucose fermentation.

ADH test was negative for all the strains and hemolysis test in the sheet blood agar. 94% of cultures at 41°C were positive (Table 1). For the identified species, 94% were *A baumanii-calcoaceticus* complex and 6% of *Acinetobacter johsonii*. 62% *A. baumanii–calcoaceticus complex* were of human origin.

The antibiogram for beta-lactam revealed that, 39.4, 52.8 and 05.6% of tested strains were resistant to the combination of respectively Ticarcillin-clavulanic acid, and preview Ceftazidime, Imipenem. Regarding aminoglycosides, 21.2 and 10.3% of the strains tested were resistant to Gentamicin and Amikacin. 35.2% of tested strains were resistant to ciprofloxacin regarding quinolones. The high proportion of antibiotic resistance was found in strains of human origin and in *A. baumanii* complex species.

Thus, 100% of resistant strains were imipenem essentially *A. baumanii* species. 05.2% of tested strains were resistant to three antibiotics families tested. This was the case of human stem *A. baumanii*. Resistance to two different antibiotics family was not observed with

Table 1. Distribution of frequency of positive biochemical tests.

Test	Number of positive (N = 110)	Percentage (%)
Positive culture at 44°C	73	66
Positive culture at 38°C	104	94
Positive culture on Simmons citrate	110	100
Fermentation of glucose	00	00
Oxidation of lactose	00	00
Production of ODC	30	28
Production of ADH	00	00
Hemolysis on GSF	00	00

ODH: Ornithine decarboxylase, ADH: arginine dihydrolase, GSF: In fresh blood agar

environmental strains, except the three strains resistant to imipenem.

DISCUSSION

In this study, in humans, *Acinetobacter* was found in suppuration from blood and urine. This is consistent with some data from the literature (Van Looveren et al., 2004). Predominantly, the suppuration is linked to the fact that *Acinetobacter* is from commensal skin bacterium (Nordmann, 2004).

The environments handled by the nursing staff and visitors (sinks, beds, door handle) were colonized with *Acinetobacter*. This presents a risk of nosocomial infection (Oie et al., 2002). This is also the problem of hygiene in our hospitals.

A. baumanii-calcoaceticus complex is a predominance species in this study. These results are in agreement with those of literature. Indeed, several studies in different countries have shown a predominance of *A. baumanii* species. It is found in two thirds of infections caused by *Acinetobacter* spp. In a study conducted in Kosovo (Raka et al., 2009), *A. baumanni* represented 81.2% of *Acinetobacter* spp. In 1993, Seifert et al. (1993) identified 73% of *Acinetobacter* strains from clinical isolates as *A. baumannii*.

Acinetobacter lwoffi, *Acinetobacter hemolyticus* and *Acinetobacter junni* species were not isolated in this study. These bacteria are rarely clinically isolated; the isolation of *A. lwoffi* suggests a port rather than an infection. For resistance to beta-lactam antibiotics, the rate of resistance to Ceftazidime and Cefsulodin was respectively, 52.8 and 89.6%. This resistance was greater with *A. baumanii* complex strains of human origin. The rate of resistance to Ceftazidime was higher than that observed in our previous studies (Meite et al., 2010, 2011). This is in favor of a steady increase in antibiotic resistance of *Acinetobacter* in our health facility. These results corroborate those of Balkhy et al. (2014) in Saudi Arabia and Rahbar et al. (2010) Iran, who found 84.1 and 99% rates of resistance to ceftazidime. Regarding

resistance to carbapenems, it was 5.6%, including 75% of the strains of environmental origin. *A baumannii* complex resistance to carbapenems has appeared in many parts of the world and is clearly increasing (Balkhy et al., 2014; Chaisathaphol and Chayakulkeeree, 2014; Gootz and Marra, 2008; Perez et al., 2008).

The main mechanism of resistance is the acquisition of carbapenemases Class B and D (Gootz and Marra, 2008). Low levels of resistance (3 and 4.5%) of *A. baumannii* to imipenem were reported in Saudi Arabia by Jaffar et al. (2007) and in Iran by Rahbar et al. (2010). However, more recent studies in these countries have shown a marked increase in resistance to carbapenems by *Acinetobacter* spp., especially in Saudi Arabia where it is grown to over 80% (Sameera et al., 2010). Asencio et al. (2010)'s study carried out in Spain showed a resistance rate of 83% by *A. baumanii* to Imipenem, while that of Cisneros et al. (2005) was 43% resistance rate. These high rates of resistance to Imipenem are found in hospitals. The variations in time of the resistance of *A. baumanii* to Imipenem are related to increased use of the molecule and use not mastered by some prescribers, sometimes. Monotherapy is not recommended and should be used in combination synergistically, taking into account the bioavailability of each molecule, the site of action, the causative organism and risk factors related to patients. Strains tested were sensitive enough to aminoglycosides. Resistance was 21.2% for Gentamicin and 10.3% for Amikacin. These results are quite close to that of Asencio et al. (2012), in which the rates were very low (12% for Gentamicin and 2% for Amikacin). These low levels of resistance are due to the fact that aminoglycosides are rarely used in treating infections in our country, generally. Oral forms are rare and injectable forms available are used only in association with other molecules and in hospitals.

In this study, the rate of resistance to ciprofloxacin was 49%. Our results corroborate those of Ben Haj et al. (2010) in Tunisia who regained a resistance rate of 50%. These results, however, are contrary to those found in Iran by Rahbar et al. (2010) which was 90.9%. Described for the first time in Taiwan in 1998 and defined as being resistance to more than three classes of antibiotics, the incidence of strains of multi-resistant *A. baumannii* (MDR) continues to grow in recent years (Appleman et al., 2000). In a study realized in the United States by Dent et al. (2010), it involved almost 72% of *A. baumannii* studied. In our present study, it was 5.2% for all strains and reached 12.5% for strains isolated from humans.

A. baumannii with P. aeruginosa are frequently pan-resistant bacteria of antibiotic. In effect, these bacteria may be resistant to all antibiotics, including the aminoglycosides, cephalosporins, carbapenems, carboxypénicillines and fluoroquinolones. Dent et al. (2010) in USA found about 46% in their series. This profile was not found in our study. However, the phenotypes of resistance to at least two families of antibiotics have been observed. This resistance was generally observed as regards the lactams and amino-glycosides. Strains involved were the cases of *A. baumanii* especially of human origin. The proportion of resistance to these two antibiotic families is linked to therapeutic habits of our health facilities. The combination of these two antibiotic families is the first therapeutic choice in many empirical studies in Côte d'Ivoire.

Conclusion

Acinetobacter remains an environmental bacterium whose species *A. baumanii* complex is the most involved in human infections in hospitals. It presents a profile of increasingly resistant to conventional antibiotics in our health facility and its main concern remains the appearance of resistant strains to imipenem. This multidrug-resistant *Acinetobacter* to imipenem requires the implementation of policy microbiological monitoring and control to limit dissemination.

Conflict of interests

The authors did not declare any conflict of interest.

REFERENCES

vitro activities of non-traditional antimicrobials against multiresistant Acinetobacter baumannii strains isolated in an intensive care unit outbreak. Antimicrob. Agents Chemother. 44:1035-1040. http://dx.doi.org/10.1128/AAC.44.4.1035-1040.2000

Asencio MÁ, Carranza R, Huertas M (2012). Resistencia a antimicrobianos de los microorganismos más frecuentemente aislados en el Hospital General La Mancha Centro entre junio de 2009 y mayo de 2010. Rev Esp Quimioter. 25:183-188.

Balkhy HH, El-Saed A, Maghraby R, Al-Dorzi HM, Khan R, Rishu AH, Arabi YM (2014). Drug-resistant ventilator associated pneumonia in a tertiary care hospital in Saudi Arabia. Ann. Thorac. Med. 9(2):104-111. http://dx.doi.org/10.4103/1817-1737.128858

Ben Haj KA, Khedher K M (2010). Frequency and antibiotic susceptibility profile of bacteria isolated blood cultures CHU Mahida. Tunis. J. Infect. Dis. 3:92-95.

Bergogne-Bérézin E, Towner KJ (1996). Acinetobacter spp. as Nosocomial Pathogens: Microbiological, Clinical and Epidemiological Features. Clin. Microbiol. Rev. 9:148-165. PMid:8964033 PMCid:PMC172888

Boni-Cissé C, Méité S, N'douba Kacou A, Koffi P, Faye Ketté H, Dosso M (2011). Colonization of catheters by bacterial in the intensive care unit in Abidjan. Sch. J. Med. 2(3):38-42.

Chaisathaphol T, Chayakulkeeree M (2014). Epidemiology of infections caused by multidrug-resistant gram-negative bacteria in adult hospitalized patients at Siriraj Hospital. J. Med Assoc. Thai. 97(3):35-45.

Cisneros JM, Rodrvguez-BJ, Fernandez-Cuenca F, Ribera A, Vila J, Pascual A, et al. (2005). Risk-factors for the acquisition of imipenem-resistant Acinetobacter baumannii in Spain: a nationwide study. Clin. Microbiol. Infect. 11(11):874-879. http://dx.doi.org/10.1111/j.1469-0691.2005.01256.x

Dent LL, Marshall DR, Pratap S, Hulette RB (2010). Multidrug resistant Acinetobacter baumannii: a descriptive study in a city hospital. BMC Infect. Dis. 10:196. http://dx.doi.org/10.1186/1471-2334-10-196

French GL, Casewell MW, Roncoroni AJ, Knight S, Phillips I (1980). A hospital outbreak of antibiotic-resistant Acinetobacter anitratus: Epidemiology and control. J. Hosp. Infect.1:125-131. http://dx.doi.org/10.1016/0195-6701(80)90044-4

Gootz TD, Marra A (2008). Acinetobacter baumannii: Multidrug resistant emerging threat. Expert Rev. Anti Infect. Ther. 6:309-325. http://dx.doi.org/10.1586/14787210.6.3.309

Halstead DC, Abid J, Dowzicky MJ (2007). Antimicrobial susceptibility among Acinetobacter calcoaceticus-baumannii complex and Enterobacteriaceae collected as part of the tigecycline evaluation and surveillance trial. J. Infect. 55:49-57. http://dx.doi.org/10.1016/j.jinf.2006.11.018

Jaffar A, Al-Tawfiq MD, Thangiah X, Mohandhas BS (2007). Prevalence of antimicrobial resistance in Acinetobacter calcoaceticus-baumannii complex in a Saudi Arabian Hospital. Hosp. Infect. Control 28:870-872. http://dx.doi.org/10.1086/518842

Meite S, Boni-Cisse C, Monemo P, Babo JC, Mlan Tanoa AP, Faye-Ketté H, Dosso M, Amonkou A (2011) . Resistance patterns of bacteria isolated from bronchial samples protected in patients receiving mechanical ventilation in the ICU CHU Yopougon. Méd Afr Noire. 58 (8/9):416-422.

Meite S, Boni-Cisse C, Monemo P, Mlan Tanoa AP, Faye-Ketté H, Dosso M (2010). Microbiological monitoring of surfaces at a tertiary hospital: CHU Yopougon, Abidjan, Côte d'Ivoire. J. Sci. Pharm. Biol.11 (1):73-81.

Munoz-Price LS, Weinstein RA (2008). Acinetobacter infection. N. Engl. J. Med. 358:1271-1281. http://dx.doi.org/10.1056/NEJMra070741

Nordmann P (2004). Acinetobacter baumannii, the nosocomial pathogen par excellence. Pathol. Biol. 52(6): 301-303. http://dx.doi.org/10.1016/j.patbio.2004.03.001

Oie S, Hosokawa I, Kamiya A (2002). Contamination of room door handles by methicillin-sensitive / methicillin resistant Staphylococcus aureus. J. Hosp. Infect. 51:140-143. http://dx.doi.org/10.1053/jhin.2002.1221

Perez F, Endimiani A, Bonomo RA (2008). Why are we afraid of Acinetobacter baumannii? Expert Rev. Anti Infect. Ther. 6:269-271. http://dx.doi.org/10.1586/14787210.6.3.269

Rahbar M, Mehrgan H, Aliakbari NH (2010). Prevalence of Acinetobacter baumannii antibiotic-resistance in bed tertiary care hospital in Tehran 1000. Iran J. Pathol. Microbial. 53:290-293. http://dx.doi.org/10.4103/0377-4929.64333

Raka L, Kalenc S, Bošnjak Z, Budimir A, Katic S, Šijak D, Mulliqi-Osmani G, Zoutman D, Jaka A (2009). Molecular Epidemiology of Acinetobacter baumannii in Central Intensive Care Unit in Kosova

Teaching Hospital. Braz. J. Infect. Dis.13:408-413.

Sameera M, Johani AI, Javed A, Hanan B, Ayman ES, Mousaad Y, Ziad M (2010). Prevalence of antimicrobial resistance among gram negative isolates in an adult intensive care unit at a tertiary care center in Saudi Arabia. Ann. Saudi Med. 30:364-369.

Schreckenberger PC, Daneshvar MI, Weyant RS, Hollis DG (2007). Acinetobacter, Achromobacter, Chryseobacterium, Moraxella,and other non-fermentative Gram-negative rods. In: Murray PR, Baron EJ, Jorgensen JH, Landry ML, Pfaller MA. Manual of Clinical Microbiology, 9th ed. Washington, DC: ASM Press. pp. 770-802.

Seifert H, Baginsky R, Schulze A, Pulverer G (1993). The distribution of Acinetobacter species in clinical culture materials. Zentralbl. Bakteriol. 279:544-552. http://dx.doi.org/10.1016/S0934-8840(11)80427-5

Unal S, Garcia-Rodriguez JA (2005). Activity of meropenem and comparators against Pseudomonas aeruginosa and Acinetobacter spp. isolated in the MYSTIC Program, 2002-2004. Diagn. Microbiol. Infect. Dis. 53:265-271. http://dx.doi.org/10.1016/j.diagmicrobio.2005.10.002

Van Looveren M, Goossens H, the ARPAC Steering Group (2004). Antimicrobial resistance of Acinetobacter spp in Europe. Clin. Microbiol. Infect. 10(8):684-704. http://dx.doi.org/10.1111/j.1469-0691.2004.00942.x

28

Potential biodegradation of low density polyethylene (LDPE) by *Acinetobacter baumannii*

R. Pramila and K. Vijaya Ramesh

Quaid-E-Millath Government College for Women, Annasalai, Tamilnadu, Chennai-00 002, Tamilnadu, India.

Acinetobacter baumannii **was isolated from municipal landfill area, Pallikaranai, Chennai, Tamilnadu. The degradation ability of the bacteria was determined by performing Fourier Transform Infrared Spectroscopy (FTIR). The by-products of polyethylene degradation were monitored by gas chromatography-mass spectrometer (GC-MS) analysis. The toxicity of degradation by-products of low density polyethylene (LDPE) was tested on the plant** *Vigna radiata* **by determining the morphological parameters such as root length, shoot length and chlorophyll content. After 30 days of degradation process, the FTIR results revealed an increase in carbonyl index and formation of peaks and occurrence of stretches. Alkane compounds were analyzed in GC-MS analysis. Determination of toxicity level of intermediate degraded products showed no changes in morphological characters.**

Key words: Biodegradation, Low density polyethylene (LDPE), *Acinetobacter baumannii*, Fourier transform infrared spectroscopy (FTIR), gas chromatography-mass spectrometer (GC-MS), *Vigna radiata*.

INTRODUCTION

Polyethylene plays an important role in our everyday life. It is a synthetic polymer, made of long chain of monomers of ethylene. Its density ranges from 0.915-0.9359 gcm^3. Polyethylene is classified into different types such as low density polyethylene (LDPE), high density polyethylene (HDPE), linear low density polyethylene (LLDPE), etc. Among these, LDPE has been used for various purposes such as packaging, making carry bags, disposable cups etc. In contrast, when considering disadvantages of polyethylene it poses one of the worst environmental problems. Polyethylene products tend to accumulate in the land areas and remain inert for several decades. This reduces the fertility of the soil, water percolating capacity into the plants and it also threatens animal life. On burning, it produces toxic

chemicals polluting the environment, leading to diseases affecting the lungs and skin.

Numerous activities are carried out to reduce the usage of polyethylene and plastic, however less attention is focused on the degradation of polyethylene. Recent research focuses on biodegradation of polyethylene. Biodegradation is the process by which organic substances are broken down by living organisms like bacteria and fungi. During biodegradation process of polymers, two categories of enzymes are actively involved; extracellular and intracellular depolymerases. During degradation, exoenzymes from microorganisms break down complex polymers into smaller molecules, for example oligomers, dimers, and monomers that are small enough to pass the semi-permeable outer bacterial membranes, and then utilized as carbon and energy sources and release end products such as CO_2 and H_2O.

Biodegradation of LDPE was studied earlier (Albertsson et al., 1987; Shah, 2007; Suresh et al., 2011; Negi et al., 2011) however the results of these reports were based on pre-treating the LDPE with UV irradiation, thermally oxidized fragments and pro-oxidant additives containing LDPE and starch blended polyethylene.

Gilan et al. (2004) and Hadad et al. (2005) have reported the degradation of LDPE by pretreatment with UV-irradiation and subsequent incubation with *Rhodococcus ruber* and thermophilic bacteria *Brevibacillus parabrevis*.

Sudhakar et al. (2008) and Harshavardhan and Jha (2013) have isolated marine bacteria and utilized them for degradation study of thermally pretreated and starch blended LDPE. Mahalashmi et al. (2012) and Kyaw et al. (2012) have reported the degradation of untreated LDPE by *Pseudomonas* species.

A bacterial culture was isolated from a municipal land fill area and identified as *Acinetobacter baumannii* during previous study (Pramila et al., 2012). The preliminary degradation ability of *A. baumannii* was studied by measuring CO_2 evolution, calculation of generation time, protein estimation, and Bacterial adhesion to hydrocarbon (BATH) test. The significance of chosen municipal dump soil for isolation of bacteria was associated with the fact, that the cultures already had stressful conditions and could develop tolerance towards such environmental conditions.

The current study was focused on determina-tion of physical changes by tensile strength and chemical changes in LDPE by FTIR analysis to measure carbonyl Index (CI). Measuring the carbonyl index (CI) is necessary to elucidate the mechanism of biodegradation process where the initial step involves oxidation of the polymer chain and leads to the formation of carbonyl groups, since these groups undergo β-oxidation and are totally degraded via citric acid cycle resulting in formation of CO_2 and H_2O (Albertsson et al., 1987). Additionally, the current study also aimed to study the formation of intermediate by-products by GC-MS analysis and to test the toxicity level of the degraded by-products on plants by *A. baumannii*.

MATERIALS AND METHODS

Preparation of LDPE powder

LDPE sheets were cut into bits and immersed in xylene. It was boiled for 15 min as xylene dissolves the LDPE film and the residue was crushed while it was warm by using band gloves. The LDPE powder so obtained was washed with ethanol to remove residual xylene and allowed to evaporate to remove ethanol. The powder was dried in hot air oven at 60°C over night.

Isolation of microorganism

Bacterial culture was isolated by spread plate method in sterilized

synthetic medium (SM) at 37°C for 24 h. SM contains the following constitutions in 1000 ml distilled water (K_2HPO_4, 1 g; KH_2PO4, 0.2 g; NaCl, 1 g; $CaCl_2.2H_2O$, 0.002 g; $(NH_4)_2SO$, 1 g; $MgSO_4.7H_2O$, 0.5 g; $CuSO_4.5H_2O$, 0.001 g; $ZnSO_4.7H_2O$,0.001 g; $MnSO_4.H2O$, 0.001 g and $FeSO_4.7H_2O$, 0.01 g, amended with 500 mg LDPE powder. Synthetic mineral medium had LDPE as the sole carbon source.

Degradation study

The degradation study was carried out in synthetic medium broth. LDPE films were cut into 2×2 cm. The films were disinfected with 95% ethanol and washed with sterile distilled water. One full inoculation loop of isolated culture were inoculated in 5 ml SM broth and incubated at 37°C for 24 h. After 24 h, the broth was compared with McFarland scale ($CFU×10^9$/ml) and poured into 45 ml of SM broth in a 100 ml conical flask. Four pieces of equally weighing LDPE films were placed in SM broth. The flasks were incubated at 37°C for 30 days with shaking at 100 rpm. SM broth with LDPE films without culture was maintained as control.

Tensile strength

For tensile strength measurement, test strips were retrieved after 30 days of incubation, washed with 2% sodium dodecyl sulphate (SDS) followed by distilled water and dried in oven overnight at 50°C. The strips were subjected to tensile strength tests as per ASTM A.370 (2012).

FTIR study

After 30 days of incubation, the LDPE sheets were taken and washed with 2% SDS followed by sterile distilled water. The LDPE sheets were dried in oven overnight at 50°C. The films were subjected to FTIR analysis to calculate carbonyl index, presence or absence of functional groups, stretches. The carbonyl index is a measure of the concentration of carbonyl group (acids, aldehydes, ketones) (Albertsson et al., 1987).

$$\text{Carbonyl index (CI)} = \frac{\text{Absorbance at 1715 cm}^{-1}\text{(Peak wavelength)}}{\text{Absorbance at 1465 cm}^{-1}\text{(Peak wavelength)}}$$

GC-MS study

After 30 days of incubation, 10 ml broth was centrifuged at 1000 rpm for 10 min. Supernatant was extracted with 10 ml dichloromethane using a separating funnel. Simultaneously, LDPE films were extracted with 5 ml dichloromethane. Both the extracts were determined by GC-MS (JOEL GCMATE II GC-MASS SPECTROMETER IIT CHENNAI) using HP5 column, helium gas, temperature from 70 to 200°C, injection liquid 1 µl. By retention time the compounds were identified by NIST library.

Toxicity study

Culture broth was analyzed for its toxicity after 30 days, towards plant *V. radiata*. 10 g of garden soil was placed in a pot. Seeds were sown and the soil was wetted regularly with 5 ml of the culture broth. The pots were kept in room temperature with normal condition. After 7 days, the seedlings were harvested and morphological parameters such as root length, shoot length and chlorophyll content of the plant were estimated by Arnon (1949) method. SM with LDPE without culture and SM alone served as controls.

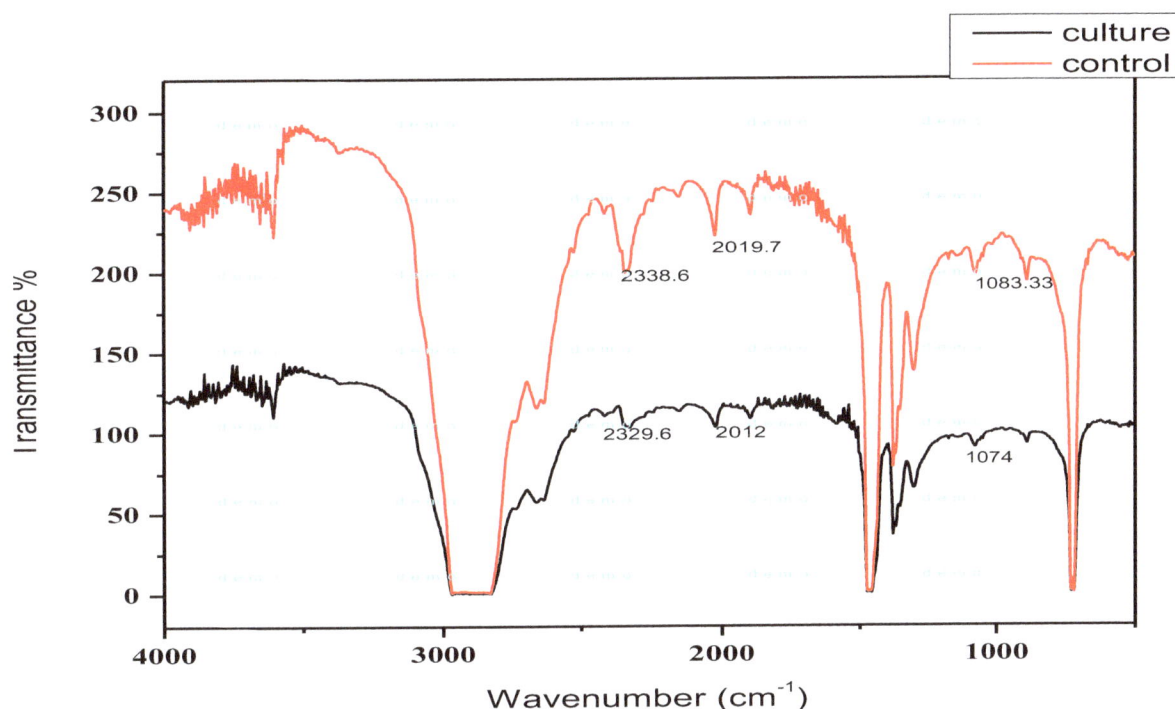

Figure 1. FTIR study of LDPE treated and untreated with *Acinetobacter baumannii* after 30 days of incubation.

RESULTS

FTIR study

Increase in carbonyl index (CI) of LDPE treated with *A. baumannii* after 30 days of incubation indicates the formation of carbonyl groups (Figure 1).

GC-MS study

Figures 2 and 3 indicate the formation of new peaks and compounds in 7.464- as 2-butene, 2-methyl, 8.250- Acetone and 17.288- ethene.

Toxicity test

Table 1 shows the toxicity results of LDPE biodegraded by-products after 30 days of incubation with *A. baumannii*. There are no changes in germination percentage as well as root length and shoot length when compared to control.

DISCUSSION

Biodegradation of polyethylene has been known for several years. In the previous study, the LDPE degradation ability of *A. baumannii* was reported (Pramila et al., 2012). The current study focused on monitoring the chemical changes of LDPE by FTIR analysis by measuring the carbonyl index (CI). The obtained results indicate the CI was increased by 0.1% after 30 days of incubation without pretreating the LDPE film.

Previous reports on polyethylene degradation utilized UV-irradated LDPE films and showed increase in CI after 30 days of incubation (Gilan et al., 2004; Hadad et al., 2005). Albertsson et al. (1987) has reported the 0.3% increase in CI after 10 years of incubation in soil burial method by pretreating with UV. Sudhakar et al. (2008) and Harshavardhan and Jha (2013) revealed the result of 0.15% increase in CI by incubating with marine bacteria for 30 days.

Suresh et al. (2011) and Negi et al. (2011) have reported the FTIR results by monitoring the changes in peaks such as formation or disappearance of peaks of LDPE film containing pro-oxidant additives by incubation with *Bacillus cereus* and soil burial method for 3 months. Mahalakshmi et al. (2012), studied the degradation of unblended or untreated LDPE using *Pseudomonas* spp. after two months of incubation and reported slight changes in peak wave numbers.

Kyaw et al. (2012) reported the result of 16-80% decrease in CI after 120 days of incubation in mineral based medium by *Pseudomonas* spp. The decrease was presumably due to the prolonged incubation time where the culture entered the Norrish II type mechanism (Albertsson et al., 1987). No changes were observed in tensile strength.

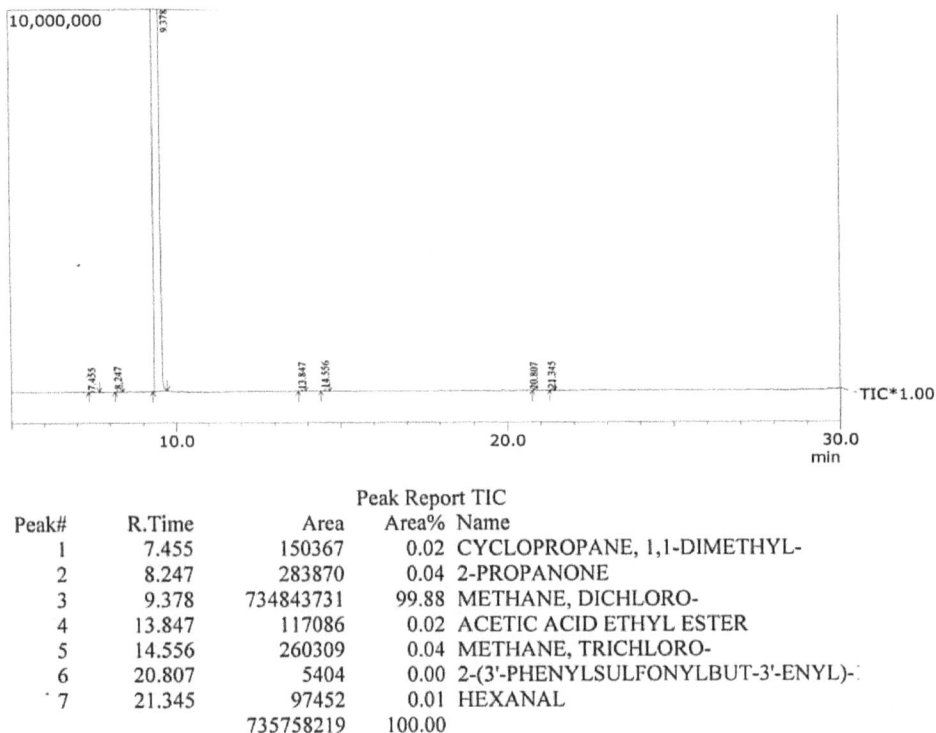

Peak Report TIC

Peak#	R.Time	Area	Area%	Name
1	7.455	150367	0.02	CYCLOPROPANE, 1,1-DIMETHYL-
2	8.247	283870	0.04	2-PROPANONE
3	9.378	734843731	99.88	METHANE, DICHLORO-
4	13.847	117086	0.02	ACETIC ACID ETHYL ESTER
5	14.556	260309	0.04	METHANE, TRICHLORO-
6	20.807	5404	0.00	2-(3'-PHENYLSULFONYLBUT-3'-ENYL)-:
7	21.345	97452	0.01	HEXANAL
		735758219	100.00	

Figure 2. Compounds detected after GC-MS study of untreated LDPE after 30 days.

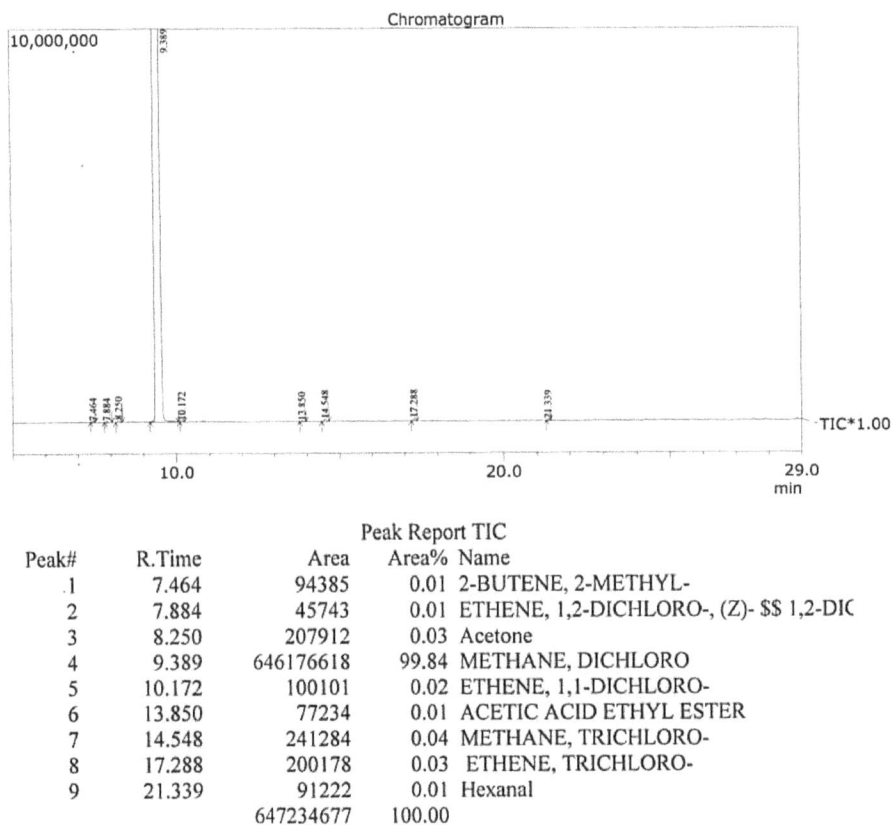

Peak Report TIC

Peak#	R.Time	Area	Area%	Name
1	7.464	94385	0.01	2-BUTENE, 2-METHYL-
2	7.884	45743	0.01	ETHENE, 1,2-DICHLORO-, (Z)- $$ 1,2-DIC
3	8.250	207912	0.03	Acetone
4	9.389	646176618	99.84	METHANE, DICHLORO
5	10.172	100101	0.02	ETHENE, 1,1-DICHLORO-
6	13.850	77234	0.01	ACETIC ACID ETHYL ESTER
7	14.548	241284	0.04	METHANE, TRICHLORO-
8	17.288	200178	0.03	ETHENE, TRICHLORO-
9	21.339	91222	0.01	Hexanal
		647234677	100.00	

Figure 3. Compounds detected after GC-MS analysis of LDPE after 30 days of incubation with *Acinetobacter baumannii.* Figures 2 and 3. Indicates the formation of new peaks and compounds in 7.464- as 2-butene, 2-Methyl, 8.250- acetone, 17.288- ethene.

Table 1. Toxicity results of LDPE biodegraded by-products after 30 days of incubation with *Acinetobacter baumannii*.

Culture	Germination percentage %	Root length (cm)	Shoot length(cm)	Chlorophyll content (mg/g)
Control	80	2.05±0.59	10.07±1.54	0.152
Culture Broth	80	2.52±0.45	11.47±2.16	0.177

Mean ± S.D n=3. There are no changes in germination percentage as well as root length and shoot length when compared to control.

GC-MS results presented in the framework of this study reveals the presence of compounds such as 2-butene, 2-methyl-, acetone, ethene. Presence of acetone indicates the formation of carbonyl groups. Kyaw et al. (2012) has reported GC-MS result of formation of alkanes, aromatic compounds and fatty acid such as hexadecanoic acid and octanoic acid after 120 days incubation.

The byproducts did not reveal any toxicity towards the tested plant characteristics.

Conclusion

Accumulation of polyethylene is becoming a serious environmental issue. Biodegradation of polyethylene process can be viewed as one of the strategic studies to overcome this problem. The current study focused on degradation of LDPE by *A. baumannii*. This isolate grows by utilizing LDPE as a sole carbon source. The bacteria are able to degrade LDPE without any additives and pretreatment in short time duration. This is the first report on degradation of non-pretreated LDPE by *A. baumannii*.

Conflict of interests

The author(s) did not declare any conflict of interest.

ACKNOWLEDGEMENT

The authors are grateful to the University Grants Commission (F.no. 42-480/2013 SR dated March 2013) for providing financial assistance for the completion of this work.

REFERENCES

Albertsson AC, Andersson SO, Karlsson S (1987). The mechanism of biodegradation of polyethylene. Polym. Degrad. Stabil. 18:73-87. http://dx.doi.org/10.1016/0141-3910(87)90084-X

Arnon DI (1949). Copper enzymes in isolated chloroplasts, polyphenoxidase in Beta vulgaris. Plant Physiol. 24:1-15. http://dx.doi.org/10.1104/pp.24.1.1

Gilan I, Hadar Y, Sivan A (2004). Colonization, biofilm formation and biodegradation of polyethylene by a strain of Rhodococcus ruber. Appl. Microbiol. Biotechnol. 65:97-104.

Hadad D, Geresh S, Sivan A (2005). Biodegradation of polyethylene by the thermoplhilic bacterium Brevibacillus borstelensis. J. Appl. Microbiol. 98:1093-1100. http://dx.doi.org/10.1111/j.1365-2672.2005.02553.x

Harshavardhan K, Jha B (2013). Biodegradation of Low Density Polyethylene (LDPE) by marine bacteria from pelagic water, Arabian Sea, India. Mar. Pollut. Bull. 77:100-106. http://dx.doi.org/10.1016/j.marpolbul.2013.10.025

Kyaw BM, Champakalakshmi R, Sarkharkar MK, Lim CS, Sarkharkar KR (2012) Biodegradation of LDPE by Pseudomonas species. Indian J. Microbiol. 52(3):411-419. http://dx.doi.org/10.1007/s12088-012-0250-6

Mahalakshmi V, Abubakker S, Niran AS (2012). Analysis of polyethylene Degrading potentials of microorganisms isolated from compost soil. Int. J. Pharm. Biol. Arch. 3(5):1190-1196

Negi H, Gupta S, Zaidi MGH, Goel R (2011). Studies on biodegradation of LDPE film in the presence of potential bacterial consortia enriched soil. Biologia 57:141-147. http://dx.doi.org/10.6001/biologija.v57i4.1925

Pramila R, Padmavathy K, Vjaya Ramesh K, Mahalakshmi K (2012) Brevibacillus parabrevis, Acinetobacter baumannii and Psuedomonas citronellolis- Potential candidates for biodegradation of Low Density Polyethylene (LDPE). J. Bacteriol. Res. 4(1):9-14. http://dx.doi.org/10.5897/JBR12.003

Shah AA (2007). Role of micro-organisms in biodegradation of plastics. Ph.D. Thesis, Quaid-I-Azam University, Islamabad, Pakistan.

Sudhakar M, Doble M, Sriyutha Murthy P, Venkatesan R (2008). Marine microbe-mediated biodegradation of low and high density polyethylene. Int. Biodeterior. Biodegradation 61:203-213. http://dx.doi.org/10.1016/j.ibiod.2007.07.011

Suresh B, Maruthamuthu S, Palanisamy N, Ragunathan R, Navaneetha Pandiyaraj K, Muralidharan VS (2011). Investigation of biodegradability of polyethylene by Bacillus cereus strain Ma-Su isolated from compost soil. Int. Res. J. Microbiol. 2(8):292-302.

29

Monitoring of enteric fever and diarrhea causing bacteria in a rural setting in Nigeria

Paulinus Osarodion Uyigue[1]* and Kingsley Anukam[2]

[1]Department of Environment and Natural Resources, Kabale University, Kabale, Uganda.
[2]Department of Medical Laboratory Science, University of Benin, Benin City, Nigeria.

Blood and stool samples of patients attending the General Hospital Abudu, Edo State, Nigeria were analyzed to know the prevalence of enteric fever and diarrhea causing bacteria in the area in 2006, 2007 and 2008. Blood sample was collected in Robertson cooked medium and glucose broth; then subcultured on blood agar, macConkey agar, *salmonella/shigella* agar and nutrient agar. Widal agglutination test was also carried out on the blood samples. Stool sample was inoculated into thiosulfate bile sucrose medium, seleniteF medium and later subcultured on macConkey agar and salmonella/*shigella* agar. Of the patients screened, the percentage incidence of *Salmonella typhi* was between 17.5 and 56.5% in 2007; *Salmonella paratyphi* C was between 2.0 and 26.7%; *Salmonella paratyphi* A was between 0 and 9.4% and *Salmonella paratyphi* B was between 0 and 0.7%; enteropathogenic *Escherichia coli* was between 0 and 0.6%; neither *shigella* nor *Vibrio cholera* was isolated. In 2008 of the *salmonella* organisms, the incidence of *S. typhi* was highest with frequency of 19.7 to 54.5%, followed by *S. paratyphi* C: 1.0 to 12.6%; *S. paratyphi* A: 0 to 3.9%, and enteropathogenic *E. coli* was 0 to 0.8%, and in 2009, the incidence or *Salmonella typhi* was highest with a frequency of 2.7 to 68.3%. There was no significant difference (p>o.05) between *S. typhi* incidence throughout the study period. However, there was a significant difference (p<.05) between the incidence of *S. typhi* and other isolates. This project revealed a high rate of typhoid fever (enteric fever) caused by *S. typhi* in Abudu (study area). Further work should be done to identify the source or sources of infection especially their water supply as typhoid fever is a water-borne disease.

Key words: Enteric fever, incidence, prevalence, subcultured, typhoid fever.

INTRODUCTION

Enteric fever and diarrhea are diseases caused by microorganisms of the genera *Salmonella, Shigella, Vibrio, Campylobacter, Yersinia* and *Escherichia*. In some communities, these diseases have become endemic especially where public health consideration is low. Incidence of these diseases can result in high mortality if not properly controlled. Typhoid fever, an enteric fever is characterized by headache, abdominal pain, fever (greater than 38°C) and general lethargy. The mode of transmission of typhoid fever is through drinking water

contaminated with faecal matter (WHO, 2004). It is estimated that more than five million people die from diarrhoeal illnesses yearly. According to WHO (2009), in developing countries, it is predominantly children under the age of five years who suffer from diarrhea. Most people die due to pathogenic microorganisms, such as bacteria or viruses, which were ingested into the gastro-intestinal tract through contaminated drinking water or food. Determining which bacterium is causing the illness in those cases is sometimes very complex. Diarrhoeal illnesses can cause long-term damage to the development of a country; micro-economically due to financial pressures for medical assistance and the physical deterioration of individuals, and macro-economically

*Corresponding author. E-mail: paulinusuyigue@yahoo.com

Table 1. Percentage incidence of enteric fever and diarrhea causing bacteria in patients attending General Hospital, Abudu in 2006.

Month	Total no. of patient screened	Salmonella paratyphi A	Salmonella paratyphi B	Salmonella paratyphi C	Salmonella typhi	Enteropathogenic Escherichia coli	Shigella sp.	Vibrio cholera
Jan	300	1.0	-	5.0	20.0	-	-	-
Feb	225	1.3	-	9.8	42.2	-	-	-
March	106	-	-	6.6	53.8	-	-	-
April	85	9.4	-	16.5	56.5	-	-	-
May	285	1.8	-	2.8	17.5	-	-	-
June	223	1.8	0.7	7.2	31.4	-	-	-
July	235	0.9	-	2.6	33.2	-	-	-
Aug	150	2.7	-	8.0	43.3	-	-	-
Sept	227	0.4	-	2.0	26.0	0.6	-	-
Oct	195	0.5	0.4	7.7	46.0	-	-	-
Nov	60	0.7	-	16.7	47.0	-	-	-
Dec	30	-	-	26.7	36.7	-	-	-

due to absences from work and the corresponding consequences for the economy. Diarrhoeal illnesses can be prevented easily by expanding the water/waste water and public health systems, enhancing hygiene and sanitary measures. Abudu, a rural settlement in Edo state, Nigeria has a population of 11,271 (Nigerian National Commission, 1991). It is a local government headquarters and is situated in a valley. Inhabitants who are mainly peasant farmers make use of the river, River Orhionmwon and untreated borehole water as their sources of water supply for their domestic activities as well as for drinking. People leaving along the river bank defecate directly into the river as well as using it for bathing, washing and cooking (Ajayi and Osibanjo, 1981). The water table is high (at depth of 10 m an aquifer is struck). This makes the aquifer prone to pollution from septic soak-away pit (APHA, 1971). The main health institution here is the general hospital which has high patient attendance. Majority of cases treated here is typhoid fever.

Justification

Majority of patients attending the general hospital Abudu present with signs and symptoms of enteric fever such as headache, fever, abdominal pains, nausea. diarrhea and general lethargy. Knowing the causative agents of this condition and monitoring the pattern of distribution of these organisms will help in achieving proper treatment for these patients.

MATERIALS AND METHODS

All media except Salmonella/shigella agar and selenite F, and glass wares were sterilized by using an autoclave at 121°C for 15 min. They were however constituted and boiled according to manufacturer's specification. Sterilized nutrient agar was cooled to

42°C before adding human blood and allowed to set to form blood agar (Cheesbrough, 2000). Loop wire was sterilized by flaming using the Bunsen burner.

Stool sample

Stool sample was inoculated into Selenite F broth for enrichment, incubated at 37°C for 24 h. This was later subcultured onto salmonella/shigella agar and incubated at 37°C for 24 h.

Blood sample

Two milliliter of blood sample was collected into Robertson cooked meat medium and glucose broth and then incubated at 37°C for 24 h. Subculturing was done on blood agar, macconkey agar, salmonella/shigella agar and nutrient agar (Ogbulie et al., 1988).

Widal agglutination test was carried out on sera using commercial stained antigens. They were for Salmonella paratyphi A, Salmonella parptyphi B, Salmonella paratyphi C and Salmonella typhi.

Identification of bacterial isolates

Bacterial isolates were identified using cultural characteristics and biochemical reactions as recommended by Baker and Breach (1980).

Typing of isolates

Commercial sera were used to type isolates using tile method (Cruickshank, 1976).

RESULTS

In Table 1, three hundred patients were screened in January 2006. Of this number, the percentage incidence of Salmonella A, S. paratyphi B, S. paratyphi C and S. typhi were 1, 0, 5 and 20%, respectively. No Escherichia

Table 2. Percentage incidence of enteric fever and diarrhoea causing bacteria in patients attending General Hospital, Abudu in 2007.

Month	Total no. of patient screened	S. paratyphi A	S. paratyphi B	S. paratyphi C	S. typhi	Enteropathogenic E. coli	Shigella sp.	V. cholera
Jan	304	0.7	-	1.3	25.7	-	-	-
Feb	247	0.8	-	1.2	36.0	0.8	-	-
March	206	2.9	-	12.6	36.4	-	-	-
April	198	1.5	1.0	3.5	54.5	-	-	-
May	200	1.5	-	2.5	30.0	-	-	-
June	301	0.3	-	1.0	27.2	-	-	-
July	284	0.4	-	2.5	19.7	-	-	-
Aug	198	-	-	1.5	39.4	-	-	-
Sept	209	0.5	-	8.0	30.0	0.6	-	-
Oct	156	1.3	-	3.8	53.2	-	-	-
Nov	108	1.9	0.9	3.7	50.9	-	-	-
Dec	51	3.9	-	2.9	39.2	-	-	-

Table 3. Percentange incidence of enteric fever and diarrhoea causing bacteria in patients attending General Hospital, Abudu in 2008.

Month	Total no of patient screened	S.] paratyphi A	S. paratyphi B	S. paratyphi C	S. typhi	Enteropathogenic E. coli	Shigella sp.	V. cholera
Jan	198	-	-	3.1	38.5	-	-	-
Feb	188	1.1	-	3.7	2.7	-	-	-
March	253	0.8	0.4	3.2	39.7	-	-	-
April	175	1.7	-	4.0	40.0	-	-	-
May	203	1.0	0.5	4.4	26.6	-	-	-
June	310	-	-	1.0	31.3	-	-	-
July	277	-	-	0.7	32.0	-	-	-
Aug	185	3.2	-	2.2	34.6	-	-	-
Sept	197	0.5	-	4.1	30.5	1.0	-	-
Oct	120	0.8	0.8	5.0	68.3	-	-	-
Nov	90	1.1	-	3.3	62.2	-	-	-
Dec	37	2.7	-	2.7	27.0	-		

coli, *Shigella* spp. and *V. cholera* were isolated. In April 2006, the highest incidence of 56.5% was recorded for *S. typhi*. *S. paratyphi* C and *S. typhi* were isolated throughout the year while *Shigella* spp. and *V. cholera* were not isolated during this period (Table 1). Also in Table 1, total number of patients screened (the lowest) in December while the highest number of patients screened was 300 in January. In 2007 (Table 2), the pattern of distribution of isolates were similar. For example, *S. paratyphi* C and were isolated throughout the year. Highest number of patients screened was in January while the lowest was in December.

However, in Table 3 the highest number of patients screened was 310 in June while the lowest was 37 in December. *S. paratyphi* C and *S. typhi* were isolated throughout the year. *S. typhi* had the highest incidence which peaked at 68.3% in October. Neither *Shigella* nor *V. cholera* was isolated.

Percentage incidence of *S. typhi* was 17.5 to 56.5%, 19.7 to 54.5% and 2.7 to 68.3% in 2006, 2007 and 2008,

respectively. Incidence of *S. paratyphi* C was 2 to 26.7%, 1 to 12.6% and 0.7 to 5% in 2006, 2007, and 2008, respectively.

Patients attending General Hospital, Abudu

Blood and stool samples of patients attending General Hospital, Abudu were screened for enteric bacteria on monthly basis from January 2006 to December 2008. *S. typhi* had the highest incidence followed by *S. paratyphi* C throughout the study period. Incidence of *S. paratyphi* A and B were low. Neither *Shigella* spp. nor *V. cholera* was isolated. Enteropathogenic *E. coli* were virtually absent; where they occurred, incidence was very low.

DISCUSSION

From the results, *S. paratyphi* C and *S. typhi* were isolated throughout the study. These organisms are the major causes of paratyphoid and typhoid fever in Nigeria. Their constant isolation in this study area implies that there is active pollution of their sources of water supply with faecal matter containing these organisms among other microorganisms. There was no significant difference ($p < 0.05$) between *S. typhi* incidence throughout the three years study period. *S. typhi* had the highest incidence as an indication that typhoid fever caused by this organism is the most common. The low patient turn out especially in December is due to the festivities (Christmas/New year celebrations). Many inhabitants travel from the study area (Abudu) to cities to celebrate Christmas thereby depleting the population here. Another reason for low patient turnout is that people save to make merriment at this time that they managed their health at home to reduce cost of going to the hospital. However, in January, patient's turnout was very high due to the effect of heavy festivities; the previous month (December) when eating wining and drinking possibly contaminated water was much. There was a significant difference ($p < 0.05$) between the incidence of *S. typhi* and other isolates throughout the study period. This is due to the high incidence of typhoid fever when compared to other enteric fever in this locality.

CONCLUSION AND RECOMMENDATION

The high incidence of typhoid and paratyphoid fever in Abudu has been revealed within the study period. This finding is alarming as tracing the source or sources of these infections becomes urgent and inevitable. More research work is recommended to trace carriers. Meanwhile water for drinking in Abudu should be boiled before drinking (APHA, 1971).

REFERENCES

Ajayi SO, Osibanjo O (1981). Pollution Studies on Nigerian Rivers 11: Water quality of some Nigerian Rivers. Environ. Pollut. Ser., B2: 87–95.

American Public Health Association (1971). Standard method for the examination of water and waste water, 13th ed., APHA, pp. 874.

Baker FJ, Breach WR (1980). Bacterial identification tests. In Medical Microbiological Techniques, Butterworth's and Co. Ltd., London, pp. 85–91.

Cheesbrough M (2000). Water and Sanitation, Decade, Bacteriological testing of water supplies. In Medical Laboratory Manual for tropical countries. Vol 11. Microbiology ELBS / Tropical Health Technology, Butterworths, pp. 206–221.

Cruickshank R, Duguid JP, Marmion BP, Swain RHA (1976). Bacteriological examination of water, Milk and Air. In Medical Microbiology, Twelfth edition, Churchill Livingstone, Edinburgh, 11: 273–300.

Nigeria National population commission (1991).

Ogbulie JN, Uwaezuoke JC, Ogiebor S1 (1988). Introductory practical Microbiology. Spring field Publishers, Owerri, pp. 160.

World Health Organization (WHO) (2004). Guidelines for drinking water Quality. Supporting Documentation to the Guidelines (3rd edition), 2: 552.

World Health Organization (WHO) (2009). Guidelines for drinking water quality. Supporting Documentation to the Guidelines (4th edition), 4: 350.

Permissions

The contributors of this book come from diverse backgrounds, making this book a truly international effort. This book will bring forth new frontiers with its revolutionizing research information and detailed analysis of the nascent developments around the world.

We would like to thank all the contributing authors for lending their expertise to make the book truly unique. They have played a crucial role in the development of this book. Without their invaluable contributions this book wouldn't have been possible. They have made vital efforts to compile up to date information on the varied aspects of this subject to make this book a valuable addition to the collection of many professionals and students.

This book was conceptualized with the vision of imparting up-to-date information and advanced data in this field. To ensure the same, a matchless editorial board was set up. Every individual on the board went through rigorous rounds of assessment to prove their worth. After which they invested a large part of their time researching and compiling the most relevant data for our readers.

The editorial board has been involved in producing this book since its inception. They have spent rigorous hours researching and exploring the diverse topics which have resulted in the successful publishing of this book. They have passed on their knowledge of decades through this book. To expedite this challenging task, the publisher supported the team at every step. A small team of assistant editors was also appointed to further simplify the editing procedure and attain best results for the readers.

Apart from the editorial board, the designing team has also invested a significant amount of their time in understanding the subject and creating the most relevant covers. They scrutinized every image to scout for the most suitable representation of the subject and create an appropriate cover for the book.

The publishing team has been an ardent support to the editorial, designing and production team. Their endless efforts to recruit the best for this project, has resulted in the accomplishment of this book. They are a veteran in the field of academics and their pool of knowledge is as vast as their experience in printing. Their expertise and guidance has proved useful at every step. Their uncompromising quality standards have made this book an exceptional effort. Their encouragement from time to time has been an inspiration for everyone.

The publisher and the editorial board hope that this book will prove to be a valuable piece of knowledge for researchers, students, practitioners and scholars across the globe.

List of Contributors

I. J. Mbuko
Faculty of Veterinary Medicine, Ahmadu Bello University, P. O. Box 1044, Samaru-Zaria, Kaduna State-Nigeria

M. A. Raji
Faculty of Veterinary Medicine, Ahmadu Bello University, P. O. Box 1044, Samaru-Zaria, Kaduna State-Nigeria

J. Ameh
Faculty of Veterinary Medicine, Ahmadu Bello University, P. O. Box 1044, Samaru-Zaria, Kaduna State-Nigeria

L. Saidu
Faculty of Veterinary Medicine, Ahmadu Bello University, P. O. Box 1044, Samaru-Zaria, Kaduna State-Nigeria

W. I. Musa
Faculty of Veterinary Medicine, Ahmadu Bello University, P. O. Box 1044, Samaru-Zaria, Kaduna State-Nigeria

P. A. Abdul
Faculty of Veterinary Medicine, Ahmadu Bello University, P. O. Box 1044, Samaru-Zaria, Kaduna State-Nigeria

O. J. Akinjogunla
Department of Microbiology, Faculty of Sciences, University of Uyo, P.M.B 1017, Uyo, Akwa Ibom State, Nigeria

N. O. Eghafona
Department of Microbiology, Faculty of Life Sciences, University of Benin, Benin City, Edo State, Nigeria

O. H. Ekoi
Department of Microbiology, Faculty of Sciences, University of Uyo, P.M.B 1017, Uyo, Akwa Ibom State, Nigeria

S. E. Baidoo
Department of Clinical Microbiology, School of Medical Sciences, Kwame Nkrumah University of Science and Technology, Kumasi, Ghana
Department of Laboratory Technology, University of Cape Coast, Cape Coast, Ghana

S. C. K. Tay
Department of Clinical Microbiology, School of Medical Sciences, Kwame Nkrumah University of Science and Technology, Kumasi, Ghana

K. Obiri-Danso
Department of Theoretical and Applied Biology, Kwame Nkrumah University of Science and Technology, Kumasi, Ghana

H. H. Abruquah
Division of Microbiology and Infectious Diseases, University Hospital, Kwame Nkrumah University of Science and Technology, Kumasi, Ghana

Allma Koçinaj
University Clinical Center of Kosova, Prishtina, Kosova

Dardan Koçinaj
University Clinical Center of Kosova, Prishtina, Kosova

Merita Berisha
National Institute of Public Health of Kosova, Prishtina, Kosova

S. Umesha
Department of Studies in Biotechnology, University of Mysore, Manasagangotri, Mysore 570 006, Karnataka, India

R. Kavitha
Department of Studies in Biotechnology, University of Mysore, Manasagangotri, Mysore 570 006, Karnataka, India

F. A. Sebastião
Aquaculture Center of São Paulo State University - CAUNESP, Rod. Paulo Donato Castellane, s/n Bairro Rural, Jaboticabal, SP CEP 14884-900, Brazil

F. Pilarski
Aquaculture Center of São Paulo State University - CAUNESP, Rod. Paulo Donato Castellane, s/n Bairro Rural, Jaboticabal, SP CEP 14884-900, Brazil

M. V. F. Lemos
Universidade Estadual Paulista, Department of Applied Biology for Agriculture, Rod. Paulo Donato Castellane, s/n Bairro Rural, Jaboticabal, SP CEP 14884-900, Brazil

Akram Hassan Mekki
Department of Microbiology, Faculty of Health Sciences, Omdurman Ahlia University, Sudan

Abdullahi Nur Hassan
Department of Clinical Microbiology and Infectious Disease, Faculty of Medicine, Alzaiem Alazhari University, Sudan

Dya Eldin M Elsayed
Department of Community Medicine, Faculty of Medicine, Alzaiem Alazhari University, Sudan

E. A. Ophori
Department of Microbiology, University of Benin, Benin City, Edo State, Nigeria

P. Imade
Department of Medical Microbiology, University of Benin Teaching Hospital, Benin City, Edo State, Nigeria

E. J. Johnny
University of Benin Teaching Hospital, Benin City, Edo State, Nigeria

Tyagi Shruti
Department of Biotechnology, Meerut Institute of Engineering and Technology, Meerut UP, India

K. Tyagi Pankaj
Department of Biotechnology, Meerut Institute of Engineering and Technology, Meerut UP, India

Panday Chandra Shekhar
Department of Biotechnology, Meerut Institute of Engineering and Technology, Meerut UP, India

Kumar Ruchica
Department of Biotechnology, N. C. College of Engineering, Israna, Panipat Haryana, India

Qurban Ali
Department of Plant Breeding and Genetics, University of Agriculture, Faisalabad, Pakistan

Muhammad Hammad Nadeem Tahir
Department of Plant Breeding and Genetics, University of Agriculture, Faisalabad, Pakistan

Hafeez Ahmad Sadaqat
Department of Plant Breeding and Genetics, University of Agriculture, Faisalabad, Pakistan

Saeed Arshad
Department of Plant Breeding and Genetics, University of Agriculture, Faisalabad, Pakistan

Jahenzeb Farooq
Department of Plant Breeding and Genetics, University of Agriculture, Faisalabad, Pakistan

Muhammad Ahsan
Department of Plant Breeding and Genetics, University of Agriculture, Faisalabad, Pakistan

Muhammad Waseem
Department of Entomology, University of Agriculture, Faisalabad, Pakistan

Amjad Iqbal
Department of Agronomy, University of Agriculture, Faisalabad, Pakistan

R. O. Ayeni
Department of Pure and Applied Mathematics, Ladoke Akintola University of Technology Ogbomoso, Nigeria

A. O. Popoola
Department of Pure and Applied Mathematics, Ladoke Akintola University of Technology Ogbomoso, Nigeria

J. K. Ogunmoyela
Department of Mathematical Sciences, Federal University of Technology, Akure Nigeria

Mohammed A. K. Al-Saadi
Department of Microbiology, College of Medicine, Babylon University, Babylon Province, Iraq

Alaa H. Al-Charrakh
Department of Microbiology, College of Medicine, Babylon University, Babylon Province, Iraq

Salim H. H. Al-Greti
Medical Institute of Karbala, Karbala Province, Iraq

Simeon Chukwuemeka Enemuor
Department of Microbiology, Kogi State University, Anyigba, Kogi State, Nigeria

James Omale
Department of Biochemistry, Kogi State University, Anyigba, Kogi State, Nigeria

Ekpa Matthew Joseph
Department of Microbiology, Kogi State University, Anyigba, Kogi State, Nigeria

Ahmed
Central Laboratory, Ministry of Science and Technology, Sudan

A. Alkando
Central Laboratory, Ministry of Science and Technology, Sudan

Hanan Moawia Ibrahim
Central Laboratory, Ministry of Science and Technology, Sudan

Afaf I. Shehata
Department of Botany and Microbiology, College of Science, King Saud University, P. O. Box 2455, Riyadh 11451, Saudi Arabia

Amal Abdulaziz Al-Hazani
Department of Botany and Microbiology, College of Science, King Saud University, P. O. Box 2455, Riyadh 11451, Saudi Arabia

Hesham Al-Aglaan
Department of Botany and Microbiology, College of Science, King Saud University, P. O. Box 2455, Riyadh 11451, Saudi Arabia

Hanan O. Al Shammari
Department of Botany and Microbiology, College of Science, King Saud University, P. O. Box 2455, Riyadh 11451, Saudi Arabia

Cesar Pedroza-Roldan
Instituto de Ciencias Biomédicas, Universidad Autónoma de Ciudad Juárez. Anillo Envolvente del Pronaf y Estocolmo s/n. C.P.32300 A.P. 1595-D. Cd. Juárez, Chihuahua, México
Instituto de Investigaciones Biomédicas, Universidad Nacional Autónoma de México, AP 70228, Ciudad Universitaria, México, D.F. 04510, México

Oscar Zavala-Tapia
Instituto de Ciencias Biomédicas, Universidad Autónoma de Ciudad Juárez. Anillo Envolvente del Pronaf y Estocolmo s/n. C.P.32300 A.P. 1595-D. Cd. Juárez, Chihuahua, México

Leny J. Alvarez-Araujo
Instituto de Ciencias Biomédicas, Universidad Autónoma de Ciudad Juárez. Anillo Envolvente del Pronaf y Estocolmo s/n. C.P.32300 A.P. 1595-D. Cd. Juárez, Chihuahua, México

Claudia Charles-Niño
Instituto de Investigaciones Biomédicas, Universidad Nacional Autónoma de México, AP 70228, Ciudad Universitaria, México, D.F. 04510, México

Angel G. Diaz-Sanchez
Departamento de Bioquímica, Facultad de Química, Universidad Nacional Autónoma de México, AP 70228, Ciudad Universitaria, México, D.F. 04510, México

Raymundo Rivas-Caceres
Instituto de Ciencias Biomédicas, Universidad Autónoma de Ciudad Juárez. Anillo Envolvente del Pronaf y Estocolmo s/n. C.P.32300 A.P. 1595-D. Cd. Juárez, Chihuahua, México

M. E. Ismail
Department of Plant Pathology, Faculty of Agriculture, El-Minia University, Minia, Egypt

M. F. Abdel-Monaim
Plant Pathology Research Institute, Agricultural Research Center, Giza, Egypt

Y. M. Mostafa
Department of Horticulture, Faculty of Agriculture, El-Minia University, Minia, Egypt

I. W. Musa
Department of Veterinary Surgery and Medicine, Ahmadu Bello University, Zaria, Kaduna State, Nigeria

M Bisalla
Department of Veterinary Pathology and Microbiology, Ahmadu Bello University, Zaria, Kaduna State, Nigeria

B. Mohammed
Department of Veterinary Public Health and Preventive Medicine, Ahmadu Bello University, Zaria, Kaduna State, Nigeria

L. Saídu
Veterinary Teaching Hospital, Ahmadu Bello University, Zaria, Kaduna State, Nigeria

P. A. Abdu
Department of Veterinary Surgery and Medicine, Ahmadu Bello University, Zaria, Kaduna State, Nigeria

Esmaeil Saberi
Department of Plant Pathology, Faculty of Agriculture, Tarbiat Modares University, PO Box 14115-336, Tehan, Iran

Naser Safaie
Department of Plant Pathology, Faculty of Agriculture, Tarbiat Modares University, PO Box 14115-336, Tehan, Iran

Heshmatollah Rahimian
Department of Plant Protection, Faculty of Agriculture, Mazandaran Agricultural Sciences, Sari, Iran

Abas Bahador
Department of Medical Microbiology, Faculty of Medicine, Tehran University of Medical Sciences, Tehran, Iran

Noormohamad Mansoori
Department of Medical Microbiology, Faculty of Medicine, Tehran University of Medical Sciences, Tehran, Iran

Davood Esmaeili
Bacteriology Department, Applied Microbiology Research Center, Baqiyatallah University

Reza Amini Sabri
Bacteriology Department, Applied Microbiology Research Center, Baqiyatallah University

Najla Haddaji
Laboratoire d'Analyse, Traitement et Valorisation des Polluants de l'Environnement et des Produits, Faculté de Pharmacie Rue Avicenne, Monastir 5000, Tunisie

Boubaker Krifi
Laboratoire d'Analyse, Traitement et Valorisation des Polluants de l'Environnement et des Produits, Faculté de Pharmacie Rue Avicenne, Monastir 5000, Tunisie

Rihab Lagha
Laboratoire d'Analyse, Traitement et Valorisation des Polluants de l'Environnement et des Produits, Faculté de Pharmacie Rue Avicenne, Monastir 5000, Tunisie

Sadok Khouadja
Laboratoire d'Analyse, Traitement et Valorisation des Polluants de l'Environnement et des Produits, Faculté de Pharmacie Rue Avicenne, Monastir 5000, Tunisie

Amina Bakhrouf
Laboratoire d'Analyse, Traitement et Valorisation des Polluants de l'Environnement et des Produits, Faculté de Pharmacie Rue Avicenne, Monastir 5000, Tunisie

Ghiadaa Abass Jassem
College of Veterinary Medicine, AL-Qadisiyia University, Iraq

Chijioke A. Nsofor
Department of Biotechnology, Federal University of Technology, Owerri, Imo State, Nigeria

Samuel T. Kemajou
Department of Medical Laboratory Science, Madonna University, Elele, Rivers State, Nigeria

Chibuzor M. Nsofor
Department of Medical Laboratory Science, Madonna University, Elele, Rivers State, Nigeria

GHELLAI Lotfi
Laboratory of Applied Microbiology in Food, Biomedical and Environment (LAMAABE), Department of Biology, University of Tlemcen, 13000 Tlemcen, Algeria

HASSAINE Hafida
Laboratory of Applied Microbiology in Food, Biomedical and Environment (LAMAABE), Department of Biology, University of Tlemcen, 13000 Tlemcen, Algeria

KLOUCHE Nihel
Laboratory of Applied Microbiology in Food, Biomedical and Environment (LAMAABE), Department of Biology, University of Tlemcen, 13000 Tlemcen, Algeria

KHADIR Abdelmonaim
Laboratory of Applied Microbiology in Food, Biomedical and Environment (LAMAABE), Department of Biology, University of Tlemcen, 13000 Tlemcen, Algeria

AISSAOUI Nadia
Laboratory of Applied Microbiology in Food, Biomedical and Environment (LAMAABE), Department of Biology, University of Tlemcen, 13000 Tlemcen, Algeria

NAS Fatima
Laboratory of Applied Microbiology in Food, Biomedical and Environment (LAMAABE), Department of Biology, University of Tlemcen, 13000 Tlemcen, Algeria

ZINGG Walter
Service de Prévention et de Contrôle de l'Infection. Hôpitaux Universitaires de Genève (HUG) Suisse

J. Leimkugel
Swiss Tropical Institute, Socinstrasse 57, Basel, Switzerland

V. Racloz
Swiss Tropical Institute, Socinstrasse 57, Basel, Switzerland

L. Jacintho da Silva
Novartis Vaccines and Diagnostics, Sidney Street 45, Cambridge, U.S.A

G. Pluschke
Swiss Tropical Institute, Socinstrasse 57, Basel, Switzerland

Nadjia BENHAMED
Laboratory of Applied Microbiology, Department of Biology, Faculty of Science, Oran University, BP16, Essenia 31000 Oran, Algeria

Mebrouk Kihal
Laboratory of Applied Microbiology, Department of Biology, Faculty of Science, Oran University, BP16, Essenia 31000 Oran, Algeria

S. Méité
Laboratory of Bacteriology-Virology, University Hospital (CHU) of Yopougon, 22 BP 539, Abidjan, Côte d'Ivoire Department of Microbiology, Faculty of Medical Sciences, University Hospital (CHU) of Yopougon, 22 BP 539, Abidjan, Côte d'Ivoire

C. Boni-Cissé
Laboratory of Bacteriology-Virology, University Hospital (CHU) of Yopougon, 22 BP 539, Abidjan, Côte d'Ivoire Department of Microbiology, Faculty of Medical Sciences, University Hospital (CHU) of Yopougon, 22 BP 539, Abidjan, Côte d'Ivoire

A. P. Mlan Tanoa
Laboratory of Bacteriology-Virology, University Hospital (CHU) of Yopougon, 22 BP 539, Abidjan, Côte d'Ivoire

F. S. Zaba
Laboratory of Bacteriology-Virology, University Hospital (CHU) of Yopougon, 22 BP 539, Abidjan, Côte d'Ivoire

H. Faye-Ketté
Laboratory of Bacteriology-Virology, University Hospital (CHU) of Yopougon, 22 BP 539, Abidjan, Côte d'Ivoire
Department of Microbiology, Faculty of Medical Sciences, University Hospital (CHU) of Yopougon, 22 BP 539, Abidjan, Côte d'Ivoire

M. Dosso
Laboratory of Bacteriology-Virology, University Hospital (CHU) of Yopougon, 22 BP 539, Abidjan, Côte d'Ivoire
Department of Microbiology, Faculty of Medical Sciences, University Hospital (CHU) of Yopougon, 22 BP 539, Abidjan, Côte d'Ivoire

R. Pramila
Quaid-E-Millath Government College for Women, Annasalai, Tamilnadu, Chennai-00 002, Tamilnadu, India

K. Vijaya Ramesh
Quaid-E-Millath Government College for Women, Annasalai, Tamilnadu, Chennai-00 002, Tamilnadu, India
Paulinus Osarodion Uyigue
Department of Environment and Natural Resources, Kabale University, Kabale, Uganda

Kingsley Anukam
Department of Medical Laboratory Science, University of Benin, Benin City, Nigeria

www.ingramcontent.com/pod-product-compliance
Lightning Source LLC
Chambersburg PA
CBHW050437200326
41458CB00014B/4977